DATE DUE

NOV 11 2015		

Governors State University
Library
Hours:
Monday thru Thursday 8:30 to 10:30
Friday and Saturday 8:30 to 5:00
Sunday 1:00 to 5:00 (Fall and Winter Trimester Only)

DEMCO

The Natural History of
Weasels and Stoats

The Natural History of
Weasels and Stoats

Ecology, Behavior, and Management

Second Edition

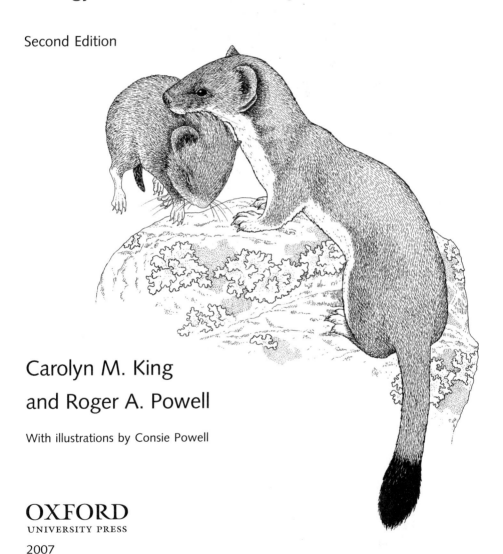

Carolyn M. King
and Roger A. Powell

With illustrations by Consie Powell

OXFORD
UNIVERSITY PRESS

2007

OXFORD
UNIVERSITY PRESS

Oxford University Press, Inc., publishes works that further
Oxford University's objective of excellence
in research, scholarship, and education.

Oxford   New York
Auckland   Cape Town   Dar es Salaam   Hong Kong   Karachi
Kuala Lumpur   Madrid   Melbourne   Mexico City   Nairobi
New Delhi   Shanghai   Taipei   Toronto

With offices in
Argentina   Austria   Brazil   Chile   Czech Republic   France   Greece
Guatemala   Hungary   Italy   Japan   Poland   Portugal   Singapore
South Korea   Switzerland   Thailand   Turkey   Ukraine   Vietnam

Published by Oxford University Press, Inc.
198 Madison Avenue, New York, New York 10016

www.oup.com

Oxford is a registered trademark of Oxford University Press.

Library of Congress Cataloging-in-Publication Data
King, C. M. (Carolyn M.)
The natural history of weasels and stoats : ecology, behavior, and management /
Carolyn M. King and Roger A Powell ; with illustrations by Consie Powell.—2nd ed.
p. cm.
ISBN-13 978-0-19-530056-7
ISBN 0-19-530056-4
ISBN-13 978-0-19-532271-2 (pbk)
ISBN 0-19-532271-1 (pbk)
1. Weasels.   2. Ermine.   I. Powell, Roger A.   II. Title.
QL737.C25K58 2006
599.76'62—dc22   2005029909

9   8   7   6   5   4   3   2   1

Printed in the United States of America
on acid-free paper

Preface

WHEN THIS BOOK was first published in 1989, it was a story produced in Britain and based mainly on research done in the Palaearctic (Britain and Eurasia), plus smaller contributions from the United States and New Zealand. British interest in natural history for its own sake goes back to Gilbert White and his clerical colleagues in the middle 1700s, who loved to observe and report on anything they saw wild animals doing as they rode around their country parishes. Later, "organized natural history" became the basis for Darwin's understanding of evolution via natural selection, and later still it became one of the roots of the science of ecology, which developed in the old-established universities in the United States and Europe from the 1920s onward.

Britain also pioneered the tradition of shooting wild-bred game birds, which required an army of professional gamekeepers to implement rigorous control of predators, including stoats and weasels. In Russia, stoats in ermine had been an important source of fur, and state-funded research programs dating back to the 1930s attempted to work out why the fur harvests varied so much from year to year. British colonists in New Zealand attempted to deal with their rabbit problem by introducing mustelids—thereby implementing a large-scale, uncontrolled practical experiment in biological control, whose consequences have since cost millions, measured in both dollars and lives. All these traditions were well represented in the first edition of this book.

Much has changed in the world since then, in the ecological sciences as much as everywhere else. One of the most astonishing changes has been the geographic shift in the sources of research papers, which in turn reflects profound changes in the way research is organized and funded. In Europe, about the same number of papers on stoats and weasels has been published every decade since the 1970s, although the subjects of interest have changed with time. But in North America, the number has been slowly declining, and in New Zealand, it has been rapidly accelerating (King 2005c).

The difference shows where the funding is. Research on native species tends to ask theoretical questions, such as (the hot topic in Europe for the last few years) the role of predation by weasels and stoats in causing the multiannual population cycles in lemmings and other small rodents. Research on introduced species tends to ask practical questions, most urgently about how the damage caused by invasive aliens to native biodiversity can be prevented or controlled. New Zealand is the only country where stoats and weasels are now regarded as important introduced pests, and the increase in state funding to support research on their biology and control since the middle 1990s has stimulated intensive research programs on these species, produced huge amounts of new information, and changed a lot of previous ideas.

The first edition of this book was produced in Britain and sold in North America as a coproduction, but this new edition of the book is produced in the United States and copublished in Britain and New Zealand. While it still owes a great deal to its British foundations, it is now much more consciously a North American book. The core zoological descriptions remain the same, as they must, but to illustrate them, stories about British gamekeepers are supplemented with early North American observations and recent studies from New Zealand. Basic facts about the species are included wherever they came from, so we have formulated the questions that arise from the facts in terms that should appeal to readers everywhere, but we hope the new version might stimulate more research on these fascinating animals in North America.

In the end, we have tried to write a book that is not only a thorough scientific reference about weasels, but also a book that provides the general public with a way to learn about these marvellous, intriguing critters.

Acknowledgments

WE THANK MANY people for their help over the long years of our joint and separate fascination with weasels. We thank our graduate supervisors, H.N. (Mick) Southern at Oxford and Monte Lloyd at Chicago. We thank all the fellow students and researchers, gamekeepers, wildlife, forestry, and national forest and national parks rangers who helped us to collect material or to make sense of it. We thank the many librarians who tracked down obscure references (especially Amie Oxley, Deidre Garland, and Ann Roth), and Ken Ayers who converted three filing cabinets' worth of reprints into a bibliographic database. We thank all those who lent us original data or illustrations or granted permission to reproduce copyright material: Cornell University Press, Charles Scribner's Sons, Farrar Straus & Giroux, LLC, The Game Conservancy, the Zoological Society of London, the British Museum of Natural History, the New Zealand Science Information Publishing Centre (DSIR), G. Caughley, T. Holmes, H. Grue, C. Magrini, P. Morris, G. Proulx, P. Sleeman, C. St-Pierre, S. Tapper, C. Vispo, and M. Weber.

It is a daunting task to read the entire manuscript of a book with sufficient attention to make constructive criticisms of it, and we are very grateful to Elaine Murphy and two anonymous publishers' reviewers for finding the time to undertake this valuable service. We thank Katharine Lenz and Peter Prescott for their encouragement.

Finally, we thank Lew, Iago, Kim, Sylvia, Carlo, Raja, Minimus, and especially Samantha (Sammy) who, through sharing their lives with us, provided personal glimpses of life as a weasel.

Contents

The Natural History of
Weasels and Stoats

1 | Weaselly Distinguished, Stoatally Different

THERE IS SOMETHING enormously satisfactory about a weasel. It has the perfection, grace, and efficiency of a well-designed tool in the hands of an expert. Just as people love to watch and applaud a craftsman or artist at work or a display of the skills of any top sportsman, so have we, for years, loved to study and applaud the lives of weasels. They are among the purest of carnivores, perfectly adapted in every feature of their bodies and behavior to live exclusively as hunters. These adaptations add up to a design for an effective mouse-harvesting machine that humans can only envy.

In Europe, North America, and especially New Zealand, farmers and foresters have even tried to make use of the powers of weasels for their own ends, particularly to control various "pest" species such as voles (small mouse-like rodents with short tails and small eyes and ears) and rabbits. But the philosophies of weasels and of people toward such situations are quite different: What for us is a problem, requiring elimination, is for weasels an opportunity, to be exploited. That is why biological control by weasels seldom works (Chapters 12, 13). One of the characteristics of a perfect tool is that it cannot be made to do a job other than the one for which it was designed.

All weasels have the sinuous physique common to the Mustelidae, the family of carnivores of which they are the smallest members. They have long, slender bodies, long necks, and short legs; their heads are rather flattish and smoothly pointed, exactly suitable instruments for poking into every possible small hole. Indeed, their Latin name is said to be derived from their small stature and long, pointed shape; *Mustela* means a "mus" (mouse) as long as a "telum" (spear). They have no apparent shoulders or hips, so the general impression is of a slim, furry tube ending in an excitable, bottlebrush tail. They have large rounded ears lying almost flat among the fur; bright, beady, black eyes; and very long, sensitive whiskers on their faces, and, like cats, short ones on their elbows. Their paws are furred between the pads (five on each foot); the claws are sharp, and not retractable. They swim well, climb trees easily, and run like small bolts of brown lightning.

WEASELS IN LITERATURE

Few animals, except maybe snakes and spiders, are as widely misunderstood as are weasels. An early description, phrased with all the vivid imagery that was still allowed in scientific writing in the last century, gives an idea of their reputation:

> A glance at the physiognomy of the weasels would suffice to betray their character. The teeth are almost of the highest known raptorial character; the jaws are worked by enormous masses of muscles covering all the side of the skull. The forehead is low and the nose is sharp; the eyes are small, penetrating, cunning, and glitter with an angry green light. There is something peculiar, moreover, in the way that this fierce face surmounts a body extraordinarily wiry, lithe and muscular. It ends in a remarkably long and slender neck in such a way that it may be held at a right angle with the axis of the latter. When the creature is glancing around, with the neck stretched up, and the flat triangular head bent forward, swaying from one side to the other, we catch the likeness in a moment—it is the image of a serpent. (Coues 1877)

At least since the mid-nineteenth century, weasels have had a bad image that invites all sorts of negative responses. As Wood (1946) put it,

> Most of us have an instinctive hatred of weasels. There is something decidedly sinister about their looks; something serpentine that makes us shudder. . . . Weasels look like killers and they are killers, and many of us get a self-righteous feeling when we shoot one.

Predators such as bears and foxes are favorites of children's literature, and they often take positive roles (Powell & Powell 1982). But weasels appear only rarely in animal fiction stories for children, and they are usually treated as thieves, robbers, or other bad characters. They are the villains of traditional nature stories—as in, for example, *The Wind in the Willows* (Figure 1.1).

The character assassination of weasels extends even to their common name. "Weasel words" are those whose meaning is bent or twisted for unearned gain; to "weasel out" of some responsibility or tight situation is to escape by cunning but unfair means. Such uses of the word "weasel" have a long history. Shakespeare noted that weasels suck eggs, and Dickens used the phrase "cunning as a weasel" in *The Old Curiosity Shop*. And in New Zealand, "weasel pee" is a derogatory description of weak beer (hence the sign once seen on the door of a local pub: "100% weaselless"!).

In addition to this moralizing and emotional reaction to the physical appearance of a weasel, the rise of game management in North America and Europe in the late-nineteenth century initiated a war against all predators that eat birds or their eggs and ensured that weasels were generally labeled as "vermin." Gamekeepers systematically persecuted them for the next hundred years. The

Figure 1.1 The traditional, popular, "bad guy" image of weasels is captured in this drawing by Ernest Shepard from *The Wind in the Willows* showing stoats armed with guns jeering at Mole dressed as a washerwoman. (Reproduced with the permission of Atheneum Books for Young Readers, an imprint of Simon & Schuster Children's Publishing Division from *The Wind in the Willows* by Kenneth Graham, illustrated by Ernest H. Shepard. Copyright © 1933, 1953 Charles Scribner's Sons; copyright renewed © 1961 Ernest H. Shepard.)

observations of early naturalists were liberally mixed with prejudice, fiction, and misinterpretation, and led to uncritical conclusions. Accounts of weasels in the older natural history books—and even in some modern ones—tell us more about the writers and the attitudes of their times than about their subjects.

Where weasels of any species have been introduced into countries outside their natural range, such as New Zealand, they enter natural communities as destructive aliens. Native species that had never previously met a weasel or a stoat are very vulnerable to them, and in these places negative attitudes to exotic species are warranted (Chapter 13). Even so, conservationists battling against the depredations of introduced stoats in New Zealand often admit a grudging admiration for the energy, speed, and skill of their little adversary.

ANOTHER VIEW OF THE WEASELS

Weasels have not always been viewed entirely negatively. In early Europe and in cultures not concerned with the preservation of game birds, the weasels have been much appreciated for their willowy elegance and for their services in killing rodents. Many of the traditional vernacular names for the common weasel in Italy and Spain are complimentary, such as *donina* (little lady or little graceful one), *bonuca-mona-muca* (pretty little one), and *comadreja* (little godmother). The ancient Egyptians apparently used to keep tame weasels (species unknown) in the period before the domestication of the cat, and the modern common weasel in Egypt is still so often found in houses that it is described as "almost completely commensal" (Osborn & Helmy 1980).

The indigenous people of the Chugache tribe of Cook's Inlet in Alaska regarded the capture of a least weasel as a piece of great good fortune, and they told F. L. Osgood that, since he had caught one, he must be destined to have great wealth and power (Hall 1951). Members of the plains tribes (Cheyenne, Lakota, and others) held weasels in high regard, especially those that were captured in their white winter coats. They decorated dress clothing, head dresses, pipes, and even their hair with ermine skins (see Figure 3.7).

Weasels certainly do kill to live, but they are neither vicious nor insatiable. At least in their native range in the northern hemisphere, the old attitudes toward weasels are simply wrong: Death is not a tragedy in nature; it is part of life. Unfortunately, the "instinctive hatred" that Wood talked about has filtered down the generations so subtly that the average person's knowledge of weasels is still influenced by old folk-stories of dubious accuracy. The effect of cultural attitudes can be demonstrated when very young children (in early primary school), whose parents and teachers have not yet indoctrinated them with negative ideas about predators in general and weasels in particular, react more positively than do older children when introduced to weasels (Powell & Powell 1982).

More recently, weasels as literary characters have been somewhat better treated. Occasionally, they have starred as heroic figures, as in the novels *Kine* and its sequel *Witchwood* (Lloyd 1982; Lloyd 1989). In *Kine*, set in rural Britain, where weasels are native and minks are introduced, it is the *minks* that are cast as the villains, not the weasels. (After being rebuked for this insult by mink biologists, Lloyd transferred to rats the role of the villains of *Witchwood*.) Modern nonfiction books for children that portray the elegance and intelligence of weasels, such as *A Bold Carnivore* (Powell 1995) and the Native American and related stories, such as *Crow and Weasel* (Lopez 1990), are well received and widely read (Figure 1.2).

Trappers in North America sometimes shared their camps with weasels. One fur trapper who accidentally caught a live long-tailed weasel took it home and kept it for a week. "If lightning is any quicker than a weasel, the margin is of microscopic breadth," he concluded, after watching it rocketing around its cage and eagerly fronting up to a terrier through the bars. When he released it, it hopped off in a leisurely way, so sure of itself that "its calm . . . movements gave no suggestion of the electric potentialities embodied within the elongated anatomy of this testy little carnivore" (Edson 1933).

Manly Hardy reported in the early 1900s of sharing his trapping camp with a weasel, showing that even hard-bitten trappers were not beyond enjoying the company of a small companion (Seton 1926:605).

Figure 1.2 Native American peoples honored wildlife, including weasels, and frequently incorporated animals into their names (e.g., Standing Bear, Little Wolf). The story of Crow and Weasel is told in the Native American tradition by Barry Lopez and illustrated with paintings by Tom Pohrt. (From *Crow and Weasel* by Barry Lopez, with illustrations by Tom Pohrt. Text copyright © 1990 by Barry Holstun Lopez. Illustration copyright © 1990 by Tom Pohrt. Reprinted by permission of North Point Press, a division of Farrar, Straus and Giroux, LLC.)

October 30, Sunday: Still blows and spits snow. Although alone in the camp, I have company; as, soon after we got settled, a Weasel came to us. At first his back was 'malty'—as we often call a blue-gray—but he soon changed to pure white. He is very tame and seems to like company, as, on our coming home, we sometimes see him running toward the camp. Often when I am alone cooking, he will come halfway out of a knot-hole in the floor, and look me in the face, letting me talk to him as I would to a Cat or a Dog. We have not seen a Mouse, Mole, or Squirrel near the camp.

One day, Hardy began to set a trap for minks:

This was some 50 yards from our camp, but our weasel must have seen me go with the trap and bait from our camp, as I had hardly begun to set the trap behind a large spruce before the Weasel came. When I was on my knees fixing the trap, he would look me in the eyes, not 2 feet from my face. As I knew he would be at the bait as soon as I left, I purposely set it too hard for him to spring, for I would not have caught him for the price of several Mink.

R.A. Powell (unpubl.), while doing research with his wife Consie in Upper Peninsula Michigan, had a stoat join his household. The basement of the house was reached either via internal stairs from the living room or stairs outside that were usually covered by a large door laid flat to the ground. Roger stored snowshoe hares and other bait for live-traps for fishers in that outside stairwell during winter. One day, Consie entered the stairwell from the basement and noted a perfectly round nest made of white hare hair, and the perfect size for a stoat. A few evenings later, she looked again at the nest and found the fellow sleeping, a white weasel curled in a nest of soft, downy, white hair. He could leave the stairwell at the top by sneaking through a crack between the door and the ground. He could also enter the basement because the door to the stairwell from the basement did not fit well. That winter, the Powells were raising lab mice to feed their fur-farm ferret, and mice had escaped into the basement. The resident stoat kept the mouse population under control in the basement.

Another winter, this time in northern Minnesota, a stoat took up residence in the big log cabin at the Kawishiwi Field Lab, a research lab for the U.S. Forest Service where Roger was working. The stoat would venture into the living room in the evening, while researchers sat around the fire in the fireplace. He would take strips of venison from their fingers. Roger observed that,

If I did not release a strip of meat when he pulled with his teeth, he would push against my hand with his feet. If I still held firm, he would let go of the meat and feint an attack on my hand, sometimes hitting my fingers with his teeth but never biting. When he first tried this trick, I dropped the meat. Thereafter, I held firm. Ultimately, he would eat the meat while I held it.

Allan Brooks reported a similar experience when he shared his camp with a stoat (Seton 1926:606):

> I have brought one to eat out of my hand within three hours of making its acquaintance, and this without confining it in any way. This was a female, and later she became a great nuisance. She generally showed up a little before midday and left about three o'clock to continue her rounds. If I happened to be skinning birds, she became greatly excited, and would rush in and try to drag the body from my fingers. A male which used to visit my cabin in the early morning never became so tame. He was a fine specimen of his kind and amazingly strong. He could drag a grouse several times his own weight a long distance over the snow.

Val Geist (1975) also shared life with a weasel while living in a remote cabin in Alaska, studying mountain sheep. He wrote (in the third person) of his daily routine in winter:

> On a winter day . . . he would take the saw and cut a slab off the frozen quarter of moose that he took from the meat house. The meat would rest and melt on a plate on the window all day from where [he] would pick it up in the evening, unless of course the weasel came first, in which case [he] would pick it up from underneath his bed—the meat, that is, not the weasel. As [he] sawed through the meat, the whiskey jacks would come and land on his shoulders or head, or wait beside him, then dart forward to pick up a bill full of meat meal, or a piece of meat the man handed to them. By tradition, the meat trimmings were theirs, as well as the scraps of fur and sinew or whatever else [he] handed them. A few scraps of melted meat were kept for the weasel, if and when it chose to show up and run to him for a free handout. It came often.

Anyone fortunate enough to know a weasel at close quarters soon becomes captivated by its charm (Figure 1.3). In England, Phil Drabble (1977) described his common weasel as "a sprite . . . a golden leaf on the tongue of a whirlwind"; and David Stephen (1969) had two that were an "intriguing, entrancing, entertaining pair of will o' the wisps . . . doing a wall-of-death act round my sitting room." Weasels are difficult to observe in the wild, and although with luck one may witness some fascinating glimpses of their activities, seldom can one get so

Figure 1.3 One never knows quite what will happen next when sharing one's home and life with a captive weasel. Roger and Consie Powell with Lew (Leweasel), a male long-tailed weasel, who treated them like part of the furniture.

good a view as Drabble had of his tame Teasy. His description conveys a vivid impression of the delightful character and restless energy of weasels:

> From whichever retreat hid him for the moment, a wedge-shaped head and wicked pair of eyes would appear. Then out he'd roll, turning cartwheel after cartwheel like an acrobat going round the

circus ring. He moved so fast that it was impossible to distinguish where his head began and his tail finished. He was like a tiny inflated rubber tyre bowling round the room. Sometimes we thought this game was purely for exercise, since we could distinguish no pattern. . . . Sometimes the weasel used his dance as a cloak for attack. . . . He usually chose me for his victim, and his cartwheeling twisted this way and that, over the carpet and up on to the settee beside me. The fabric . . . was cut and ragged. . . . It made a perfect foothold for the weasel, who could run up and down the perpendicular arms of the furniture with the ease of a squirrel. . . . When his gyrations fetched him up on the seat beside me, I always knew what the next act would be. . . . The tiny scratching of the pen and the movements of my fingers were irresistible. From the cover of his dynamic camouflage, he could dive on to my hand, grasp my first finger in his forepaws with the strength of a tiny bear, and bite the fingertip with mock ferocity but, in reality, as gently as a kitten. . . . If I tickled his belly he'd roll on his back and attack as if his very life depended on it. Then he'd gradually relax, until he was licking the tips of my fingers and croon his high-pitched little purring love-song.

Modern research has rapidly increased our knowledge of the lives of weasels, especially since the 1960s. Rigorous, systematic observations have replaced the more casual, subjective accounts, and experiments have replaced the verbal descriptions, which used to be all we had to go on. Something will be lost if the objective approach is taken too far, but it does help to interpret the old stories, field observations, and anecdotes collected by lucky chance, and to eliminate longstanding fallacies. The modern weasel-watcher needs to move in both worlds, that of systematic, critical analysis of data and that of hours of patient fieldwork, waiting for the occasional, exciting glimpses of undisturbed weasels behaving naturally in the wild. We have tried to integrate both in this book.

THE WEASEL FAMILY

The Mustelidae, the weasel family, is large (some 70 species) and includes the weasels, minks, ferrets and polecats, martens, badgers, and otters. They all have, to some degree, the long, thin body and short legs that are characteristic of weasels. The genus *Mustela* is the largest in the family and contains some 13 to 16 species, depending on which taxonomist one asks. Besides weasels, this genus includes the ferrets or polecats, and the minks. These groups are generally divided by size and coloration, which match evolutionary relatedness only in part. Weasels are usually considered to be those that are small, brown on the back, and white to cream to yellow on the belly, and that specialize in hunting the smallest of the rodents. Ferrets and polecats are generally larger than weasels and have distinct, black masks on their faces and black legs and tails. Minks are dark brown all over, except for light patches on their chins or chests, and are adapted to foraging in streams and wetlands. They live mostly on aquatic and semi-aquatic prey such as fish and frogs, water voles, and water birds.

Weasels live on all continents except Australia and, of course, Antarctica. The three species that are most common throughout the north temperate and boreal zones are by far the best known weasels: the short-tailed weasel or stoat, *Mustela erminea*; the long-tailed weasel, *M. frenata*; and the least weasel or common weasel, *M. nivalis*. These three species are the subjects of this book. At least six other species may have lifestyles similar to the three best known ones, and are sometimes called weasels (Macdonald 2001): the tropical, *M. africana*, and the Columbian, *M. felipei*, weasels of South America; the mountain, *M. altaica*, yellow-bellied, *M. kathiah*, back-striped, *M. strigidorsa*, and barefoot, *M. nudipes*, weasels of Asia; and the Egyptian weasel, *M. subpalmata*, if it is truly a species distinct from *M. nivalis* (van Zyll de Jong 1992; Reig 1997; Abramov 2000; Abramov & Baryshnikov 2000). An undescribed, small weasel has been reported from the island of Taiwan recently, but may be a form of *M. nivalis* (Hosoda et al. 2000).

COMMON NAMES

Weasels of all three species are distinct and easily recognized but, unfortunately, the common name "weasel" can be confusing unless carefully defined. At one extreme, the name can be applied to all members of the weasel family; at the other, it can mean only the smallest member of the genus *Mustela*, *M. nivalis*. In North America, "weasel" generally refers to any of the three small, native *Mustela* species, while in England and New Zealand, "weasel" is reserved for *M. nivalis*, and *M. erminea* is called "stoat." In Europe it might be correct and acceptable to apply the American term "ermine" to all European *M. erminea*, a word that is similar to the French common name "hermine" or the German "Hermelin," but it would not be correct to apply the American name "least weasel" to European *M. nivalis* except in the far north. So when someone refers to a "weasel," one needs to know the nationality (or accent) of the speaker to know whether the subject is *M. nivalis* only or any of these small, skinny, graceful critters. Equally confusing, the common name "ermine" is used by some to mean *M. erminea* only, and by others to mean a weasel of any species when it is in white winter coat.

Using Latin names avoids all confusion, of course; that is, after all, their function. But Latin names can be stultifying. In this book, we use the word "weasel" alone to mean any of the small *Mustela* species or all of them in general. We refer to *M. frenata* as the "long-tailed weasel" or "longtail." In North

America, *M. erminea* is the "short-tailed weasel" or "short-tail," but in the rest of the world, and in most of the scientific literature, it is called the "stoat," so we use "stoat" here. We use "least weasel" to refer to *M. nivalis* in North America, Asia, and eastern and far northern Europe, and "common weasel" for *M. nivalis* in Britain, New Zealand (where they were introduced from Britain), western and southern Europe, and northern Africa.

In parts of England, country folk will swear that there are two species of common weasels, the normal one plus a smaller one known as a "grass weasel," "finger weasel," "mouse hunter," or "miniver." Museum biologists have tried to collect specimens of these small weasels, but without success. The best explanation is that people easily become confused by the large difference in size between males and females of the same species, exaggerated by the slow growth rate of young born late in the year, and by imprecise use of common names.

THE EVOLUTIONARY ORIGINS OF THE WEASELS

Weasels belong to the canoid group of placental carnivores, which originated in the New World (Flynn & Wesley-Hunt 2005). The first predatory mammals with characteristics of the weasel family, and clearly different from their closest relatives the Procyonids (Bininda-Edmonds et al. 1999), appeared in North America in the early Miocene, some 28 to 30 million years ago (Figure 1.4). Throughout the Miocene period, these animals were forest-dwelling hunters, probably somewhat like martens. Some mustelids must have also have reached Eurasia by the beginning of the Miocene, because diverse mustelids were by then living in both North America and Eurasia. The oldest fossil of a mammal clearly belonging to the genus *Mustela* comes from eastern Eurasia, dated to late Miocene, and an independent lineage of mustelids very similar to the first weasels appeared in North America at roughly the same time. Extensive movement back and forth between North America and Eurasia (Wederlin & Turner 1996) set the scene for an active evolutionary radiation within the Mustelidae over the next 15 to 20 million years. Among the early (by mid-Miocene at least) migrants to travel west from America must have been the far-distant ancestors of the stoat and common/least weasel (Hosoda et al. 2000), while the distant ancestors of the longtail remained at home.

By the early Pliocene (5 million years ago), at least three separate lines of true *Martes* were well established, as well as other species intermediate between *Martes* and *Mustela*. Throughout the Pliocene, extensive open savannahs began to develop in both North America and Eurasia, as the climate cooled toward the approaching glacial periods and the grasses evolved and progressively replaced the forests. The grasslands were soon populated by the early species of voles and lemmings, and then by the ancestors of the weasels. It seems likely (King 1983a; King 1984a) that those early weasels, descended from the larger,

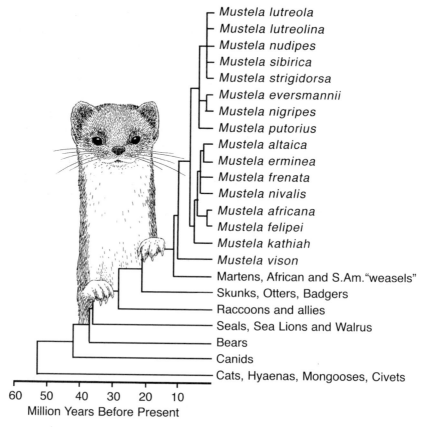

Mustela lutreola
Mustela lutreolina
Mustela nudipes
Mustela sibirica
Mustela strigidorsa
Mustela eversmannii
Mustela nigripes
Mustela putorius
Mustela altaica
Mustela erminea
Mustela frenata
Mustela nivalis
Mustela africana
Mustela felipei
Mustela kathiah
Mustela vison
Martens, African and S.Am. "weasels"
Skunks, Otters, Badgers
Raccoons and allies
Seals, Sea Lions and Walrus
Bears
Canids
Cats, Hyaenas, Mongooses, Civets

60 50 40 30 20 10
Million Years Before Present

Figure 1.4 A phylogenetic tree for the weasels developed from morphological, fossil, and biochemical data. This and most other phylogenetic trees made using different data agree consistently that *Mustela erminea* is most closely related to *M. altaica* and not most closely related to either *M. nivalis* or *M. frenata*. (Redrawn from Bininda-Edmonds et al. 1999.)

marten-like mustelids already existing, discovered the advantage in becoming small enough to exploit the new niche for predators able to get into the burrows and runways of voles, mice, and lemmings.

The living forms most similar to ancestral weasels are probably either the modern stoat or the Asian mountain weasel (*M. altaica*). An ancestral stoat, *M. plioerminea*, appeared in Eurasia in the Pliocene, some 4 million years ago (Kurtén 1968). During the long transition between the end of the Pliocene era and the beginning of the Pleistocene, and throughout the first of the four great glacial epochs conventionally recognized and the first (Cromerian) interglacial epoch, an intermediate species called *M. palerminea* was common. Around that time, the evolutionary lineages leading to stoats and mountain weasels separated

(Bininda-Edmonds et al. 1999: Kurose et al. 2000). Definite specimens of the modern species *M. erminea* date only from the time of the third major glaciation, which started about 0.6 million years ago, and are common in European deposits dated to the last full glaciation (Sommer & Benecke 2004). *M. erminea* returned to its ancestral homeland via the Bering Land Bridge into North America roughly half a million years ago, and now occupies about 13 million km² of the United States and Canada (Fagerstone 1987).

The earliest form of the common weasel was *Mustela praenivalis* (probably derived from *M. pliocaenica*) (Kurtén 1968), which appeared in Eurasia some 2.6 million years ago (Bininda-Edmonds et al. 1999) and survived for nearly 2 million years alongside *M. palerminea*, until the Cromerian interglacial epoch. The transition to *M. nivalis* was gradual, but was completed by the time of the second major glaciation (less than half million years ago). The modern form is found among a forest fauna of that age at West Runton, Norfolk, eastern England, and is common in European cave deposits dating from the last glaciation (Yalden 1999; Sommer & Benecke 2004). In the late Pleistocene, during the last glaciation (van Zyll de Jong 1992; Abramov & Baryshnikov 2000), *M. nivalis* also crossed the Bering Bridge, and followed *M. erminea* back into North America. It now occupies about 7.5 million km², mostly in Canada.

M. frenata appeared quite abruptly in North America more than 2 million years ago, before the first glaciation. It has therefore survived in its present form much longer than either of the other two modern species, and it has the longest stratigraphic record and widest fossil distribution (>30 locations) of any North American weasel. Its present range is about 8.5 million km², mostly in the United States. Its ancestor is unknown: The only suggested candidate so far is *M. rexroadensis*, a medium-sized weasel, known only from two fossil sites containing mammal faunas dating from around 3 million years ago (Kurtén & Anderson 1980). The immediate origins of *M. rexroadensis* are unknown.

During the cold phases of the Pleistocene, the weasels found themselves already adapted to live under snow, finding there both food and shelter from the killing cold above. They were already the right shape to burrow through soft powder snow on the surface, dive through the deeper layers, and follow the tunnels made by rodents on the ground underneath, just as they do today (Chapter 2). Most of the huge ranges of the stoat and least weasel, and the northern end of that of the longtail, still lie within climatic zones having a severe winter with prolonged snow cover (Figure 1.5). Weasels still live all year round at elevations of 2,000 to 3,000 m or more in the snows of the high mountain ranges of, for example, the Sierra Nevada of California, the Alps of Europe, and the Caucasus, Altai, and Tien Shan of Asia.

The interesting implication of these family histories is that, although the stoat and the longtail look and behave alike, they not closely related. Their far distant common ancestor came from America more than 4 million years ago, before *M. plioerminea* existed. The modern forms evolved separately, one in Eurasia

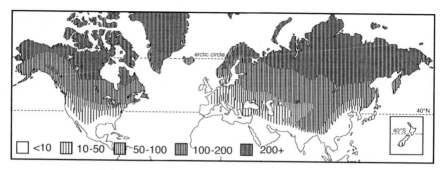

Figure 1.5 The approximate distribution of snow cover in countries inhabited by weasels, shown as the average number of days per year with snow lying on the ground in the morning. (Note: New Zealand, inset, is slightly enlarged relative to the Northern Hemisphere.)

and the other in North America. Combined phylogenetic analyses of the contemporary species show that the most closely related pair is the stoat and the Asian mountain weasel (Bininda-Edmonds et al. 1999); the longtail is about equally distant from these two as from the common/least weasel. The next closest relatives are the New World tropical and Colombian weasels. Yellow-bellied weasels are more distant relatives of the central group of four, and the back-striped and barefoot weasels appear to be closer to minks and polecats than to the rest of the weasels. Other authors (Sato et al. 2003; Abramov 2000) have produced phylogenies differing slightly from this one, because the details depend on the time scale of interest, the methods used, and which species are included, but all agree on the main outlines of mustelid history.

A Little Bit About Snow

Because snow has been so important in the evolutionary history of weasels, it deserves some special attention. Snow provides, in a sense, another world, the subnivean world. Just the name "subnivean" sounds a bit like Middle Earth or Narnia, and perhaps with good cause. The subnivean world is a chilly but safe haven from the bitter cold and fierce storms that rage through the northern winter.

Snow is like a duvet or down comforter, because both hold much air. That is why both are such good insulators (LaChapelle 1969), and why the ground under a stable snow blanket is relatively warm. In temperate climates, ground temperature under snow may be up to 10°C, although in northern boreal forests and tundra it is closer to zero. Nonetheless, the ground is never as cold as the subfreezing air temperatures of mid-winter. Snow holds in warmth, protecting the small mammals and insects that remain active at the ground surface throughout the winter. Snow also transmits light, though the attenuation is great. Animals emerging out of their thickly insulated nests and underground burrows to move about under shallow snow can still tell the difference between day and night.

Most important, a snow blanket is not the same all through from top to bottom. As one snowfall follows another, the warmth of the ground and the accumulating pressure changes the structure of the original six-sided snow crystals. At the ground surface, what might have fallen as fine powder snow becomes granular, much like sugar, so that deep layer is called depth hoar or sugar snow. Voles, mice, and weasels can push through it easily, forming tunnels and small rooms of crystal between the ground surface and the more solid blanket above, and forging new pathways (see Figure 6.2).

Weasels evolved their superb tunnel-hunting abilities in response to the opportunity to hunt small mammals that formed burrows and runways in grass, and for them this subnivean world is no different than life under a thick layer of grass. But if they had not been able to make use of the insulation supplied by snow, they might have joined the many other species of animals that did not survive the savage cold of the glacial periods. Least weasels and stoats are completely at home in snow, and common weasels and longtails can burrow through it when they need to, although most common weasels and longtails live in more southerly latitudes where snow falls less often and lies less thickly than in the far north.

PATTERNS OF POSTGLACIAL COLONIZATION

As each new glacial period took hold and the volume of ice built up in the north, the world sea level fell by various amounts, at worst by more than 120 m. Wherever rain or snow was insufficient to build up glaciers, vast ice-free plains formed on land abandoned by the falling seas, such as across Beringia, the exposed land between northeastern Asia and Alaska, and over much of central and eastern Eurasia (Kurtén 1966) (Figure 1.6).

Entire ecosystems moved southward, along with the climatic zones to which they were adapted. The periglacial environment supporting the tundra vegetation and lemmings, which we now think of as typical of the far north, became established where oaks and mice had previously flourished. Weasels of all three species were often included in these displaced communities. Conard Fissure, Arkansas, has fossils of both stoats and long-tailed weasels of glacial age (Kurtén & Anderson 1980) and stoats lived in southeastern New Mexico during terminal Pleistocene times (Harris 1993), even though their nearest living relatives are now found hundreds of kilometers north of these places. The broad expanses of what is now the southern United States provided ample refuge for the displaced faunas of glacial times; other populations of stoats survived in Beringia and in ice-free patches of northern Canada (mapped in Eger 1990).

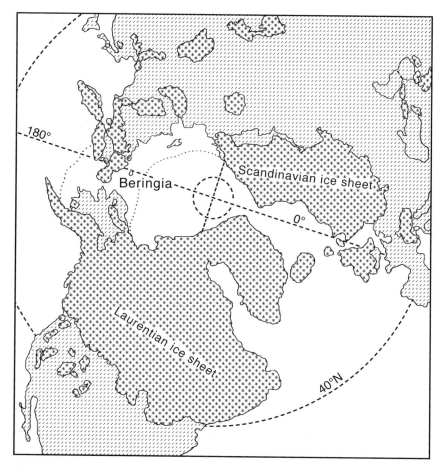

Figure 1.6 The approximate extent of the last major glacial advance, north polar view. The extent of Beringia is shown by dotted lines. (Redrawn from Ehlers & Gibbard 2004.)

The story was the same in Europe, except that the east-west chain of mountain ranges stretching from the Pyrenees to the Caucasus divided the available refuges south of the permafrost zone into three distinct areas, represented by the present Iberian, Italian, and Balkan/Grecian/Turkish peninsulas (mapped in Hewitt 1999). Stoats must have lived in Britain, and on the exposed continental shelf south of the present coast, during the last interglacial, because remains dating from 35,000 years ago have been found in Castlepook Cave in Ireland (Stuart 1982). During the last glaciation, large populations of stoats and common weasels would have hunted voles and lemmings across the cold and windswept plains immediately south of the ice front (Sommer & Benecke 2004) and in the extensive southern refuges.

As the glaciers retreated, entire temperate ecosystems slowly migrated north-ward again. First the tundra, then the coniferous forests, then the temperate forests gradually covered the denuded landscape. Small mammals and birds, accompanied by their predators, moved north with the returning vegetation. But the population characteristics of the weasels that returned to the far north after the glaciations were not quite the same as those of their ancestors that had moved southward. When populations of any species are isolated by geographic barriers for many generations, they diverge both physically and genetically. Populations of animals that had sat out the glaciations in different places and returned via different routes had developed differences in genetic composition, which can still be detected by analysis of their descendants. Modern techniques make it possible to reconstruct the routes they took, and, in several key places around the northern hemisphere, what happened when they met.

For example, small rodents are the favorite prey of weasels, especially the various species of voles and lemmings. In North America, stoats and least wea-sels must have recolonized the newly reestablished forests as members of the community of small mammals and birds on which they depended. The Pacific Northwest has been a crossroads for postglacial colonists arriving from differ-ent directions. The early-succession generalist herbivores came first, along with their predators, from three main refuges—from Beringia via Alaska, and from the south along both eastern and western sides of the continent (Figure 1.7). In southwest Alaska, the postglacial histories of the long-tailed vole (*Microtus longicaudus*) and the stoat have been analyzed by Conroy and Cook (2000) and Cook et al. (2001), with fascinating results. For both species, there is evidence of two routes of migration from the south, one inland and one along the west coast, plus a separate migration from Beringia in the north. The differences between contemporary populations of stoats in northwestern North America do appear to reflect these different origins and routes of recolonization. By con-trast, genetic differentiation among the stoats of the Palaearctic appears to be relatively low (Kurose et al. 2005).

Not only is genetic evidence strong for at least three quite separate lineages of stoats now living in the Alexander Archipelago, but also there are morpho-logical differences visible even between adjacent populations elsewhere in the Northwest. For example, the skulls of stoats on Vancouver Island differ from those on the adjacent mainland (Eger 1990), and stoats from coastal Oregon never turn white in winter, even when translocated to Alaska, whereas those from inland Oregon always do (Feder 1990).

In northern Europe, geographical patterns of genetic relatedness among populations of voles and weasels and other species are nicely documented. Jaarola and Tegelström (1995) found a distinct difference between the field voles (*Microtus agrestis*) living in Finland and northern Sweden and those in south-ern Sweden, Denmark, and Britain. They concluded that the two groups are derived from populations that survived the glaciations in different places and

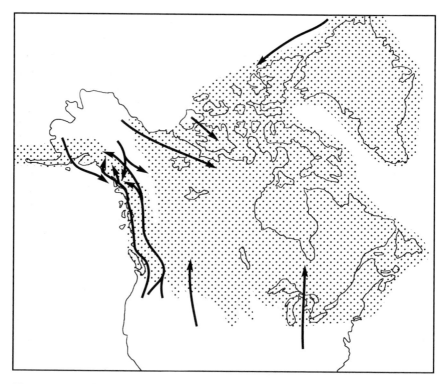

Figure 1.7 Possible routes (*arrows*) taken by flora and fauna recolonizing northern North America after the most recent retreat of the glaciers (*stippled area*). (Redrawn from Eger 1990: fig 1 and Conroy and Cook 2000: fig 4b.)

then recolonized Scandinavia from two different directions, the south and the east. The southern migration moved north via the late glacial and postglacial land bridges connecting southern Sweden with the Danish islands and Germany. The eastern migration approaching from Russia swung through Finland, around the northern Gulf of Bothnia, and south again into Sweden. Where the two advancing fronts met, they established a hybrid zone that still persists at the genetic level, although in voles the differences between them are not visible to the eye and raise no taxonomic questions.

Very similar postglacial histories have been established for other members of the same assemblage of glacial fauna, which presumably all returned northward in the same two separate streams. Examples include the brown bear, the shrew, the bank vole, and the water vole (Hewitt 1999; Jaarola et al. 1999). Naturally, weasels were part of this community, and they recolonized Europe along the same two routes (Figure 1.8). One group, which we now recognize as *Mustela nivalis vulgaris*, colonized all the southern and western countries of Europe; the other, now called *M. n. nivalis*, came from the east and north, en-

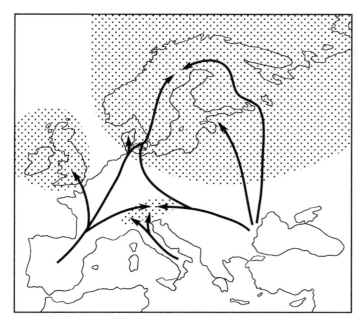

Figure 1.8 Possible routes (*arrows*) taken by flora and fauna recolonizing northern Europe after the most recent retreat of the glaciers (*stippled area*). (Redrawn from Hewitt 1999 and Jaarola et al. 1999: fig. 5.)

tering Scandinavia via Russia and eventually turning south again through Norway and Sweden. The hybrid zone marking where they met in southern Sweden is still the best place to observe the visible differences between these two subspecies that developed during their separation, including, most obviously, that one turns white in winter while the other does not (Stolt 1979). Despite this high level of intraspecific variation, the different forms of *nivalis* did not develop any reproductive isolating mechanisms. The two lineages are still genetically distinguishable (Kurose et al. 2005), but as taxonomic units they remain distinct only at the subspecies level. Common weasels in Poland and the former Czechoslovakia are classed as *M. n. vulgaris*, but with so much geographical variation that they seem to combine the characteristics of both subspecies (Schmidt 1992; Zima & Cenevova 2002).

The weasels that now live in Morocco, Algiers, and Tunisia are similar to those of southern Mediterranean Europe, so are classed as *M. n. vulgaris*, but they have probably not been there very long. The Straits of Gibraltar have maintained the geographic separation of Iberia from North Africa throughout the Pleistocene and ever since, so the ancestors of these weasels were probably introduced with human help (Dobson 1998). By contrast, the weasels in the lower Nile Valley of Egypt belong to a different subspecies (*M. n. subpalmata*). They have certainly been isolated for a long time, and are so different from other weasels classed as *M. nivalis* that they should perhaps be promoted from a sub-species to a separate full species (van Zyll de Jong 1992; Reig 1997; Abramov 2000; Abramov & Baryshnikov 2000). They may represent one of the refuge populations that survived the glacial periods in the south, whose descendants have stayed put rather than migrate north. As Egypt dried out, they escaped the encroaching desert by adapting to live in towns and gardens, and even in houses (Osborn & Helmy 1980).

The idea that the morphological variation we see today can be interpreted in terms of glacial history is an old one (Macpherson 1965). Others have argued that although stoats, for example, recolonized the deglaciated areas of North America from different stocks, the present differences among their descendants might just as easily be explained as adaptations to local environmental conditions. Eger (1990) used a sophisticated analysis of 13 craniometric characters of stoats collected from across North America to distinguish between these possibilities. She concluded that present variation in skull *size*, which reflects body size, one of the key characters of life for any animal (Chapter 4), is better explained by local climatic conditions than by ancestry. Natural selection keeps a relentless scrutiny on all creatures, and those that, like weasels, maintain high levels of genetic variability and breed rapidly are able to respond quickly to changes in local conditions. That means that their most important physical characteristics, including body size, say more about their present environment than about range changes and recolonization in response to glacial advance and retreat. This process has long ago superimposed contemporary local adaptation upon any size differences that might have developed during isolation in glacial refugia.

On the other hand, the fine details of differences in skull *shape* have no such dire ecological consequences, so it might be expected to retain traces of differences in ancestry for much longer. Eger's analysis confirmed that this is so. Variation in skull shape of North American stoats is not closely correlated with local climate; rather, discontinuities in geographic pattern are, indeed, consistent with the refugium hypothesis. Eger identified distinct groups of samples that were probably descended from three different ancestral refugial populations in Beringia, in eastern North America, and on the west coast.

Glacial history helps explain a lot about modern populations, including puzzling observations that cannot sensibly be explained any other way. In Europe during glacial periods, the least weasel, the northern form now typical of colder climates, once occupied the continental plain north of the Alps and also the valleys between the high mountains of Switzerland and Austria. As the glaciers melted, the mountains provided a cold-climate refuge to which least weasels could retreat, and they survive into the present day, long after the surrounding lowlands had been reclaimed by the common weasel moving north from warmer climates. In Switzerland, the least weasel (often called the pygmy weasel there) is now found only at high elevation, while common weasels occupy the intervening valleys (Güttinger & Müller 1988).

The difference in appearance between the two hinges on genetically determined characters; the local details of their distributions confirm that these differences are determined by history rather than contemporary conditions. For example, in some places both whitening and nonwhitening races of weasels live in similar environments, and the conditions associated with whitening in stoats seem to be quite different from those associated with whitening in least weasels living nearby (Güttinger & Müller 1988). These differences seem to be more easily explained on the grounds of ancestry than of ecology (Chapter 3).

SPECIES DIAGNOSES

The general weasel appearance is common to all three species (Figures 1.9. and 1.10). The differences among them hinge mainly on size, color, and reproduction (Table 1.1). Size and color are obvious to the eye, so one might think that they would be easily defined and reliable guides. Most of the early taxonomists based their species descriptions on size and color, often using the standard practice of describing whole species—or, at any rate, whole populations—from one or a few "type" specimens. But the three weasel species between them occupy an enormous geographical range (Figure 1.11), across which climate, habitat, and prey vary equally enormously. Under these different conditions, each weasel species has evolved considerable variation in external appearance, especially in size (Chapter 4), in the details of summer pelage, in the length and color of the tail (Table 1.1), and in whether or not they turn white in winter (Chapter 3).

Until the early 1800s, the names of European weasels were applied to their North American counterparts, except that the species' distinctions were not made consistently. From then into the twentieth century, many local variants of North American weasels, and sometimes even brown and white individuals of the same species in the same place, were described as separate species. The "type-specimen" method of nineteenth-century taxonomists led to glorious confusion and to dozens of new species and subspecies descriptions. Hosts of new names confound the literature of that period. As more and more specimens were collected, however, taxonomists realized that *M. erminea* is a circumboreal species and should have the same scientific name in both North America and in Eurasia. During the second half of the twentieth century, least weasels in North America have also become generally accepted (but not by Reig 1997) as belonging to *M. nivalis*, instead of to the separate species *M. rixosa*.

The color patterns of weasels are not much help in distinguishing the species, because they vary so widely. In summer coat, the head, back, all or most of the tail, and the outsides of the legs of almost any weasel are rich chestnut in color, but range from sandy tan to dark chocolate. The throat, chest, and belly range from pure white through creams and yellows almost to apricot in some longtails. The feet often have white markings, and the outer edges of the ears

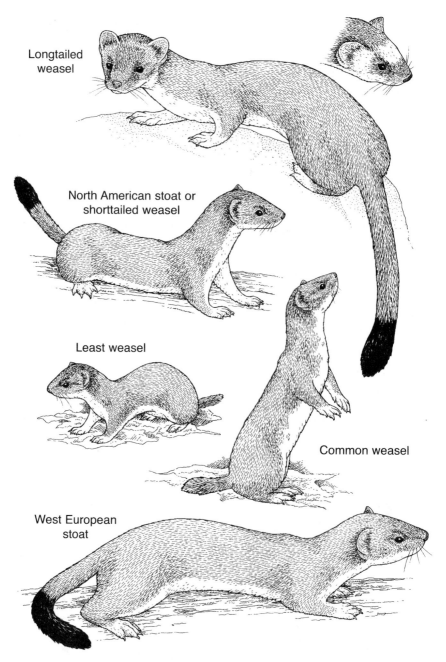

Figure 1.9 Least weasel (*Mustela nivalis nivalis*), common weasel (*M. nivalis vulgaris*), North American stoat and west European stoat (*M. erminea*), and long-tailed weasel (*M. frenata*) in summer coat. The inset shows an example of the facial patterns of long-tailed weasels, often called bridled weasels, in the North American Southwest. European stoats are about the same size as longtails, distinctly larger than North American stoats. The five weasels are drawn to scale.

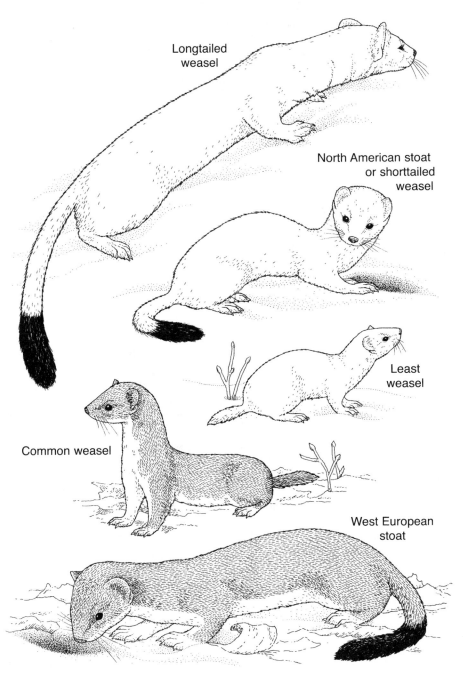

Figure 1.10 Least weasel (*Mustela nivalis nivalis*), common weasel (*M. nivalis vulgaris*), North American stoat and west European stoat (*M. erminea*), and long-tailed weasel (*M. frenata*) in winter coat. Common weasels never turn completely white in winter, while stoats, least weasels, and longtails do not turn white in the southern parts of their ranges where winters are mild. The five weasels are drawn to scale.

Table 1.1 Distinctions between the Species of Weasels[1]

	M. nivalis		M. erminea		M. frenata
	N. NIVALIS	N. VULGARIS	E. HIBERNICA	OTHERS	
Authority and date	Linnaeus 1766	Erxleben 1777	Thomas & Barrett Hamilton 1895	Linnaeus 1758	Lichtenstein 1832
English name used here	least weasel	common weasel	Irish stoat[2]	stoat	long-tailed weasel
Other names	grass weasel, miniver		Irish weasel	ermine, short-tailed weasel	
Summer coat demarcation line[3]	straight	wavy	straight or wavy	straight (Eurasia), wavy (N. America)	straight or wavy
Tail tip	brown	brown	black	black	black
White in winter	yes	no	no	maybe	maybe
Tail length[4]	<25%	<25%	30–45%	30–45%	40–70%
Diapause	no	no	yes	yes	yes

1. For species distributions, see Figure 1.11; for sizes, see Chapter 4.

2. This form is also present on the Isle of Man, Islay, and Jura.

3. Demarcation line: the line separating the brown from the white summer fur.

4. Tail length expressed as a percentage of head–body length.

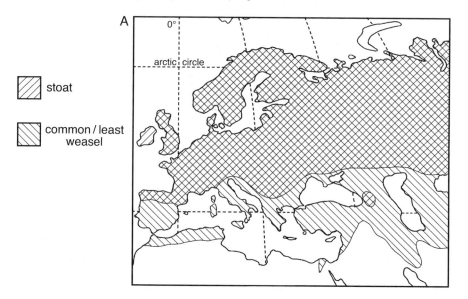

Figure 1.11 Global distributions of the three northern weasel species *Mustela nivalis* (common and least weasels), *M. erminea* (stoat or short-tailed weasel, ermine), and *M. frenata* (longtailed weasel). (A) Europe. (B) Asia. (C) North America.

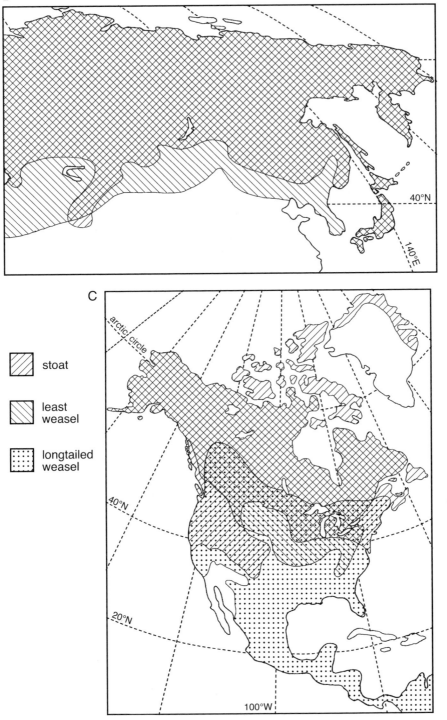

B

C

stoat

least
weasel

longtailed
weasel

40°N

140°E

arctic circle

40°N

20°N

100°W

Figure 1.11 (continued)

are often fringed with white in stoats. The boundary between the brown and white fur is irregular in common weasels, often with odd spots or large patches of brown among the white (Figure 1.12). This pattern is also found in many longtails, in many stoats and least weasels in North America, and in some 80% of Irish stoats.

Common weasels usually have a brown spot, called the gular spot, behind the angle of the mouth (sometimes joined by an isthmus to the rest of the brown area), though least weasels do not. On common weasels, and on other weasels with crooked belly lines, the details of the belly pattern are individually unique and constant through successive molts (Linn & Day 1966); on one common weasel live-captured 54 times (on the right in Figure 1.12), the pattern remained unchanged through two molts (King 1979). Stoats, longtails, and least weasels living in cold climates turn white in winter; stoats, longtails, and common weasels living in mild climates stay brown.

Every stoat and longtail has a distinct, bushy pencil of long, black hairs on the tip of its tail. This mark distinguishes the members of these species reliably

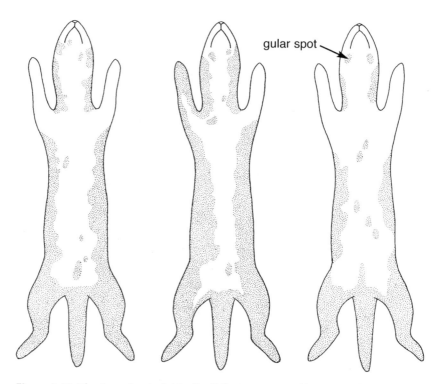

Figure 1.12 The irregular, individually distinct patterns of brown and white blotches on the bellies of three English common weasels. The unique patterns can be used to identify individuals. (Redrawn from sketches made by C. M. King of anesthetized, live animals in the field.)

from common and least weasels, though individual common and least weasels may have a few black hairs at the tips of their tails. Longtails in southwest North America have attractive patterns of white and brown markings on their faces, which explains why their alternative names in these parts of the country is "bridled weasels" (Fagerstone 1987).

Common and least weasels in Europe are very different in appearance, even in the few places such as Sweden and Switzerland where they live close together (Stolt 1979; Güttinger & Müller 1988). Only the least weasel is the true winter-white "snow-weasel" described and aptly named as *M. nivalis* by Linnaeus in 1766; evidently he did not distinguish the southern one, so that was named *M. vulgaris* by Erxleben in 1777. Later, *vulgaris* was reduced from full species to subspecies status, although Erxleben's name remains attached to the subspecific epithet. The two are now regarded as sibling subspecies of one polytypic species, *M. nivalis* Linnaeus. Yet in color patterns and winter whitening, in reproductive cycles, and consequently in population dynamics, they are more different from each other than stoats are from longtails, suggesting that the idea of recognizing them as two species has some merit (Reig 1997).

Two strong counterarguments to the two-species idea, however, hold greater weight. First, if size and color pattern are indeed critical characteristics separating the two "potential" species, they should vary together, but they do not. Mediterranean common weasels are all very large, but in some populations they have straight belly lines (Type I), in others they have crooked lines (Type II), and some populations have both (Figure 1.13). The two patterns are controlled by the two alleles of a single gene, of which the one producing the Type I straight line pattern typical of *nivalis nivalis* is dominant over the one controlling the Type II crooked line pattern of *nivalis vulgaris* (Frank 1985). Second, both forms have the same number of chromosomes ($n = 42$), are fully interfertile, and inherit size and belly pattern independently (Frank 1974; Mandahl & Fredga 1980; Frank 1985; Zima & Cenevova 2002). *M. nivalis* in Sicily have the same straight belly line as those in northern Scandinavia, but they weigh several times more.

If common and least weasels are clearly not separate species, yet still obviously are different, the only alternative policy is to recognize them as separate subspecies. In fact, the most recent review of the intraspecific taxonomy of *Mustela nivalis* recognizes 17 subspecies in three groups: a small *nivalis* type, a large *vulgaris* type, and large *nivalis* (*boccamela*) type (Abramov & Baryshnikov 2000). Ideally, it is best not to recognize subspecies as valid unless they represent incipient speciation (Whitaker 1970), so here we ignore most of the named subspecies of all three species.. But the differences between the *nivalis nivalis* group and the *nivalis vulgaris* group, which presumably developed while their ancestral populations were isolated in separate glacial refuges, may represent an advanced stage of the speciation process, interrupted just before it was completed. The modern common and least weasels are different in many ways but definitely still interfertile, as Frank (1985) proved, so under the conventional

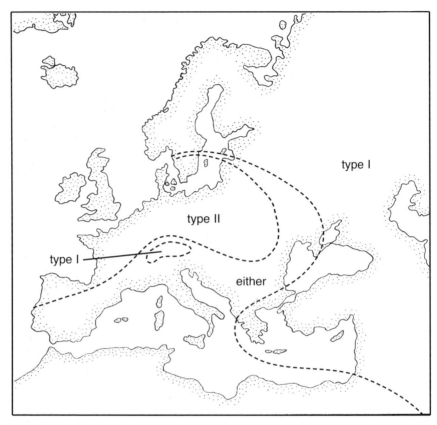

Figure 1.13 The approximate distributions of the two patterns of coat color of *Mustela nivalis* in Europe, designated type I and type II by Frank (1985). Figure 1.9 shows the type I pattern on the least weasel and the type II pattern on the common weasel. (Distributions deduced by King [unpublished] from the collection of skins in the Natural History Museum, London. Güttinger and Müller [1988] provided local detail for Switzerland. For a more detailed version of this map extending across the Palaearctic, see Abramov 2000b.)

definition of a biological species they both still belong to *M. nivalis*—unless they are found to maintain reproductive isolation in the wild. On the other hand, the differences between them are too strong to ignore.

We, therefore, continue to follow van Zyll de Jong (1992) in distinguishing two subspecies (or subspecies groups) of *M. nivalis*, labeled as *M. n. nivalis* and *M. n. vulgaris*. We do not reject the idea of a third (*M. n. boccamela*) group, but there is not enough information about the ecology of these large southern weasels to treat them separately here. This is not a very satisfactory solution, but the jury is still out on what might constitute the most useful definition of such a variable species, so at the moment we cannot suggest a better one. Recent analyses of morphological variation across the entire geographic range of *M. erminea* and *M. nivalis* suggests that both vary extensively across the entire northern hemisphere (Chapter 4), and that describing this variation is difficult even for taxonomists (Meia & Mermod 1992; van Zyll de Jong 1992; Abramov & Baryshnikov 2000). At present, the best we can do in this book is to discuss three species but four distinct forms.

2 | Hair-Trigger Mouse Traps with Teeth

THE WEASELS, THE smallest of the true carnivores, are among nature's more recent inventions. Both their relatively recent origin and their small size are due to their evolutionary profession, or "niche," as supreme specialists in the difficult art of hunting rodents in confined spaces. All the details of their ecology and behavior can be explained by the idea that they have evolved to exploit a particular kind of high-risk, high-reward resource—the boom-and-bust, unstable populations of voles and lemmings. They developed this skill well before the ice ages began, so that now they are among the very few predators that can overwinter in the simple and savagely inhospitable (to us) environments of the contemporary snowy north (Figure 2.1). They are outstandingly successful in this profession, and their natural distributions are among the largest of any mammals in the world.

The weasels have also spread south to milder climates, where they can find more stable populations of larger prey. There, the greatest problems are likely to arise not from fierce cold and unpredictable food supplies, but from the more constant dangers of competition and predation from other species. Nevertheless, their primary home is the circumpolar boreal region, and that is the key to understanding everything about them because it is where they display their marvellous adaptations to best advantage.

THE GENERAL ANATOMY OF WEASELS

In general body plan, the weasels are pretty much standard mammals. Weasels have the ancestral five-toed limbs and the usual arrangement of internal organs. Their specializations are mainly in the tools they need to hunt and to kill rodents in confined spaces—in the elongation and extraordinary flexibility of their spines, in their short limbs, and in the strength and shape of their skulls and teeth. These features put the weasels among nature's most interesting examples of the coevolution between predators and prey.

Weasels are much less conspicuous than the better-known big cats and wolves, and their predatory exploits are conducted on a scale that is less dramatic, at least to us; but in their own way they are just as spectacular as the big carnivores, or more so. They are much more widespread than large carnivores and easier to handle, and when common they are much more available and

Figure 2.1 The ability of weasels to forage under the snow is the key to their survival in the far north.

amenable as subjects for teaching and research. When their populations crash, which can happen at any time—a drastic problem if it happens in the middle of a graduate student's research—they epitomize the problems inherent in research on wild carnivores.

THE SKELETON

Everything about a weasel is attuned to the profession of hunting for small prey in dark, confined spaces. In motion, weasels appear almost boneless. We have seen weasels leap into a hole and then look out again in a single, fluid action so fast that the tail was not in before the nose came out again. A weasel can do this because the articulations between its vertebrae are so flexible that it can turn over and walk back over its own hindquarters. Living and working in tunnels are normal and natural activities for weasels; the short legs (less than half the length of the body) (Holmes 1980) swing through their normal arc in a space only twice the depth of the animal's head (Figure 2.2). In an "average-shaped"

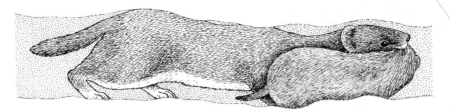

Figure 2.2 Weasels are perfectly adapted for hunting in tunnels. Their short legs allow a normal gait, and their long necks allow them to carry prey without tripping.

mammal, the legs and the body are about the same length, and in a confined space the legs are cramped, folded, and hampered in movement.

A weasel's neck is so long that prey can be carried in its mouth without tripping its front feet. No part of the skeleton of a weasel is wider than the skull, so that anywhere the head can go, the rest of the body can follow. The limb girdles at the shoulders and hips do not bear any great weight, so they are small; they fit with the sinuous lines of the vertebral column and allow great freedom of movement to the limbs. It is said that a common weasel can pass through a man's wedding ring. This tale may be an exaggeration for a living weasel, but it is not far off the mark for a skull.

The vertebrae of the neck are very strong and provide anchorage for the large muscles of the neck (Figure 2.3). The joint between the skull and the atlas (the first vertebra in the neck) is wide with smoothly curved surfaces. Each side of the atlas has a large flat wing, which receives muscles from the skull that move the head up and down. The axis (the second vertebra) is a completely different shape, compressed sideways rather than horizontally, because it receives a different set of muscles from the skull that allow a weasel to tilt its head.

The rest of the vertebrae are much smaller than these two, and number five more in the neck (total seven cervical vertebrae), 14 to 15 in the chest (the thoracic), 5 to 6 in the abdomen (the lumbar), 2 to 4 in the hips (sacral), and 11 to

Figure 2.3 The skeleton of a stoat. Weasels have elongate heads, necks, and vertebral columns and short legs. Otherwise, their skeletons are generalized and typical of mammals.

33 in the tail (caudal). The tails of least and common weasels are short (11 to 16 vertebrae, <25% of body length; see Table 1.1), for several reasons. One is that an important function of a tail, maintaining balance while jumping and turning (as in tree-dwelling mammals), is not needed in burrows; another is that a tail may be a source of serious heat loss, especially in very cold climates. On the other hand, a long thin tail carrying a deflection mark can decrease the risk of a weasel being caught by other predators (see Figure 11.6). The long-tailed weasel, living up to its name, has 19 to 33 caudal vertebrae, stretching its tail to 40% to 70% of its body length. The tails of stoats are intermediate in length (30% to 45% of body length) and in numbers of vertebrae.

Other than their very long necks and spines, weasels have a generalized type of mammalian body, with no particular specializations (Ondrias 1960, 1962; Holmes 1980). They have five toes on each foot, and walk with their heels on the ground in a plantigrade posture, much as the earliest mammals did. Their paws seem large for their body size—good for grabbing a fleeing vole—and, like those of their close relatives the martens, their claws are semiretractable but not sheathed (Powell 1993). Weasels are good climbers, another ancient characteristic of mammals in general and mustelids in particular (Holmes 1980), and they frequently climb trees and shrubs in their search for food (Chapter 6). Like tree squirrels, they can rotate their ankles and hang or climb head-downward as naturally as they can climb straight up (see Figure 6.5).

SKULL AND TEETH

The skulls of weasels are where their specialized life style is best demonstrated. Weasel skulls are unusually long and narrow, have exceptionally large areas for jaw and neck muscles, and enclose large brains (Radinsky 1981). The bones of an adult weasel's skull are fully fused, forming a single, box-like unit of massive strength. Sturdy, boney crests along the top (the sagittal crest) and back (the nuchal crest) of a full-grown skull provide extra space to anchor the temporalis muscles, which are responsible for a weasel's powerful bite.

This feature is unusual among small mammals: In general, only large mammals have crests on their skulls (Hildebrand 1974). For example, the large cats must have huge jaw muscles and need large crests to anchor them. Most small mammals need no crests. Even a wild cat the size of a house cat, which may weigh 10 times more than a weasel, can function well with no crests on its skull at all. Similarly, the skulls of adult red foxes show no large sagittal crests for jaw muscles. By contrast, the exceptionally well-developed temporalis muscles and tremendously powerful jaws of the large weasels need to be anchored to crests on the skull, just as in the large carnivores.

The zygomatic arches (cheekbones) on a weasel's skull are widely separated, to leave room for the temporalis muscles to pass through from the sagittal crest to the large coronoid process on the back of the jawbone (Figure 2.4). The zygomatic arches anchor the masseter muscles, which in herbivorous animals, such as voles or horses, provide a grinding motion to the molars as well as closing the jaws. Because the jaws of weasels can move only up and down, and because the masseter also prevents the jaws from opening very wide, the masseter muscles of a weasel are small. The zygomatic arches are therefore slim and weak, which shows that there are no strong pressures on them (Ewer 1973).

A short thick jaw is a better lever than a long slender one, and it can exert a stronger force (Biknevicius & van Valkenburgh 1996). The jaws of weasels are short, and their teeth are specialized for a diet of flesh, to a degree matched only in the cat family. The four carnassials, the last upper premolar and the first lower molar on each side of the jaw, are critically important. They are strategically

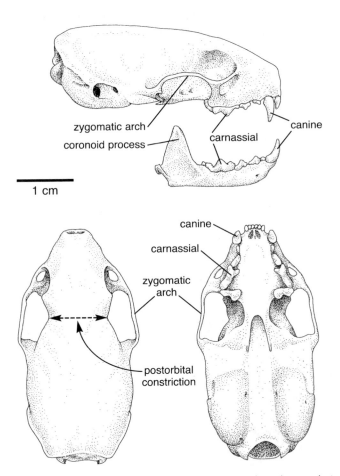

1 cm

Figure 2.4 The skull of a stoat, in lateral, dorsal, and ventral views. For the significance of the postorbital constriction, see p. 76.

placed at the back of the jaw so as to take the utmost advantage of the leverage of the jawbone and the huge strength of the temporalis muscles. This arrangement explains why weasels, like dogs, chew bones at the corners of their mouths (Figure 2.5)

The upper and lower carnassials on each side are parallel and slightly offset, and they shear closely past each other, to slice flesh and bones from a carcass into swallowable lumps. Weasels do not grind up their food finely; indeed, they physically cannot and do not need to, because meat is digestible in chunks without grinding. In the upper jaw, the first molar is set a little behind and crosswise, with a deep recess in its center (Figure 2.4). The rearmost blade of the lower carnassial slots exactly into it to form a slicing guillotine. Not only is this arrangement devastatingly effective, it is also self-sharpening and self-tightening (Mellett 1981). The joint between the lower jaw and the skull is cylindrical, and flanges from the skull wrap around the mandibular condyle on the lower jaw to prevent any sideways motion or dislocation of the jaw. Sideways movement would make the carnassials ineffective, like scissors held together with a loose screw. When the temporalis muscles contract, they pull

Figure 2.5 Because the carnassial teeth are at the back of the jaw, weasels always chew muscle, sinew, and bones at the corner of the mouth.

on the coronoid process, causing the jaw to rotate around the jaw joint and creating torque (Biknevicius & van Valkenburgh 1996).

The next most highly specialized teeth are the long, slim canines, whose function is not to slice food but to catch it. The canines act together as a trap and humane killer, grabbing hold of a fleeing mouse and dispatching it with a piercing bite through the bones of the neck and skull. The crunch can be audible several meters away.

The rest of the teeth of a weasel have little to do, and most of them are small or have been permanently lost. The incisors are jammed into such a tiny space at the narrow front of the mouth that, in the lower jaw, two of the six have been squeezed out into a second row behind. Between the incisors and canines in the upper jaw is a large gap on each side, which, when the mouth is closed, is filled by the canines of the lower jaw. The first premolars and the third molars are absent in both jaws, and so are the second molars in the upper jaw.

This reduction in the number and size of the nonessential teeth has permitted the front part of the face to become shorter, forming the characteristic wedge-shaped head. The dental formula is I3/3, C1/1, P3/3, M1/2 = 34. This means that weasels have three incisors, one canine, three premolars and one molar on each side of the upper jaw, and the same in the lower jaw except for an extra molar on each side. Most of the work is done by only 10 of the 34 teeth.

With this equipment, a weasel can kill a mouse or bird in seconds or less. The big cats of Africa are usually regarded as the ultimate in predatory power but, in relation to their size, the weasels are equally formidable predators.

THE SENSES

To find and to catch its prey, a predator depends absolutely upon its acute senses. Prey must often be located from a distance, and an actual attack requires accurate orientation, stealthy approach, and judgment of distance. Weasels, especially the smaller ones, hunt often in burrows or dense cover or in the dim silence under the snow, so hearing and smell are important in finding prey. But weasels also have the sharp eyes, with binocular vision to judge distance, needed to catch and kill prey in broad daylight. As weasels are often themselves prey for other, larger predators (Chapter 11), their keen senses also alert them when they are in danger of becoming a meal rather than finding one.

Ears

The ears of weasels are well adapted for the profession of hunting rodents in confined spaces. Each pinna, or direction-finding flap, is large and flat to the head, since a high pinna, like that of a bear or horse, would collect soil and get in the way underground. The external ear canal is also very large. As in many burrowing mammals, the sensitivity of the ear is magnified by the greatly enlarged tympanic bulla (cavity) of the middle ear, which is braced inside with trabeculae, or bony struts.

One would expect all weasels to be very sensitive to the high-pitched squeaks of mice, which range up through the ultrasonic to 92 KHz. Tests in captivity show that the range of hearing in the least weasel runs from 52 Hz to at least 60 KHz, and is best at 1 to 16 KHz (Heffner & Heffner 1985). Wild, long-tailed weasels, and presumably the others, are able to hear clearly the ultrasonic sounds of voles and mice (Powell & Zielinski 1989) and insects (Willey 1970). At the other end of the scale, the least weasel's sensitivity to low-frequency sounds is among the best of all mammals, and unusually good for a small animal.

Gillingham (1986) established a 3×3 m indoor grassland into which he could release least weasels and prairie and meadow voles to watch hunting behavior of the weasels. By lighting the grassland with red light only, he could eliminate use of sight; by flooding the area with white noise, he could eliminate use of hearing; and by flooding the area with odors, he could eliminate use of smell. The responses of his five least weasels to various deprivations of senses suggest strongly that they hunted primarily by sound, less so by smell, and least by sight. Longtails moved so fast in the grassland that data were difficult to collect. Nonetheless, longtails, too, appeared to hunt predominantly by sound.

High-frequency sounds are strongly directional, and the value for a small hunter of a keen ability to locate sounds is obvious. In this case, a weasel's ability to hear high-frequency sounds is not so much an adaptation as the natural consequence of having small, close-set ears (Heffner & Heffner 1985), but all weasels certainly make the most of it. A weasel sitting up with alert, sparkling eyes could be actually *hearing* around, not looking around.

Smell

The muzzles of weasels are pointed, providing ample space for the complex turbinal bones inside the nasal cavity and suggesting that weasels have an excellent sense of smell. Weasels use smell more than vision when hunting, although less than hearing (Gillingham 1986), and they mark their home ranges with complex, informative scent signals (Chapter 8). Herman (1973) tested three least weasels in a Y-maze, 30 times each, to see whether they could distinguish which arm of the maze had previously been travelled by a rodent, and found (to no one's surprise) that they could. Herman made sure that the weasels were responding to scent by running tests both in daylight and in the dark and by offering sound cues at the ends of both arms of the Y-maze but scent in only one. The time each weasel spent in the maze varied from around 140 to 180 seconds at first, whether they got a reward or not. When a correct choice was reinforced by gaining access to the rodent, they learned after about 15 trials to follow the scent trail more quickly to the right door, and after 60 trials their running time was down to less than 30 seconds.

The difference between the results reported by Gillingham and Herman are not necessarily as contradictory as they appear. It may be simply that all the experimental animals could perceive both sounds and smells but reacted to whichever was the most informative in the circumstances.

Eyes

The short, pointed face of a weasel allows both binocular vision forward and a wide arc of monocular vision on each side. The eye of a weasel, like that of a cat, is constructed so that it can see well in both bright and dim light. Since the requirements for good vision by day and by night are not the same, the eyes of an animal active at both times have to achieve something of a compromise. The retina of a mammalian eye has three kinds of receptor cells, of which two, called (from their appearance in histological sections) cones and rods, are part of the visual imaging system. The cones are used for perception of bright light and colors, and are found in diurnal species. The rods are particularly sensitive to low-intensity light and are most numerous in the eyes of nocturnal animals. The balance of rods and cones gives a fair idea of how sensitive and acute an eye is, and whether or not it can see in color. The third type registers only the intensity of light, and its function is to direct a mammal's internal clock (Hattar et al. 2002).

All species of weasels have duplex retinas (including both rods and cones), suggesting some degree of color vision. Unfortunately, although Herter (1939) and Gewalt (1959) set out to test the color discrimination abilities of common weasels, neither got past preliminary training before their animals died. Nonetheless, the structure of the retinas in common weasels is histologically very similar to that in stoats, which have been shown in behavioral tests to be able to see at least red, and perhaps also yellow, green, and blue (Gewalt 1959).

On the other hand, the presence of cones need only show that the eye is adapted for use in bright light. Since all weasels and most of their prey are shades of brown, white, or grey and important markings on them are usually black or white, weasels may not need color perception for identification of food, potential rivals, or mates. Even if they can see it, color may not mean much to weasels, anyway.

In most carnivores, the sensitivity of eyes in dim light is increased by the tapetum lucidum, a reflective layer behind the retina. The eyes of an animal that has a tapetum, such as a cat, have very obvious "eyeshine" when caught in a flashlight, compared with the dull red glow of human eyes. Weasels have a vivid green eyeshine, a fact well known to naturalists: Wood (1946) wrote of "their eyes . . . glowing with a strange green fire."

Eyes sensitive enough for night hunting must also be well protected if their owners are to be about during the day. The brown iris of the eye, which automatically closes in bright light, protects the retina from damage: The more sensitive the retina is, the more protection it needs. A slit pupil can close more tightly than a round pupil, so that even very sensitive eyes are not blinded in sunshine. Weasels have slit pupils, but the slits are horizontal rather than vertical as in cats.

Eyes that are good at seeing in dim light often achieve this ability at the expense of sharp acuity in bright light. As every photographer knows, photos taken on film of 400 ASA, or digital images saved to too few pixels, should not be too greatly enlarged because they lack fine detail. Most carnivores key into movements rather than entire pictures, but that does not mean they cannot be sharp-eyed when something catches their attention. A long series of patient, form-discrimination tests with one common weasel in Germany showed that this individual could distinguish quite minor variations in shapes presented as cues for food rewards (Herter 1939). In fact, this animal eventually learned to distinguish seven letters, offered in various combinations. The letters were attached to a pair of boxes, in one of which was the reward (a mealworm). By the end of the series, the weasel could "read" the label on the box containing the reward, which was WURM (worm), and preferred this box consistently to the other one, which was labelled LEER (empty). Learning the location of a reward, and detours to it, are easy tasks for weasels, and they easily remember pathways and places around their home ranges.

VOICE

All the weasels are capable of making a range of sounds, which no doubt mean much more to other weasels than to us. Yet, some of the messages these sounds convey are quite unmistakable even to our ears (Huff & Price 1968; Svendsen 1976). A weasel that feels slightly uneasy will make a low hissing sound, often while retreating into a safe place, which indicates low-intensity fear or threat. If the danger is more pressing, the weasel will probably turn and stand its ground, making a series of sharp, explosive barks or chirps, loud enough and aggressive enough to deter all but the most steely nerved attacker. Taken unawares, 9 out of 10 people will jump out of their skins at the sudden noise. If the danger is really acute and all retreat cut off, such as when a weasel is cornered or trapped, the chirp escalates into a prolonged defensive screech, or a really pitiful wail often accompanied by a "stink-bomb" (Chapter 8).

At the other end of the scale is a low-intensity, high-pitched trilling sound, often heard during friendly encounters between mates, or when a mother is calling to her young (pp. 223, 224, 309). Occasionally it may be heard from exceptionally tame hand-reared weasels playing with a trusted human companion. According to Huff and Price (1968), young weasels give high-pitched squeaks at birth, which are replaced by the adult chirps at 4 weeks of age; trills are added at 5 to 7 weeks. Another sound, best described as a medium-pitched, medium intensity "zheep" or "zhzhzhzhp" in longtails, less harsh in stoats, is used between mates and at other times when a weasel appears interested. It is a vocalization of an alert, distinctly interested but unworried weasel, made with the mouth wide open, almost smiling (Figure 2.6). Both male and female longtails and stoats use this vocalization, but *only* when in brown, summer coat (R.A. Powell, unpubl.). One male longtail zheeped every day one summer and fall, abruptly stopped upon molting to white, and abruptly started again when his spring molt back to brown was complete. The reasons for this sudden, reversible change are among the many things we still do not understand about these enchanting critters.

PLAY

Weasels are among the most playful of mammals. Although not as well known for playing as are their cousins the otters, weasels play often, even as adults (Figure 2.7). Some examples of weasels playing with human companions are given in Chapter 1, but that may not be the same thing as weasels playing among themselves.

Play is more difficult to define than it is to recognize. It involves normal behaviors but in exaggerated form, in unnatural combinations, and often in

Figure 2.6 Lew, a male long-tailed weasel kept by the Powells, "zheeping."

Figure 2.7 Weasels, like many carnivores, play vigorously and apparently with enjoyment.

unfinished or jumbled sequences (Fagen 1981). Youngsters play more than adults, and many components of play are behaviors that youngsters will need to use as adults. Play in herbivores often involves running and other escape behaviors, and play in carnivores often involves imitation prey capture (Fagen 1981). Playing carnivores often crouch and pounce on and wrestle with siblings or inanimate objects. A leaf blowing past can stimulate a young weasel to pounce on it, wrap its body around it, and hold it with all four legs, biting it with a mock-intensive kill bite. Adult weasels will bounce off logs, tree trunks, or other objects, twisting and contorting, nearly turning themselves inside out. Their movements are quick and intense, but then they will suddenly stop, standing absolutely stock-still. Then, without any warning or perceptible reason, the bounces and contortions start again. Although intent and emotion are hard to understand in another animal, we have the distinct feeling that a weasel playing is deep-down enjoying itself. Developing the ability to zig zag and turn on a dime could also have survival value to a weasel being pursued by a raptor or owl (see Figure 11.6).

Byers and his colleagues (1995; 1998) have recently offered a good evolutionary explanation of play in young animals. The nervous system of a mammal continues to develop after birth. The growth of axons and the synaptic links between them is greatest during the juvenile period, when play is most intense. Those axons and synapses that are used most are retained to adulthood, and the rest are lost. Playing youngsters need not perform adult behaviors perfectly in form or sequence; they gain an advantage if they simply perform the adult behaviors that will be most important to them. Play is not practice, but it is critical to the full development of a young mammal's nervous system. Adult mammals play significantly less often than do young, and those that do tend to be among the more intelligent species. So the frequent reports of adult weasels playing with each other or with their young may be a sign of their intelligence.

THE CONSEQUENCES OF BEING WEASEL-SHAPED

The small size and thin shape of the weasels, so essential to the profession of burrow-hunting rodent predator and so advantageous in competition with other rodent hunters, have both advantages and disadvantages (Figure 2.8). The small size and long, thin shape of a weasel's body affect the whole of the rest of its life, and exploring these relationships is one of the recurring themes of this book. For example, the short legs, large home ranges, and tight energy budgets of weasels hinder regular, direct communication (friendly or unfriendly) between neighboring residents. This is no disadvantage to weasels, because they have large scent glands under their tails and use them as part of an advanced scent communication system (Chapter 8). Small size makes weasels almost as vulnerable to larger predators, especially raptors, as voles and lemmings are; the white winter coat of all species, and the black tail tip of the two larger, is a defense against the hunter being hunted (Chapter 11). Small size, short life span, and the capacity to make a large reproductive effort early in life tend to go together, especially in species that stand to gain a huge advantage from a rapid response to a sudden increase in food supplies (Chapters 9, 10, and 14). The consequences of the weasel body plan relevant here concern metabolism and physical strength.

Metabolism: The Cost of Being Long and Thin

The advantages and disadvantages of being small, long, and skinny illustrated in Figure 2.8 tend to cancel each other out. There are no one-for-one tradeoffs but, in general, weasels have both a substantial advantage over other, larger predators in hunting efficiency and the disadvantage of being vulnerable to attack by them. Weasels have the advantage of being able to increase production rapidly in good years but the disadvantage of risking starvation in bad years. Most important, the metabolic penalties incurred by a small animal in a cold climate are all serious. The size and shape of a weasel are hugely inefficient in physiological terms, and impose real energetic costs to living the weasel way of life.

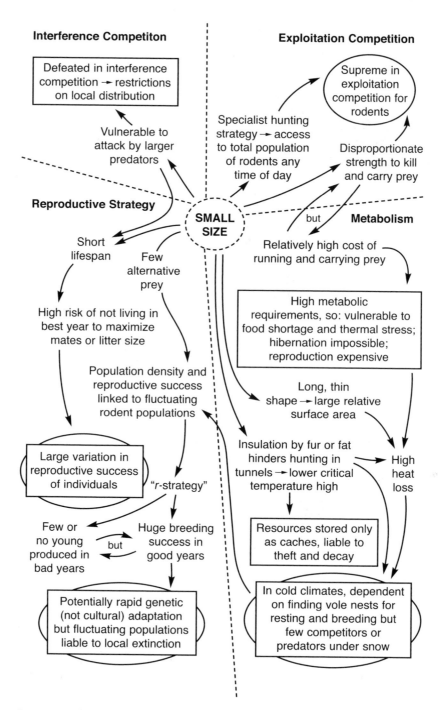

Figure 2.8 The advantages (*circled*) and disadvantages (*boxed*) of small size to the weasel way of life. Text not in boxes or circles explains steps of logic. Large arrows between text shows direction of logic. Small arrows within text mean "leads to." (Redrawn from King 1989b.)

To supply the massive energy they need, weasels have huge appetites. Captive animals eat between a quarter and a third of their body weights each day, and more active wild ones probably eat more (Table 2.1). They are tied to a life of frequent small meals, five to ten a day, and to the necessity of finding food at frequent intervals (Gillingham 1984). None of them, especially not least weasels, can endure going without food for long, which is one of the difficulties of live-trapping them. Nearly all this enormous intake of food is burnt up merely to keep warm.

The alert, rapid movements of weasels reflect their constant hunger, geared-up metabolism, and galloping pulse—measured at rest at about 360 to 390 beats per minute in stoats and 400 to 500 beats per minute in common weasels (Tumanov & Levin 1974; Segal 1975). If you put your ear to the chest of an anesthetized weasel, you can hear the heart "purr-r-r-ring" away, but you cannot count the heartbeats. This frantic pace of life is also reflected in the anatomy of organs that have to do with metabolic processes. The load on the heart and lungs is great, so the heart is large relative to the weight of the body, especially in small individuals. A weasel's food comes in large, infrequent packages, so the digestive system is adapted to deal with alternate feast and famine. The gut is short and meals pass through quickly. Dyed bait fed to a least weasel reappears in 2 to 4 hours, and the defecation rate is high, averaging 19 scats per 24 hours (Short 1961). Weasels often nap after a meal, but not for long.

Vulnerability to Cold

Every animal must balance, every day, the energy gained from food against the energy spent in getting it. Small mammals have to spend relatively more energy

Table 2.1 Food Consumption by Captive Weasels

Species	Males			Nonbreeding females			Reference
	Wt food/day	Body wt weasel	% body wt eaten/day	Wt food/day	Body wt weasel	% body wt eaten/day	
Stoat	57	—	23	33	—	14	(Day 1963)
Stoat	—	—	19–32	—	—	23–37	(Müller 1970)
Common weasel	41	133	33	28	81	36	(Moors 1977)
Least weasel	32	81	40	—	—	—	(Gillingham 1984)
Long-tailed weasel	—	—	c. 33	—	—	—	(Hamilton Jr. 1933)

to maintain a constant body temperature than large ones, because they have a large surface area, relative to their mass, from which heat radiates. For weasels, this problem is even greater than for other small mammals. Their long, thin shape is ideal for hunting through small spaces and burrows, but it has the huge disadvantage of giving them a larger surface area than "normally" shaped animals of the same weight, even when they are asleep (Brown & Lasiewski 1972).

Weasels feel the cold tremendously. Their body temperature is 39°C to 40°C (102°F to 104°F), and their body heat escapes easily, not only because of their shape, but also because they cannot afford to insulate themselves too much. Layers of fur or fat would get in the way and perhaps prevent them from entering vole runways and burrows. The length of a mammal's fur is related to the diameter of its body, not its weight; hence, weasels have shorter fur than regularly shaped mammals of their mass. Their winter fur is hardly warmer than their summer fur.

Most mammals also use fat as an insulation from the cold. Fat is a good insulator, and it produces more energy when burned to run a body. Yet, weasels appear to benefit from storing excess nutrition as muscle instead of as fat (Harlow 1994). Fat is dead weight for a weasel, requiring energy to carry around and slowing a weasel's escape from predators. Muscle, on the other hand, can be used until it must be burned as fuel. So, few weasels carry much fat, and any fat that they do have is confined to dips in the body outline to retain their sleek figures.

The cold northern winters that weasels endure over much of their range must be, therefore, periods of great energetic stress. The simple maintenance of body heat at rest requires twice as much energy as in summer, at three times the cost incurred by a lemming of similar size resting at the same temperature (Casey & Casey 1979). Arctic and alpine weasels at rest absolutely depend upon having a thickly insulated den, which they take from recent prey and improve by lining with fur, as illustrated by Sittler (1995).

However safe and attractive a warm den may be, a weasel cannot stay in it indefinitely. Hunger will eventually drive it out into the cold again, and that means using up yet more energy for running. Curiously, researchers who have studied the energy metabolism of weasels disagree as to whether weasels are restricted in their movements by subzero temperatures. In Alaska, Casey and Casey (1979) estimated that arctic least weasels outside their dens may have to generate up to six times the basic metabolic rate just to keep warm, and they may have little capacity left in reserve to provide energy for active hunting. If so, weasels should stay under the snow when the air temperature drops too far, which is exactly what they do. Stoats tracked by Kraft (1966) in Western Siberia did not emerge onto the snow surface when the air temperature got below −13°C.

On the other hand, Sandell (1989) concluded that stoats in Sweden do not need to generate more than three times the basic metabolic rate under any conditions, and that, since 75% of the energy expended during activity is released as heat, a running weasel does not need to spend any extra to keep warm whatever the temperature. If this second interpretation is correct, an active weasel is virtually independent of air temperature and is safe even in severely freezing conditions so long as it keeps moving.

Whether these results actually contradict each other is not clear. The authors used different species, methods, and materials and their results may apply to different situations. For example, weasels may avoid hunting above the snow because they cannot afford to stop moving, even when stops are necessary for foraging. Or, weasels may remain below snow in extreme cold not only because no prey venture above snow at those temperatures, but also to avoid exposure to hawks and owls. Or, the differences in body size and geographic origins between the weasels studied in Alaska and in Sweden were enough to introduce real differences in the results. In any case, it seems safe to conclude that weasels balance their winter energy budgets with little to spare, and in the far north they probably survive at the limits of their metabolic capacities.

Strength and Loading

Hunters small and thin enough to enter rodent burrows must not be too small to execute a kill, so weasels make up extra size in length rather than in girth. A weasel can wrap its long body around a catch, holding it with all four feet while it struggles (see Figure 6.7). Even so, weasels have not sacrificed muscular strength for size. On the contrary, weasels are relatively stronger than large carnivores.

The difference is a result of scaling. The apparently disproportionate strength of a weasel is one of the mechanical advantages of being small. The force that can be exerted by an individual muscle is the same, per unit of cross-sectional area, in mammals of any size (Schmidt-Neilsen 1984). As animals get larger, the strength of their muscles increase in proportion to the square of body length, but the total weight of the body to be moved increases in proportion to its volume, as the cube of body length. Consequently, the force that can be exerted by the muscles, relative to the mass of the body, is greatest in small animals.

The economics of hunting have some peculiar costs and benefits for a small predator (King 1989a). For example, the energy cost of running is relatively high in short-legged animals, because they have to take many small steps or bounds to move one unit of body mass over one unit of distance, each step requiring work in proportion to mass. Weasels never merely walk anywhere. They either glide along with a straight-backed scuttle or bound hump-backed at speed—and either method takes a lot of energy. Bounding is especially energetic, because the supple back is used as an extension of the legs, so the whole body is involved in every step. On the other hand, when a weasel climbs a tree or a steep slope, the additional energy required to work against gravity is negligible. Running is always an expensive activity to a weasel, but, as in all small animals, it makes hardly any difference whether it is going along the ground, straight up, or straight down (Schmidt-Neilsen 1984).

Small predators also have the problem of carrying prey that may be as big as themselves or bigger. The energy cost of carrying a load increases in direct proportion to its weight. If the load is 100% of the weasel's own mass, its oxygen consumption (a measure of energy expended) increases by 100% (Schmidt-Neilsen 1984). Weasels routinely carry prey that heavy. Even the smallest of them have the strength, but the cost is high. A stoat can kill a rabbit of twice its own weight, and then carry it away, looking like a terrier bounding off with a sheep. On the other hand, no lion can run at speed carrying a carcass of even half its own weight. For its size, the weasel is a fiercer, stronger, and tougher predator than any lion.

3 | Molt and Winter Whitening

ALL WEASELS MOLT twice a year, in spring and autumn. In mild climates the incoming fur is always about the same brown color as the old, so it is usually difficult to tell from the outside, on cased museum skins or on living animals, whether a particular individual is molting or not. But in the autumn in cold climates, the incoming fur is white, and then the patterns and timing of the molt process become easily observed.

The process of molting is really a compound of two other processes, which quite independently control the growth of the new hair and the color it will be. The growth of new hair is stimulated when the lengthening days of spring, or the shortening days of autumn, reach a certain critical number of hours of daylight, which acts like a trigger. The color of the new coat (always brown in summer, white or brown in winter) is controlled mainly by heredity.

Cold temperatures also affect a third process, the shedding of the old hair, the final stage in the replacement of the old coat. So in the arctic, where the shortening days of autumn always herald the rapid onset of a severe winter, the old coats are replaced quickly, within a few days. In mild climates, the shortening days of autumn set off the molt process just as predictably, but it is slower, spread over a month or 6 weeks. In spring, cold temperatures delay the shedding of the old fur, so that in changeable temperate climates, weasels do not lose their winter coats too early if the spring is particularly cold or late (Rust 1962). So there is a lot of variation possible in the molting patterns shown by individual animals in different places and years.

THE MOLT CYCLE

The Skin Follicles

Each of the two molts in a year involves a series of changes in the shape and activity of the hair follicles in the skin (Figure 3.1). First, the follicles enlarge and extend themselves deeper into the skin. There they produce the new hair, which grows out alongside the old. If the growing hair is to be brown, the follicles contain a dark pigment, melanin. So in spring, the active growth of new brown hair can be detected from the small dark flecks on the inside of the skin, which represent the enlarged, active melanin-containing follicles.

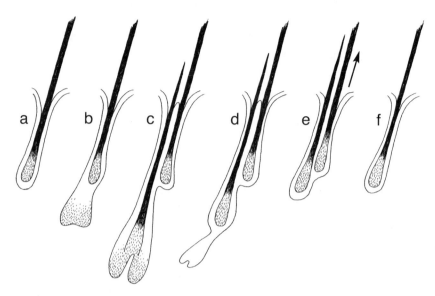

Figure 3.1 The cycle of activity in the hair follicles of the skin during the molt and regrowth of brown hair. (a) Normal condition; (b) preparation stage; (c) growth of hair and follicle and accumulation of melanin; (d) formation of root; (e) shedding of old hair; (f) process complete, normal condition again.

By the time the new hair is fully formed and anchored in the follicle, the melanin is used up, and the follicle retreats back to its smaller resting state, nearer the surface of the skin. The inside of a dried skin now appears a clear, almost translucent honey color. The old hair is shed a few days, or even weeks, later. Hair cannot grow when the follicles are not active, and if a patch on the rump of a weasel is shaved during the resting phase, it will remain naked until the beginning of the next molt cycle (Feder 1990).

The Pattern and Timing of the Molt

A complete molt cycle lasts from the beginning of the active phase of the follicles to the end of the shedding of the old fur. Since these are quite separate processes, even in one follicle, the cycles of growth and shedding can overlap. The biology of weasels offers two quite different methods of working out the timing and length of the cycle, depending on the species and location.

In places where weasels reliably turn white in winter, the simplest technique is to arrange a set of skins collected all year round, ranging from full brown through all the intermediate pied stages to full white, in order on a table. Glover (1942a) was one of the first to do this, using northern longtails, and van Soest and van Bree (1969) did it for European stoats from the collections of the Amsterdam Museum. Both studies showed that the autumn molt starts on the belly, and the new white hair with its thicker underfur appears there first. Then new fur grows on the flanks and back, and finally on the head. In spring the pattern is reversed; the old white coat is replaced first on the head, then across the body, and finally underneath. The spring molt passes as a distinct, sharp-edged wave of new brown hair across the body, whereas the autumn molt is somewhat more diffuse, producing a "salt-and-pepper" effect (Figure 3.2). Either way, the belly is the last part of the body to lose its extra insulation in spring, and the first to acquire it in autumn, which seems a practical arrangement for a short-legged animal living in a cool climate.

The disadvantage of this simple technique of describing the molt from dead weasels is that every observation is necessarily of a different individual, and if the variation between individuals, climates, and capture locations is extensive, any estimate of timing could become blurred. This problem can be resolved using another technique, watching and photographing captive weasels every day as their coats change. Here we meet a different set of disadvantages in estimating normal seasonal patterns, which could arise if the number of animals is small and their behavior affected by captivity, but at least the individual variations through each molt cycle can be documented.

Figure 3.2 The annual cycle of coat color of the northern weasels.

Feder (1990) watched the molt cycles in captive stoats collected from Oregon and from Alaska. She confirmed that their molt patterns were much the same as expected from the previous work, but the Oregon animals molted more slowly, and began later in autumn and earlier in spring. That is because the difference between day length in summer and in winter increases toward the poles, and so the rate of daily change in day length from one solstice to the other speeds up in the same direction. Therefore, the higher the latitude of a weasel's home is, the more rapid the transition is from summer to winter coats: at lower latitudes, the transition from one coat to the other is slower. Also, the longer the part of the year during which the day length is too short to maintain the summer coat is, the longer the winter coat is worn. The lower the latitude is, the slower the transition from one coat to the other.

Most studies of molt and color change in weasels have used one or other of these techniques, which means that their observations have been done only from the outside, and in climates where all or most individuals turn white every year. But these studies really only observe the last stage, shedding, and the vital early stages of preparation by the follicles and most of the phase of growth of new hair are not detected this way. Also, observing molt only by color change compounds the effects of two quite independent processes, and cannot be used in places with mild climates, where most weasels and stoats do not turn white.

Fortunately, there is an alternative technique by which the molt process can be studied quite easily, even in places where both summer and winter coats are brown, from the inside of the skin. On a flat weasel skin, which has been scraped clean of fat and dried in air without preservative, the distribution of the dark active follicles, the small black flecks along the inside of the summer-brown area, show what phase of molt that animal was in when it died. In the belly fur, the follicles do not accumulate pigment because they are growing white hairs, so the onset of the molt there cannot be detected by this method. But on the back and sides, the active follicles appear just before the new brown hairs grow, and fade as their melanin is used up, so it is possible to plot the progress of the molt by following the migration of the active follicles across the summer-brown area and estimating the period at which molt activity might have reached the white underparts (Figure 3.3).

This method has been used to show that the pattern of molt in common weasels in England is the same as in northern longtails and in Dutch stoats (King 1979). The spring molt begins in March on the head, then spreads along the spine and down across the flanks, ending in May or June on the underside. The autumn molt follows the same route in the opposite direction, beginning in about September or October and ending in about November. The spring molt ends, and the autumn molt begins, in the white fur, so the period during which the summer coat is worn is not exactly known. In King's sample, a few common weasels were still completing their winter coat in December and January: These were all young ones born late in the season, which were probably late in starting.

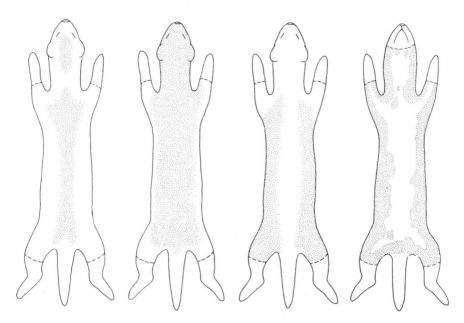

Figure 3.3 Active hair follicles on four common weasels killed at successive stages in the spring molt (left to right, dorsal views; far right, ventral view), drawn as if the follicles could be seen from outside through the skin. Active follicles (*dark*) first appear on the back; then the activity of follicles migrates down each side to the belly. (Redrawn from King 1979.)

The actual color of the fur is found mainly in the long guard hairs, whereas the underwool is a nondescript grey at all seasons. The winter fur is thicker because it contains more underwool, and by the spring the guard hairs are worn and broken and less intense in color, and more underwool shows (Powell 1985a). Hence, even in weasels that stay brown all winter, the winter fur often appears to be slightly paler in shade. The difference is best appreciated during the spring molt. Among King's common weasels were several that showed quite a marked change in color as the incoming new fur, noticeably darker, contrasted with the old, paler fur. This was obvious only on animals caught at just the right moment, when the new fur had extended along the middle of the spine as a dark dorsal stripe, but before it had widened and spread across the curve of the flanks. On over a quarter of 122 skins, the dark molt patches on the inside of the skin could be seen spreading along in front of the new fur, while the skin under the area of the new fur itself was clean again.

When the same technique was applied to a large collection of stoat skins from New Zealand, the same pattern appeared again, appropriately adjusted to the southern hemisphere seasons. In New Zealand, the spring molt begins in August and ends in October (although in the cooler parts of the country the old hair might still be hanging on into December), and the autumn molt begins in March and ends in June (King & Moody 1982).

The control over molting by day length is so powerful that stoats in captivity can be induced to molt into their summer coats in midwinter and vice versa, merely by manipulation of the lighting over their cages. For example, five Wisconsin stoats kept on 18-hour "days" molted into their summer coats at a temperature of –6°C (Rust 1962). They took longer to get started than a control group kept at 21°C, but they could not postpone the change for more than 3 weeks when their eyes were insisting that spring had arrived. Conversely, Feder (1990) induced eight stoats to start their autumn molts out of season by switching them to short days.

The same conclusion was clear from the large collection of New Zealand skins, which came from 14 study areas representing a wide range of habitats, altitudes, and latitudes. Analysis of these skins confirmed that the timing of the molt is controlled by day length. On average, stoats in the far south of New Zealand (44°S to 45°S), which is cooler and reaches a given day length later than the north (39°S to 40°S) in spring and earlier in autumn, were always the last to start molting in spring and the first to start in autumn. Stoats from high and low elevations at the same latitude, whose homes differed in temperature but not in day length, molted at the same time (King & Moody 1982).

The molt of the wild stoats in New Zealand was much slower than that of Feder's captives in Alaska, although comparison is hampered by the different observation methods. On the other hand, climate probably influences the speed of molt much more than does observation method. The molt cycles in New Zealand stoats and in the common weasels of England were both studied by the same method, and both were slower and more diffuse than that of the stoats in Alaska. New Zealand and Britain both have mild maritime climates influenced by warm ocean currents, although New Zealand is nearer the equator (35°S to 47°S) than Britain (50°N to 59°N). Both are much more temperate than Alaska.

Control of Molt and Reproduction

The seasonal cycles of molt and reproduction are closely related and are coordinated by the neuroendocrine system. The nerves receive stimuli from the outside world (Is it day or night? Is it warm or cold?); the brain then instructs the endocrine organs to secrete the appropriate hormones. Hormonal messages travel, via the bloodstream, more slowly than the electrical messages that flash along the nerves, but they reach every cell of every limb, organ, and tissue in the body. Cells for which the message is not intended will ignore it; cells sensitive to it will swing into coordinated action, even if they are spread from one end of the body to the other—as are the skin follicles.

The pivot of this system is the pituitary, a small gland under the base of the brain (see Figure 9.5), constructed so as to maximize the contact between brain and circulation. When the eyes perceive the lengthening of the days after the winter solstice, the pineal gland produces less melatonin, a substance that is released only in darkness and has an inhibitory effect on prolactin. Higher levels of prolactin in the circulation are a signal to the pituitary to release gonadotropins, which stimulate the gonads to prepare for the approaching breeding season. The same message is also received and understood by the hair follicles. So, in early spring the testes and the hair follicles of males enlarge together, and the testes begin to manufacture testosterone and sperm while the hair follicles accumulate melanin and manufacture hair (Wright 1942b).

The autumn molt takes place in the opposite conditions, with shortening days, increasing levels of melatonin and falling levels of plasma testosterone. The connection between the two systems can be demonstrated experimentally by showing that it fails in stoats in which the pituitary has been removed (Rust 1965). The same effect can be achieved by inhibiting the pituitary, for example, with melatonin. The large testes of five intact captive stoats, given a melatonin implant in spring, regressed within 6 weeks to near-winter condition, while at the same time the animals molted into their normal winter coat (Feder 1990). Conversely, there is a significant correlation between the spring rises in testis weight and in readiness to molt, well documented in wild stoats in New Zealand (King & Moody 1982).

In females the pituitary gonadotropins also have a stimulating effect on both the hair follicles and the ovaries, at least to start with. But when the ovaries begin to manufacture large quantities of the female sex hormone estrogen, things suddenly change. Among the common weasels examined by King (1979) were some that had started their spring molt and had produced a dark dorsal stripe of new brown hair, but had no molt patches on the insides of their skins. These females were all in breeding condition, and it appears that, when they came in heat, molting had stopped. In the domestic ferret, molting is inhibited when the blood contains high quantities of estrogen; possibly the same thing happens in common weasels. However, when the heat period subsides, molting is resumed, and several female weasels were found to be still molting late into spring.

Molt and reproduction are both active processes for which the small female must find energy, in addition to the already considerable energy required just for daily living. Van Soest and van Bree (1969) suggested that female stoats complete their spring molt earlier than males, before their embryos implant, because they could not provide the energy for molt and gestation simultaneously. But female common weasels are able to do this, because they resume molting after they have been fertilized; and if they can do it, there is no obvious reason why stoats could not.

If there is any difference between the two species, it is probably less to do with energy conservation than with their different reproductive cycles. Female stoats are in heat in summer, so their spring molt is not interrupted by estrogen, and there is time for them to finish growing their summer coats before their embryos implant. The winter coat is a different matter, since it is thicker and no doubt more expensive in energy to grow, but by that time both stoats and common weasels are in their winter (anestrous) period, and no energy need be diverted from the serious business of preparing for the winter.

WINTER WHITENING

The Southern Limit of Winter Whitening

Hall (1951) noted that the southern limit of regular whitening in longtails is not a sharp line, with white animals on one side and brown ones on the other, but a broad zone up to about 350 km wide in which white, brown, and pied animals could be found in various combinations. He mapped the position of this zone as running for most of its length along one side or other of the 40th parallel, but extending north on the West Coast (Figure 3.4). From Figure 1.5 it is possible to work out that this zone corresponds very roughly to the southern limit of regular snow cover at least 2.5 cm thick and lying for about 50 days a year.

Virtually all North American stoats east of the Cascades turn white, because they all live in mountain or continental climates; only the stoats of the coastal Northwest Pacific stay brown in winter. In Britain, the transition zone for stoats runs through a reverse S-shaped curve lying between 52°N[0] and 56°N: White coats are common in Wales and Scotland; brown is the rule in England. Coastal Holland straddles the transition zone at 51°N. Across Europe, however, the moderating influence of the Gulf Stream is left behind, and the southern limit of whitening follows the snowline southeastward.

The climate conditions required to set off winter whitening are apparently not the same for all species of weasels, or even for the same species in all places. Minnesotan stoats turn white much later than Alaskan ones, and return to brown earlier (Feder 1990:108). British stoats seem to whiten in relatively mild winters: The hilliest parts of Britain where white stoats may be seen are much less snowy than the country occupied by white longtails in the United States, even though the British ones are 1,800 km further north. The same is true of the stoats of British stock introduced into New Zealand. By contrast, Russian stoats behave just like longtails; in Byelorussia (50°N to 55°N) the transition zone for stoats again coincides with the southern limit of stable snow cover lasting at least 40 days a year (Gaiduk 1977).

The southernmost limit of regularly white *Mustela nivalis* in western Europe lies at a much higher latitude than for *Mustela erminea*. All-white weasels

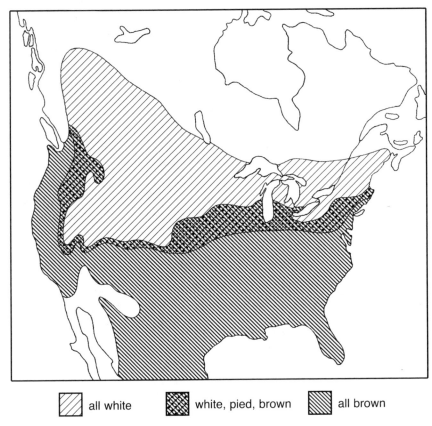

| | all white | | white, pied, brown | | all brown |

Figure 3.4 North of the transition zone, all long-tailed weasels turn white every winter; south of it, all stay brown. In the transition zone itself, all gradations can be found during winter from white through pied to brown. (Redrawn from Hall 1951.)

are a great rarity in Britain, even in the north where stoats regularly achieve full ermine. Salomonsen (1939) calculated, from the geographical distribution of white and brown winter skins from Greenland, Scandinavia, and Britain, that the critical minimum temperature associated with whitening in common/least weasels is a full 5°C lower than that for stoats. He concluded that, paradoxically, the smaller species is much more resistant to the cold. But from what we know about the metabolic stress endured by small mammals in the Arctic, this seems an unlikely explanation; and besides, there is a better one.

In southern Sweden, *nivalis* stay brown all winter as far as about 59°N to 60°N (the latitude of the Orkney Islands, in the far north of the British Isles); then there is a narrow boundary zone, about 100 km across, roughly from Norrköping through Stockholm to Uppsala, beyond which all *nivalis* turn white. But this boundary does not, as in *erminea* and *frenata*, mark a transition zone where some individuals of one species turn white or not according to the local climate. It marks the meeting of the two distinct subspecies of *nivalis* (see Figure 1.13), which have different histories and genetic makeup, and which are distinctly different in their summer coats as well as in winter (Stolt 1979). The intriguing implication is that *M. n. vulgaris* evolved at a more southerly latitude than *M. n. nivalis*, and so in the *vulgaris* group the allele for winter whitening is (or has become) rare, whereas in the *nivalis* group it is widespread or even fixed (Zima & Cenevova 2002). British common weasels all belong to the *vulgaris* group; hence, the fact that they do not turn white in Britain has more to do with genetics and evolutionary history than with climate.

The same is true of the stoats of northwestern North America that were the subjects of Feder's experiments. Those that now live on the Pacific coast probably moved north after the glacial periods from a refuge somewhere on the continental or coastal mainland, whereas those that now live in Alaska could have been living on the wide expanses of ice-free Beringian tundra throughout the Pleistocene (Eger 1990). Their different ancestry explains their different average size and winter dress.

The Mechanism of Winter Whitening

In the Far North

At landscape scale, there is a general relationship between cold winters and snow cover—the further north the latitude or the higher the elevation is, the greater the chances are that a regular covering of snow every winter can be expected. Likewise, there is also a general connection between whitening in the landscape and in weasels. In cold climates the autumn temperatures plunge quickly, and all northern weasels always turn entirely white.

The processes setting off the autumn molt, initiated by the shortening days, are the same everywhere, but the processes controlling the growth of the new hair are different in weasels living in mild compared with cold climates. The follicles can produce brown hair only if they contain melanin, which they cannot have unless the melanocytes (small cellular factories in the skin) have first been told to make it by MSH (melanocyte-stimulating hormone). In warm-climate weasels, the MSH is produced on cue by the pituitary, the follicles respond to it, and the new winter hair grows out brown. But in cold-climate weasels, an inhibitor either prevents the pituitary from producing MSH, or prevents the follicles from responding to it. The effect is to turn off the supply of MSH and of melanin, and the new hair grows out white (Rust & Meyer 1969) (Figure 3.5).

On the other hand, the winter climate alone does not provide a complete explanation of the control of winter whitening. Early evidence for why it does not came from transplant experiments across the transition zone between all-

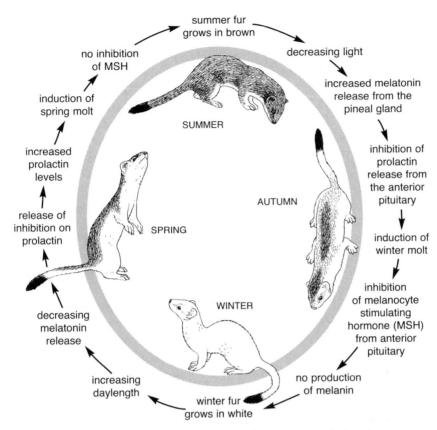

Figure 3.5 The annual molt cycle and winter whitening in North American longtails. In northern populations, the production of melanin in autumn is inhibited, so the winter fur grows white. In southern populations, the inhibition of MSH is switched off, so melanin is produced and the winter hairs grow brown. The rest of the cycle is the same in all populations. (Redrawn from Feder 1990.)

white and all-brown winter fashions. If temperature or some other purely environmental condition were the sole controlling agent, then weasels born on one side of the dividing zone and transferred to the other side should change to white or not according to the custom in their new home, but they do not.

In the summers of 1936 and 1937, a male and two female long-tailed weasels, captured on the warm, snow-free coast of California, were transported east along the latitude lines to Lake Tahoe, high up in the cold and snowy Sierra Nevada; in the other direction, a female from Lake Tahoe and a male from Salt Lake City, Utah, were moved from snowy environments to the mild Mediterranean climate of Berkeley, California (Hall 1951). All the animals were kept in outdoor cages with wire mesh sides open to the elements; all were exposed to about the same day length as at home; all molted at the usual time—but into the "wrong" winter coat. Those from east of the line turned white even in California where all weasels wear brown in winter; and those from west of the line stayed brown even in the mountains, where all the residents turn white. Two wild-caught longtails from northern climes kept in heated quarters at a constant temperature of 18°C to 24°C still turned snowy white in October–November and back to brown in March–April (Noback 1935; R.A. Powell, unpubl.). In all these cases, the animals remained at about their usual latitude.

So heredity, as well as winter climate, must be involved in the control of winter whitening. The weasel populations of a local locality must be *adapted* to the expected conditions of their homeland. Where it is always very snowy or always very mild it is easy to predict whether a white coat will be needed or not; there is no need for any individual to make a decision. This proposition was confirmed by Feder (1990) in a series of elegant experiments on eight captive stoats kept on short-day photoperiod at the University of Fairbanks, Alaska. Two stoats from Alaska and two from Oregon were kept warm (18°C to 20°C), and two from each place were kept cold (below zero). All four Alaskan stoats molted to white, and all four Oregon stoats molted to brown, regardless of temperature.

At one stage it seemed that a second argument for some degree of genetic control could be deduced from the fact that, in all temperate countries where not all individuals turn white automatically, those that do are significantly more often females. Hutchinson and Parker (1978) pointed out that a correlation between sex and tendency to turn white would necessarily follow if the gene controlling winter whitening is sex linked—that is, dominant in one sex and recessive in the other. The advantage of this arrangement would be that it is an efficient way of maintaining a genetic polymorphism, a constant readiness to meet sudden changes in the characters favored by natural selection in present conditions. With respect to winter whitening, those animals that stay brown in a mild winter are best adapted, whereas those that turn white in the occasional exceptionally long snowy winter are the lucky ones. A stable, sex-linked genetic polymorphism ensures that the population always includes a few individuals with every combination, every winter, so some will always benefit whatever happens.

The problem with this idea is that the mechanism controlling the molt cycle involves prolactin, which also plays a very important part in the reproductive cycle of females. It would take a very careful, critical experiment to disentangle the different roles played by prolactin in the control of winter whitening and reproduction in the two sexes, and that has not, so far, been done.

In the Transition Zones

Not all weasel populations are so predictable in their winter dress as are those of Alaska and Oregon. As Figure 3.4 shows, there seems to be a "zone of indecision" where winter climates are quite unpredictable and where it might be an advantage to have some sort of temperature switch, enabling a given animal to decide one way or the other—to turn white this year or not. Within this zone, white or brown animals may be found in different proportions *apparently* varying with location and temperature.

In the transition zone, the results of translocation experiments appear unpredictable. For example, in 1953 Rothschild and Lane (1957) caught a young stoat in the Swiss Alps and brought it home to England, about 5° of latitude further north. In September, the transplanted animal turned pure white, despite the mild autumn and the established custom of locally resident stoats to stay brown. In contrast, Roger and Consie Powell brought two stoats already in their white winter coats to North Carolina from Minnesota in winter (10° of latitude further south). After a summer in brown, they did not turn white in the following winter, and neither did the female's offspring, born in North Carolina.

The idea that individuals might have a "choice" of whether to turn white or not is a completely different proposition from the idea of genetic adaptation to expected climate at the population level. It implies there could be provision for some level of response to weather, rather than climate, and at the level of the individual rather than the population. How could this happen?

One early study that seemed to suggest a possible mechanism was that of Gaiduk (1977). Gaiduk calculated that in Belorussia, where he worked, the fur of the hindquarters and lower flanks of stoats would grow white if the temperature at the time was 2°C, but it had to be below −1°C before the head and back would turn white. Even though all stoats in Belorussia normally turn white every year, Gaiduk's results implied that there could be a minimum critical temperature below which new fur grows only white, which is different for different parts of the body. In milder regions the critical temperatures required for whitening could be different, probably higher, and the autumn weather more variable, so that the threshold might be exceeded for one part of the body but not for another. If true, this arrangement could amount to a "switch" that could control the supply of melanin with short-term changes in the weather. For a given individual animal, the switch could be "off" while the new fur grew on the tail and sides, but by the time the new dorsal fur was growing, it could be "on" again.

The result would be that the animal would appear to be "pied" (Figure 3.6), and in places where the weather conditions in autumn are very variable, one might observe a mixture of white, brown, and pied individuals in different proportions each winter. These animals would look as if they were caught in the middle of molting. An alternative explanation could be that they were wearing a full-grown winter coat in which the switch had turned off the supply of melanin before the cycle of hair growth was complete. If so, the proportion of pied stoats should be correlated with local and annual variation in winter temperature.

The trouble is that Gaiduk's intriguing conclusion includes a hidden assumption, that low temperatures predict the snow cover against which white coats

Figure 3.6 Pied stoats appear to have been caught halfway through molting, but are more likely to be heterozygous for the alleles that control whitening of the winter fur.

would be an advantage, which is not always true. Moreover, it was based on simple observations, not on manipulative experiments. Where such experiments have been done, using captive animals from populations that always reliably turned white or stayed brown, temperature did not control whether stoats turned white or not (Feder 1990).

We suggest another explanation. If winter whitening is controlled by a polygenetic system, then *all* individuals must have the genes that determine the response of the follicles to MSH in autumn. The difference between well-adapted populations in the far north and far south is that, in each, only the genes for *either* white or brown in winter would be expressed. The "transition zones" could represent, not places where *individuals* can exercise an option, but places where different *proportions of individuals* have different genetic origins, including hybrids. The "transition zones" could mark the places where postglacial recolonization of the northern lands (Chapter 1) has brought populations of winter-brown weasels from snow-free southern refugia into contact with populations of winter-white weasels from snowy northern refugia. The contact zone is narrow and well defined where least and common weasels meet in Scandinavia and where stoats of different origins meet in northwestern North America, but is wider and more diffuse in other temperate populations of stoats living in variable climates.

If this hypothesis is correct, then there is no need to postulate a temperature "switch." Extensive hybridization between these different stocks, plus natural selection acting on the results to produce locally adapted resident populations, would be sufficient to produce the patterns we see. For example, it would be reasonable to suppose that more of the stoats that have lived in cool northeast Scotland since soon after the ice ages might have retained their adaptation to colder periods in the past, including expression of genes for winter whitening, than stoats in the milder climate of England. Indeed, in northeast Scotland at 57°N, over 90% of stoats of both sexes seen by Hewson and Watson (1979) in each January from 1969 to 1974 had changed color, of which most were in full ermine.

By contrast, most English stoats stay brown. An inquiry conducted by questionnaire in 1931–32 in Yorkshire (54°N) found that, of 2,930 stoats killed, only 21 were in full ermine (all females) and another 175 (85 females, 90 males) were pied. Ninety percent of the informants had never seen a fully white stoat, and 40% had never seen a pied one (Flintoff 1933, 1935).

Likewise, very few stoats collected over 4 years by King and Moody (1982) in Fiordland, in the cool far south of New Zealand, during the months between the end of the autumn molt and the beginning of the spring molt (June to November), were white. In one area at 44°S to 45°S, only 71% of the 34 females collected and 47% of the 62 males collected had any white hairs at all. Of 124 females and 305 males collected from the whole country during those months, only eight were in almost full ermine, and even they still had a few brown hairs remaining around the eyes. Most were pied, at best, and many had only a few white hairs just in front of the black tail tip.

King and Moody (1982) compared the proportions of white or pied stoats collected in 14 samples taken in the southern winter and spring months of June to November with the local weather records. The results were consistent with the idea of local adaptation linked to temperature. The variation among samples in the proportion of pied stoats was significantly correlated, both with the mean daily minimum temperature in July (the coldest month) and with the number of days of ground frost per year. Further, white and pied stoats were more common at higher elevations and more southerly latitudes, although, even there, the vast majority of stoats stayed entirely or mostly brown.

Stoats from the cool end of Britain (the far north) turned white more consistently than those from the cool end of New Zealand (the far south) even though they were not exposed to a much colder or snowier climate than the New Zealand ones sampled in these studies. In both areas, the minimum temperature for the coldest month is usually within 5°C either way from zero, and in both, snow seldom lies more than 10 to 20 days in any winter. In this and other respects, the New Zealand stoats are more like English than Scottish ones. New Zealand stoats are probably descended from English (non-whitening) stock, and they have lived in New Zealand for only about 120 generations. In those that have spread into the coolest parts of their new homeland, natural selection might still be in the process of retrieving their lost or latent genes for whitening. These observations make sense if heredity, rather than temperature, controls what color of winter coat is worn by individuals.

The Advantage of Winter Whitening

Why do weasels turn white at all? Obviously, the mechanism must be maintained by natural selection; that is, it would disappear if it did not confer an advantage more often than a disadvantage. For example, if the small stoats of Northern Ireland are descended from a glacial relict population that once whitened regularly (Chapter 1), they must have lost the habit since the climate warmed. Where snow cover is too infrequent and too brief to convey any advantage, then the genes for whitening should soon be silenced or weeded out, especially if a white stoat is especially conspicuous against a dark background. Contrariwise, in northern Canada the penalty for staying brown in winter must be substantial, since no Arctic weasels ever try it—or those that do never live long enough to be observed.

At least three sorts of benefits of winter whitening have been suggested. The first is that white fur is supposed to conserve heat. According to the laws of physics, a black body is a perfect radiator of heat, so some early biologists assumed that a white body would therefore be a less perfect one. But this idea is mistaken, because the radiation of heat from an animal's body is in the far infrared regardless of the color of its coat. Consequently, all animals are "black bodies" with respect to heat loss by radiation (Hammel 1956). It is the length and thickness of the fur, not its color, that determines whether it conserves heat well, and the winter fur of weasels is a poor insulator even in areas where it is always white.

The second suggestion is that white fur is a camouflage that helps a weasel catch its prey. Predators that lie in wait for their prey to pass by unawares, or that sneak up on them quietly from some distance away, rely on camouflage and appropriately stealthy behavior to conceal themselves from their intended victim for as long as possible. This might explain the white coat of the polar bear; however, weasels, brown or white, do not hunt so much by stealth and cunning as by constant, active searching in every possible runway and hiding place (Chapter 6). Even if the match of a white weasel's coat against snow were perfect, which it is not, its movements would give it away—a weasel is hardly ever still except when asleep. The trick of melting into the background works only for animals prepared to move very slowly or to sit immobile for long periods, and that is too much to ask of a weasel.

The third suggestion is that white fur is a camouflage that helps a weasel avoid larger predators. All weasels are small enough to be in danger of attack from hawks, owls, and foxes (see Figure 11.6). Very obvious mismatches between coat color and background would invite the immediate attention of any larger predator, especially raptors. Predation by raptors can be a serious hazard for individual weasels, which explains several things about them, including the old question of why stoats and longtails have black tips on their tails and common and least weasels do not (Chapter 11). It seems most likely that the northern weasels try to match their snowy backgrounds, not to catch a meal but to avoid becoming one.

ROBES OF STATE

Northern members of the weasel family wear fine, lustrous furs that are not only beautiful in themselves, but have also become symbols of luxury and status. Native Americans used ermine skins to trim not just their ceremonial headdresses and pipes but also their clothing and hair. Headdresses were often draped with many ermine skins falling to either side of the bearer's head (Figure 3.7) and decorations on ceremonial shirts included ermine skins. Pipes sometimes boasted many ermine skins among other decorations. Not only men used ermine

Figure 3.7 Native Americans incorporated weasel skins into their ceremonial headdresses and pipes and used weasel skins for personal ornamentation. This portrait of Two Moons, a Cheyenne who fought at Little Big Horn, shows his headdress decorated with ermine skins falling to either side of his face. (Photograph by Edward S. Curtis.)

skins for decoration; women also ornamented their clothing and sometimes braided ermine skins into their hair with striking effect.

In medieval Europe, the furs of ermine, marten, and sable were reserved for the upper ranks of the nobility, while lesser folk had to make do with rabbit and cat. Ermine is traditionally worn by British justices and peers; in 1937, 50,000 ermine skins were sent from Canada to make robes for the coronation of King George VI (Haley 1975). There is an ironic contrast between the villainous reputation of living weasels (Chapter 1) and the use of their pelts to adorn the robes of the highest ranking justices.

Ermine skins were once very important in the international fur trade and still do have some value, though much less than before. In the late 1920s and early 1930s in New York State alone, Hamilton (1933) reckoned that 100,000 ermine skins were traded each year, at an average price conservatively estimated at $0.50 US each. These figures imply that, during an economic depression, stoats and longtails between them were generating an income of $50,000 a year for the rural community in that state alone. Likewise in Saskatchewan, around 60,000 to 70,000 longtails were taken every year during the 1930s. Since then, the total harvest of weasel skins in North America (all three species) has steeply declined (Figure 3.8). By the 1970s, the annual harvest of weasel pelts for the whole of the United States and Canada together was negligible, and those that

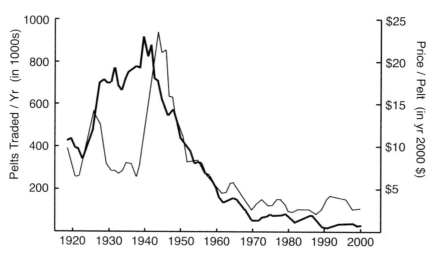

Figure 3.8 Fur buyers' records show that the steady decrease in trading of weasel pelts (*heavy line*) in North America through the twentieth century correlates with the prices paid for pelts in constant dollars (*light line*). During the Depression, many weasels were trapped for their pelts despite the low prices paid. The modest increase in prices paid for pelts in the 1990s did not lead to increased trapping. (Data for pelts traded redrawn from Proulx 2000; data for pelt prices from Statistics Canada.)

were collected comprised only a tiny proportion (under 2%) of the millions of carnivore pelts traded in those years.

The lack of interest in weasel furs nowadays is clear from the huge fur warehouse of what used to be the Hudson's Bay Company in London. In 1984 there were no tradable weasel skins among their current stock of thousands upon thousands of pelts (C. M. King unpubl.). The reasons for the decline are probably that the small ermine skins are so fiddly to prepare, and that so many are required to make even a collar for an earl's cape. Cheaper kinds of fine, white fur can now be obtained elsewhere. It is certainly not that fur trapping has had any influence on weasel populations (Chapter 11). Nevertheless, the fur trade did, in its time, stimulate some important research on the biology of weasels, especially in the former Soviet Union and the United States.

4 | Body Size

ANIMALS ARE NOT the size they are just by chance. One of the most important facts of life for any animal is its own body size. Size governs, among other things, what foods it can reach or catch; what refuges it can get into; what other animals are its predators or competitors; how successful it will be in obtaining a home range and one or several mates, and in leaving young; and, for mammals and birds, how much it costs in energy to keep warm.

THE IMPORTANCE OF BEING THE RIGHT SIZE

Because size is so important to the lives and family prospects of individuals, and is at least partially a heritable characteristic, it is a powerful agent of evolution. In any generation individuals will be of various sizes, but the ones that happen to have developed the best size (and other qualities) for the conditions of the time will be the ones that leave the most offspring, which will also be more or less that size. If the conditions change, the best size may be different, and natural selection will quickly favor another genetic lineage, with different average body size, for preferential survival. Evolution is the result of a conversation between the environment and the lives, particularly the reproductive processes, of the animals.

We may, therefore, expect to find that the average body size of each species of weasel living in a given place is the one that gives individual breeding adults the best chance of success in producing young in the local conditions. The best size to be in a particular environment need not be the same as in any other environment; hence, there is enormous local variation in size across the huge geographic range occupied by weasels (Table 4.1). Moreover, the best size to be is not the same for both male and female weasels, for reasons explained in Chapter 14. Males are always substantially larger, but not by a constant amount.

PROBLEMS OF MEASURING SIZE

Study of the geographical variation in body size of animals is a hazardous business, especially when weasels are the subject. The first problem is to decide which measure of body size to use. The two most obvious are head-and-body length

Table 4.1 Some Representative Examples of Local Variation in Mean Body Size of Weasels

	M. nivalis		*M. erminea*	*M. frenata*
	N. NIVALIS	N. VULGARIS		
Alaska, Yukon and Northwest Territories (Holmes unpubl.)				
HBL　Male	172	—	230	—
Female	148	—	203	—
Alaska, Kodiak Island (Holmes unpubl.)				
HBL　Male	—	—	236	—
Female	—	—	191	—
Alaska, mainland Southeast (Holmes unpubl.)				
HBL　Male	—	—	211	—
Female	—	—	187	—
Boreal Canada (Holmes unpubl.)				
HBL　Male	151	—	222	—
Female	136	—	182	—
Vancouver Island, British Columbia (Holmes unpubl.)				
HBL　Male	—	—	191	—
Female	—	—	170	—
British Columbia island population (Holmes unpubl.)				
HBL　Male	—	—	203	—
Female	—	—	176	—
Rocky Mountains, British Columbia, and Montana (Holmes unpubl.)				
HBL　Male	—	—	279	261
Female	—	—	204	235
Great Plains, U.S. and Canada (Holmes unpubl.)				
HBL　Male	—	—	—	270
Female	—	—	—	243
Great Basin and Rocky Mountains, U.S. (Holmes unpubl.)				
HBL　Male	—	—	166	247
Female	—	—	153	215
Northeastern Oregon (Holmes unpubl.)				
HBL　Male	—	—	—	220
Female	—	—	—	213
Coastal Oregon and northern California (Holmes unpubl.)				
HBL　Male	—	—	177	248
Female	—	—	158	216
Coastal Central California (Holmes unpubl.)				
HBL　Male	—	—	?	267
Female	—	—	?	221
Sierra Nevada (Fitzgerald 1977)				
BWT　Male	—	—	59	256
Female	—	—	45	122
Upper Midwest U.S. (Holmes unpubl.)				
HBL　Male	168	—	210	262
Female	147	—	182	245

Table 4.1 (Continued)

		M. nivalis		M. erminea	M. frenata
		N. NIVALIS	N. VULGARIS		
Southwestern Midwest, U.S. (Holmes unpubl.)					
HBL	Male	—	—	—	257
	Female	—	—	—	231
Indiana (Gehring & Swihart 2000)					
BWT	Male	—	—	—	195
	Female	—	—	—	111
New York State (Hamilton Jr. 1933)					
HBL	Male	—	—	201	270
	Female	—	—	181	218
BWT	Male	—	—	81	225
	Female	—	—	54	102
Northeastern U.S. and Southeastern Canada (Holmes unpubl.)					
HBL	Male	156	—	217	253
	Female	154	—	181	209
Québec (St-Pierre et al. in press-b)					
HBL	Male			217	274
	Female			—	—
BWT	Male			134	256
	Female			—	—
Britain (Corbet & Harris 1991)					
HBL	Male	—	202–214	297	—
	Female	—	173–181	264	—
BWT	Male	—	106–131	321	—
	Female	—	55–69	213	—
Ireland (Fairley 1981)					
HBL	Male	—	—	252–278	—
	Female	—	—	208–230	—
BWT	Male	—	—	233–334	—
	Female	—	—	123–165	—
Sweden (Stolt 1979; Erlinge 1987)					
HBL	Male	166	189	?	—
	Female	148	154	?	—
BWT	Male	54	73	184–230	—
	Female	35	36	98–137	—
Switzerland (Güttinger & Müller 1988)					
HBL	Male	167	168	?	—
	Female	144	142	?	—
BWT	Male	53	54	?	—
	Female	31	32	?	—
Italy (C. Magrini, unpubl.)					
HBL	Male	—	222	—	—
	Female	—	180	—	—
BWT	Male	—	193	—	—
	Female	—	82	—	—

(*continued*)

Table 4.1 (Continued)

| | | M. nivalis | | | |
		N. NIVALIS	N. VULGARIS	M. erminea	M. frenata
Egypt (Osborn & Helmy 1980)					
HBL	Male	—	289	—	—
	Female	—	242	—	—
BWT	Male	—	390	—	—
	Female	—	209	—	—
Siberia (Heptner et al. 1967)					
HBL	Male	160	—	260	—
	Female	?	—	212	—
BWT	Male	53	—	166	—
	Female	41	—	?	—
New Zealand (King 2005b)					
HBL	Male	—	217	285	—
	Female	—	183	256	—
BWT	Male	—	127	324	—
	Female	—	58	207	—

—, species absent or rare in that area; ?, data missing

Mean head–body lengths (HBL) in mm, weights (BWT) in g. Measurement ranges given are between, not within, sample means. For continental-scale geographical variation in skull lengths, see Figures 4.2 to 4.4. Note that sexual dimorphism is more pronounced in weight than in length. There is extensive north–south variation in Irish and Swedish stoats (larger in the south) and in British common weasels (larger in the north).

and total body weight, but records of these are scarce. Body weight is easy to measure but not very reliable, as it is affected by how recently a weasel last had a decent meal (King 1975c).

Skulls are easy to clean using a simple chemical method (McDonald & Vaughan 1999). Clean, dry skulls are small and inoffensive, and can be stored indefinitely in air. Large numbers of skulls of weasels can be kept in a few drawers; some natural history museums have collections going back for a hundred years. The condylo-basal length of a skull (defined in Figure 4.2A) is easily measured from existing material, so it is, therefore, the most readily available estimate of size. The simpler phrase "skull length" used here refers only to this measure of skull length.

Skull length has the slight disadvantage that it gives a less immediately obvious picture of body size than body weight or length, but it is easy to measure accurately, and in adults is immune to minor variations in environmental conditions. Other potential measures of body size in carnivores, such as the diameter of the canine teeth, vary more with the technique of attack, and the mechanical stresses of making a kill, than with body size itself (Biknevicius & van Valkenburgh 1996).

A second problem is that, since variation in size of a species as a whole must

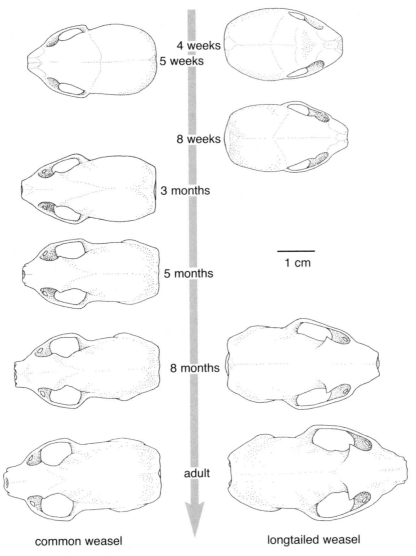

4 weeks
5 weeks

8 weeks

3 months

1 cm

5 months

8 months

adult

common weasel   longtailed weasel

Figure 4.1 Skulls of male common weasels and long-tailed weasels of known age, showing the changes in skull shape with age. Skulls are drawn at actual size. (Redrawn from King 1980e [common weasels bred in Germany by F. Frank] and Hamilton 1933 [long-tailed weasels bred in New York].)

be studied from adult animals, it is important to reduce the amount of error due to the accidental inclusion of young animals, not yet full-grown. Fortunately, this can also readily be done from skulls. The skulls of all young animals have open sutures between the growing skull bones, whereas in the skulls of older animals these suture lines are closed or even invisible (Figure 4.1).

The skulls also change in shape with age. The fully adult form is quite distinct from the juvenile form, especially in the larger species in which the sequence of changes continues through most of the first year of life. The single feature that best expresses these changes is the development of the postorbital constriction. The ratio between interorbital and postorbital widths, the postorbital ratio (King 1980e, 1991a), is a reliable method of identifying young animals and, unlike the best alternative method, tooth sectioning, it can be used on museum material, which must not be damaged.

A third problem is that of comparing measurements taken by different people at different times, perhaps in different ways, and often on very small samples. Potential errors thereby introduced are less troublesome with skulls of small animals, which can be measured with a micrometer between easily definable points. In recent years, the data on cranial variation in European and North American weasels have greatly improved. Many recent studies present much larger datasets than had ever been available before, controlled for individual variation in measurements, and use sophisticated new techniques to analyze them (Eger 1990; Meia 1990; van Zyll de Jong 1992; Reig 1997).

Yet a fourth problem arises from natural variation. This is particularly significant in weasels, in which pronounced sexual dimorphism in body size is very marked and variable both with locality and in time. Males are nearly always better represented than females in samples of weasels but, fortunately, there is a general covariation in size of the males and females of one species. If the objective is only to map the variation on the scale required to make comparisons across whole continents, it is fair to take the males as representing each local population as a whole.

SIZE VARIATION IN THE WEASELS OF THE NORTHERN HEMISPHERE CONTINENTS

Mapping the Patterns

Only two of the three North American species are represented in most places. The most widespread combinations are least weasels and stoats in the north and longtails and stoats in New England and the west. All three are found together in a broad stretch of country from Minnesota across to Pennsylvania. If longtails are present, they are always the largest, and if least weasels are present, they are always the smallest. Stoats may be either the larger or the smaller of a set of two, depending on which the other one is. In the southern United States and south into Mexico and beyond, longtails hold the stage alone.

In North America, stoats become larger in the Northwest, where they are about the same size or a little smaller than the northeastern Eurasian stoats just across the Bering Strait (Eger 1990). The smallest American stoats, in the South and especially in the Southwest, are considerably smaller than European common weasels. Stoats elsewhere range in size from much smaller to much larger than longtails (Figure 4.2, Table 4.1). By contrast, North American least weasels and longtails are no larger or smaller in any particular direction (Figures 4.3 and 4.4); one can find longtails of roughly the same size from Canada right down into South America. Most North American least weasels are very small, and the next smallest are their nearest relatives in the Old World, those inhabiting Siberia. In parts of the southwestern United States, longtails are smaller than West European stoats and stoats are smaller than West European common weasels.

Stoats and least or common weasels live together over most of Eurasia, and of the two, stoats are always the larger. Both are smallest in the Far East and north of Eurasia, and become larger toward the west and south (Figures 4.2 and 4.3). Both stoats and common/least weasels living at high elevation in the Alps, Caucasus, Altai, and Tien Shan Mountains are particularly small. Otherwise, the largest Eurasian stoats are found at the southwestern edge of their range (from the Netherlands across the North European Plain), and the largest common weasels are found south of the southern limit for stoats, particularly in Egypt, where they are much larger than stoats are in eastern Siberia. The general patterns are very obvious from skins (Figures 4.5 to 4.7), and the finer details plotted on continental-scale maps show a mosaic of local variations (Ralls & Harvey 1985; Meia 1990; van Zyll de Jong 1992; Reig 1997; Abramov & Baryshnikov 2000). The details of these local variations, however, do not affect the general trends, which are unmistakable at continental scale.

Why are the patterns of variation in the individual weasel species so confusing and contradictory? There is no obvious reason why the smallest stoats should be found in the south and west of North America but in the north and east of Eurasia. Both are harsh environments, certainly, but they are harsh in different ways—the high Rocky Mountains present different challenges to animals compared with the vast cold expanses of Siberia. Even more difficult to explain are the smaller-scale trends that contradict the general regional pattern. For example, stoats vary over a much greater range of sizes in North America than Eurasia, while common weasels vary much more in Europe than in North America. Although common weasels in Europe generally tend to become larger southward, in Britain the largest common weasels are in the north (Table 4.1). Even within Britain, the general north–south cline is interrupted by local variations.

If we view all the weasels as a single group, however, the continental-scale pattern is the same in both the Old and the New Worlds. Weasels in general are relatively small in the far north, both in North America and in Eurasia. The mean skull length in stoats, always the largest species in the north, seldom exceeds 46 mm right around the Pole, and least weasels measure 30 to 32 mm in Alaska

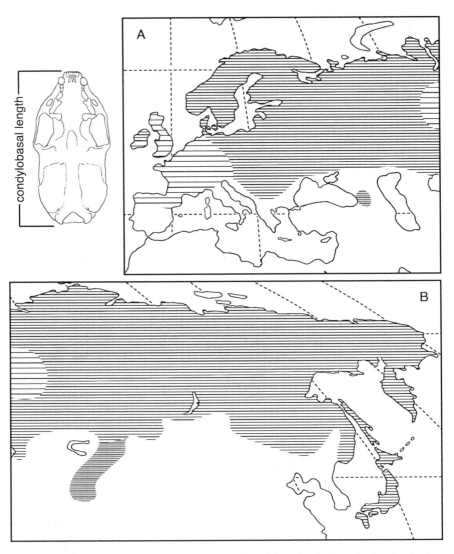

Figure 4.2 Geographical variation in condylobasal length (skull size) of male stoats in (A) Europe, (B) Asia, and (C) North America. The females generally vary in the same way, although the degree of difference between them is locally variable. The distribution of size classes is schematic only. (Data from Eger 1990, Güttunger & Müller 1988, Hall 1951, Heptner et al. 1967, Holmes 1987 and unpublished data, Kratochvil 1977, Meia 1990, Meia & Mermod 1992, Ralls & Harvey 1985, Reichstein 1957, Reig 1997, and Schmidt 1992.)

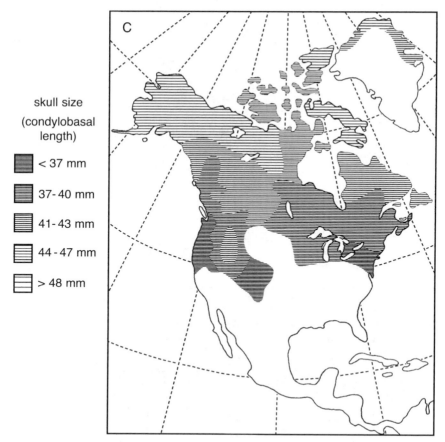

skull size
(condylobasal
length)

▨ < 37 mm

▨ 37-40 mm

▨ 41-43 mm

▤ 44-47 mm

▤ > 48 mm

Figure 4.2 (continued)

and Canada and 33 to 35 mm in Russia. The largest weasels are always found in the south: Longtails in the United States and common weasels in Egypt reach 50 to 53 mm in skull length or more.

Size and Temperature

Geographical variation in the body sizes of animals is common, and many explanations have been offered since the nineteenth century. At that time, many zoologists still believed that it should be possible to explain much of the riotous variety of life in terms of simple, formal "Rules," comparable with those of physics and chemistry and based on the same properties of energy and matter that govern inanimate things. For example, warm-blooded animals living in cold climates have to expend a huge proportion of their total energy budget on keeping warm, and the smaller they are, the greater the relative expenditure and the

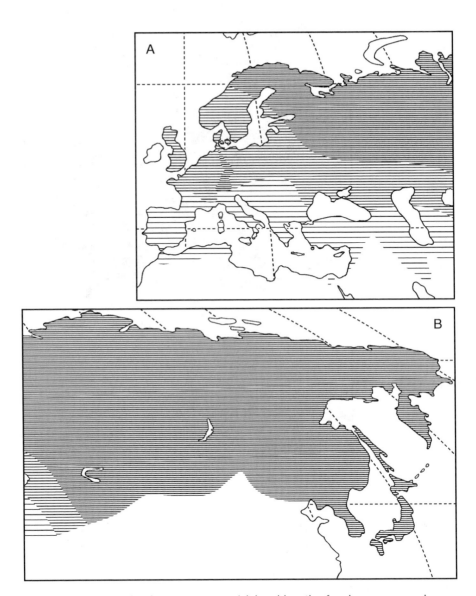

Figure 4.3 Geographical variation in condylobasal length of male common and least weasels in (A) Europe, (B) Asia, and (C) North America. Our arbitrary contour break in Scandinavia falls north of the boundary between least and common weasels, shown in Figure 1.13. See comments in Figure 4.2. (Data from Hall 1951, Holmes 1987 and unpublished data, Kratochvil 1977, Meia 1990, Meia & Mermod 1992, Morozova-Turova 1965, Ralls & Harvey 1985, Reichstein 1957, and van Zyll de Jong 1992.)

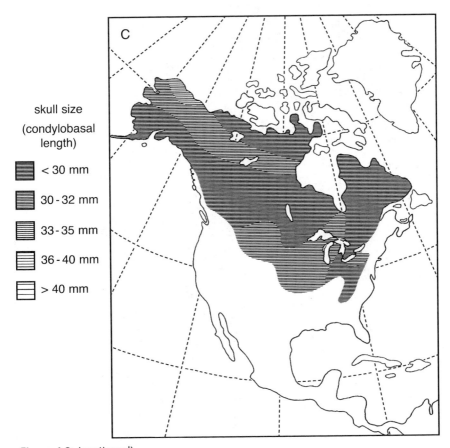

Figure 4.3 (continued)

greater the danger of irreversible chilling. In some species of mammals, individuals living in colder climates have larger bodies than their relatives in the south, and shorter ears and tails.

In 1847, Bergmann suggested that the larger size of northern mammals is a matter of energy conservation, since large mammals have relatively less surface area exposed to the cold air in relation to the mass of the body in which heat is generated. Bergmann's Rule therefore predicts a steady increase in body size northward in related mammals and birds living in habitats equally exposed to increasingly severe environmental conditions. The same logic underlies Allen's Rule, which explains the relatively shorter ears and tails of northern mammals as a means of reducing the area of vulnerable appendages from which heat may escape.

Weasels in general are particularly sensitive to thermal stress at low temperatures (Chapter 2). Metabolic inefficiency costs them very dearly in the vast, cold northern parts of their ranges, so one might expect them to be prime

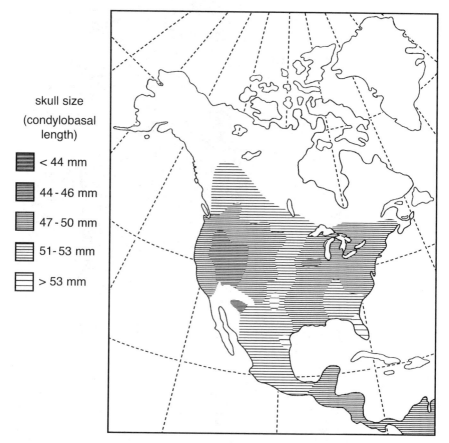

skull size
(condylobasal
length)

■ < 44 mm

■ 44 - 46 mm

■ 47 - 50 mm

■ 51 - 53 mm

■ > 53 mm

Figure 4.4 Geographical variation in condylobasal length of male long-tailed weasels. See comments in Figure 4.2. (Data from Hall 1951, Holmes 1987 and unpublished data, and Ralls & Harvey 1985.)

examples of Bergmann's Rule. A simple comparison between skull size and latitude, however, does not confirm this expectation. Only in North America are stoats substantially larger in the north of their range than in the south (Eger 1990); common weasels in Eurasia are just as substantially larger in the south, and all the rest are more or less indifferent (King 1989a). Moreover, the American stoats that are larger in the north can be said to be so only by comparison with their exceptionally small relatives further south in North America; they are not larger than their relatives at the same latitudes in eastern Eurasia. Within single regions spanning a range of climates, Bergmann's Rule does not explain the local variation of stoats in Russia, Europe, or New Zealand (Petrov 1962; King & Moody 1982; Erlinge 1987).

Figure 4.5 Skins from the collection of the Natural History Museum, London, arranged to represent the range of variation in body size of adult male Eurasian weasels with latitude (except that the Swiss specimens are out of order). Above, stoats; below, common and least weasels. Origin of specimens, left to right; stoats: Switzerland, Norway, France, Hungary; common and least weasels: Switzerland, Norway, France, Hungary, Italy, Egypt. Scale of centimeters at left.

Figure 4.6 Skins from the collection of the Natural History Museum, London, arranged to represent the range of variation in body size of adult male Eurasian weasels with longitude. Above, stoats; below, common and least weasels. Origin of specimens, left to right, both species: England, Germany, European USSR, Asian USSR.

Figure 4.7 Skins from the collection of the Natural History Museum, London, arranged to represent the range of variation in body size of adult male and female long-tailed weasels with latitude. Origin of specimens, left to right: Colombia (female right, male left), northeastern United States, Canadian Northwest Territories.

Bergmann assumed that temperature is directly correlated with latitude. At low elevation this is generally correct, and the skull lengths of weasels in North America are about as well correlated with temperature as with latitude. But temperature is also correlated with elevation. The high mountains that cross Eurasia from east to west at low latitudes (e.g., the Alps, Caucasus, and Tien Shan) therefore confuse the simple size–latitude correlation for that part of the world. For example, stoats and common weasels living high in the mountains of Switzerland are all distinctly smaller than those in the surrounding countries, and least weasels survive there as a separate, southern population isolated from their relatives in the cooler north (Güttinger & Müller 1988). This problem can be removed by comparing size directly with mean annual temperature; the result shows clearly that both sexes of both species of weasels in Eurasia are smaller in colder environments—the very opposite of what Bergmann's Rule predicts.

This result is puzzling only until we remember that weasels evolved their long, thin shape in response to the opportunity to hunt small rodents in their own burrows and under snow. This means that, for much of the time that the northern weasels are hunting, they are not exposed to the full rigors of the winter air temperatures as much as their southern relatives are. So, comparing the sizes of northern and southern weasels in terms of Bergmann's Rule is logically invalid. Besides, the relationship between size and temperature cannot be crucial for weasels, since the difference between the males and females living in one place is often at least as great as that between the northern and southern members of either sex. We need, then, to find another explanation for the general north–south variation in size of the weasels as a group.

Because the size of a body affects many of its functions, and the best size for one function is not necessarily the best for any of the rest, the actual size of the whole animal is usually a compromise. Sandell (1989) proposed a model predicting the best sizes that male and female stoats could be. Sandell designed his model to take into account the differences in the reproductive roles and in the seasonal energy requirements of males and females. He assumed that the total amount of energy available to any animal in any one day, expressed as a multiple of its basal metabolic rate, is limited—which certainly seems reasonable for weasels (Chapter 2). Then Sandell allowed for the important facts that a weasel's priorities for energy expenditure are not the same all year round, and that they are not the same in the two sexes.

In winter, the huge costs of keeping a small thin body warm in a cold climate load the loss columns of a weasel's budget with high "overheads." Consequently, during winter the most important thing for weasels of both sexes is to economize on their daily expenditure of energy as much as possible. The best way to do this is to be extremely good at hunting, so as to be able to spend the least possible time each day out of the den searching for food. For any combination of values measuring foraging efficiency and air temperature, there will be a theoretically "optimum" body size, which will be the same for both sexes. In the breeding season, however, the equations must be different. In the intense competition for mates, larger males tend to be more successful, so during the breeding season the optimum size for males is larger than in winter. Conversely, even in the breeding season, foraging efficiency is still the most important consideration for females, since it determines the number of young they can rear. The optimum size for a female in summer is, therefore, controlled by the same variables as in winter.

Real animals, of course, vary in size individually, and the biogeographical history of their species is important too (Chapter 1). Sandell's work suggests that the most successful animals in any given place should be the ones that have chanced on the best compromise between possible sizes at different seasons. Sandell's model could in theory predict the optimal body size for members of each weasel species in each place, and, thereby, predict geographical variation in body size of weasels, if only we could accurately measure foraging rates. That has not yet been done.

The Costs and Benefits of Being Small

A simple, verbal model that explains the continental-scale variation in body size of the weasels as a group (see Figure 2.8) was worked out, independently of Sandell's calculations, by King (1989a). The small size and elongated shape of weasels give them a huge advantage when hunting rodents in their burrows, but the physiological consequences of smallness are nearly all unfavorable. In years when rodents are scarce, a weasel's constant need for well-insulated shelter is a handicap, because just at the very time when food is most hard to find and energy most precious, hunting expeditions must be long and ready-made dens are few. Weasels can store surplus food as caches, liable to theft and decay, but not as extensive body fat; they are extremely vulnerable to temporary food shortages, yet hibernation and long-distance migration are impossible. The huge additional energy needed for breeding is hard to find, except when rodents are abundant.

These are all serious problems for small, short-lived, warm-blooded mammals living in cold climates. The list of considerations in Figure 2.8 is not, of course, exhaustive, and none of them can be assigned to a scale of relative values. Nonetheless, on the whole Figure 2.8 appears to confirm Sandell's assumption that body size is probably most strongly influenced by whatever is the most workable balance between hunting efficiency and energy balance under the local conditions.

In very cold climates, the metabolic toll imposed on weasels by being long and thin must be outweighed by the advantages of being able to live and to hunt under the insulating blanket provided by snow (Chapter 2). The snow pack plays a crucial role in the life of northern animals and plants, and conditions within it and under it are well studied (Chernov 1985). Like their small mammal prey, the small northern weasels can pass the entire winter under the snow; in fact, the ability to escape from the infinite heat sink of the clear night sky is the condition of existence for all small mammals in the far north (Pruitt 1978), including weasels. Weasels can find both food and shelter under the snow—two pressing reasons why northern weasels must not be too large to live for many months of the year in the subnivean runways and nests of voles and lemmings (see Figure 6.2).

Small weasels are still deadly hunters, because for them small size does not re-duce killing power (Chapter 2) as fast as it increases searching efficiency. In the far north for much of the year, that means being good at searching under snow.

When the air temperature is mild and the wind-chill factor low, weasels can emerge onto the snow surface and their tracks are often seen (Nyholm 1959b; Korpimäki & Norrdahl 1989a; Korpimäki et al. 1991), but they are still abso-lutely dependent on subnivean nests and prey. For example, Teplov (1948) tracked stoats chasing game birds and squirrels across the snow in the Pechorsk state game reserve, but subnivean species (rodents and shrews) still comprised 88% of the stomach contents of male stoats, and 98% in females. Therefore, the need to retain access to undersnow tunnels and nests imposes a distinct limit to the size of northern weasels. This limit overrides all other considerations (King 1989b), and could explain why sexual dimorphism is less pronounced in the smallest weasels that live in snowy climates than in the larger weasels of further south (Table 4.1).

By contrast, in the milder climates of western Europe, weasels have less need to avoid exposure to the winter weather, and this relaxes the restrictions op-posing larger size in males. Larger males are socially dominant (Erlinge 1977a) and probably enjoy greater breeding success, so this is a powerful reason why large males will be favored wherever larger size is permitted by energy economy. This process, sexual selection, acts only on males, but it affects females, too, since both sexes share the same gene pool. Both sexes would also respond to the op-portunities presented by the presence in the south of various large prey, such as rabbits and squirrels, that can more easily be caught by large weasels and offer a better economic return than do small rodents. The net energy value of small prey is low because a weasel has to invest energy into finding and killing each one separately and has to eat it whole, including the bones and the fur. Large prey are more risky to attack but, once secured, they provide more than one meal, and the second and later meals are all free. There will be local variations of course but, all in all, larger-sized weasels are better off in the south.

This hypothesis predicts that, even in the south, females will tend to remain nearer the "right" size for rodent hunting than will males, because in most places in the northern hemisphere rodent hunting is still the key to their breeding success even where other prey are available (Chapter 10). It would be interest-ing to know more about the relationship between body size and breeding suc-cess in females. One might expect that, in rodent peak years when hunting is easy, females of any size might be successful, but in poor years a small female might be more likely to feed a litter than a hypothetical female that was the size of a male (Moors 1980).

It is one thing to demonstrate (1) a southward increase in body size among the weasels as a group, (2) the penalties of getting too large in the north or at high elevation, and (3) the advantages of getting larger in the south, for example, the presence of larger prey. It is quite another thing to suggest that (1) is caused by (2) and (3). All three could be unrelated consequences of something else altogether. As both King (1984a) and Erlinge (1987) have pointed out, in different contexts, it is very difficult to distinguish between a causal relationship and coincident side effects of some other process of adaptation. We can, however, suggest falsifiable explanations for observed patterns that future students of weasels might test. We hypothesize that the combination of energy balance, size of available prey, and sexual selection, in unknown proportions, explains why the niche for a weasel-shaped carnivore allows for only small individuals in snowy northern and high mountain climates, but larger ones in the milder lowlands (King 1989a).

We emphasize that the idea put forward here, that it is advantageous to weasels to be relatively small in the coldest and snowiest climates, is very general and applies only to the weasels as a group or, rather, to the largest or only local species. Other considerations govern the sizes of the separate species and of the two sexes of weasels relative to each other (Chapters 1 and 14). The concept is derived from the size distribution of contemporary populations of living weasels, but it is also consistent with other evidence that small size really is advantageous to cold-climate weasels, both in the past and now.

First, Kurtén (1960) showed that fossil stoats belonging to *M. palerminea*, the direct ancestor of modern stoats, were relatively small during the cold phases of the middle Pleistocene, and larger during the warm ones. Wójcik (1974) pointed out that the common weasels found as fossils in Polish caves and dated to the Eemian, the last interglacial period (about 120,000 years ago), resemble the common weasels of modern Poland, but those from the Weichselian, the last glacial period, are smaller, like the modern boreal least weasels that no longer live so far south except at high elevation in Switzerland. Less numerous fossil stoats from the same caves and dated to the last glacial period are within the range for modern stoats but smaller than average.

Second, among more than 4,000 skulls of stoats collected over 5 years from Tjumen, at 57°N in the north of the former USSR, there was a progressive decrease with age in mean skull length in every annual cohort. To Kopein (1969), that meant that small stoats were better adapted and lived longer in that severe climate than did larger ones. Extrapolated over the very long term, the advantages of small size to weasels might explain why weasels have remained small for millennia, in defiance of Cope's Rule, which predicts a gradual increase in size with time in long-lived mammalian lineages.

SIZE VARIATION IN WEASELS ON THE NORTHERN HEMISPHERE ISLANDS

There are thousands of islands off the mainland coasts of North America and Eurasia, ranging in size from Newfoundland and Britain down to scattered rocks.

All those now separated from the mainland by water less than 100 m deep must have been joined to it during the last glaciation, so all those that were not glaciated, or were freed from the ice before the rising sea level cut them off, could presumably have been colonized across dry land by animals from the nearest continental fauna living at that time.

Some islands were no doubt colonized by weasels then and have been continuously inhabited by weasels ever since. Other islands have been colonized since their isolation, once or many times, by weasels swimming from the nearest point on the mainland or, occasionally, deliberately or accidentally carried in boats by farmers or traders, not necessarily from the nearest mainland port. Only the largest islands provide enough space and enough prey for a permanent population of weasels: Among the offshore islands of Britain the lower limit is around 60 km² (King & Moors 1979a), and, except on mainland Britain (230,000 km²), Skye (1,600 km²), and the Isle of Wight (380 km²), only one species of weasel lives on each island.

Weasels on islands are often at least slightly different in size from those on the nearest mainland, especially if they have lived there in isolation for a long time. The published data on island weasels are sparse and inadequate, and the origin, route, and date of colonization of the immigrant weasels are seldom known. Yet, in recent years we have begun to compile a reasonable general explanation of the patterns we see. Included with Eger's (1990) analysis of the distribution of size and shape of the skulls of stoats across North America were several representative populations living on islands. For example, the stoats on Newfoundland, Kodiak, and Baffin Islands are significantly different in size, but not necessarily in shape, from those on the adjacent mainlands. The skulls of the stoats living on the Alexander Archipelago of southeast Alaska are not only much smaller and distinctly different in shape compared with skulls from the adjacent mainland, but also different from skulls collected in the far north and in the Pacific Northwest (Eger 1990). In the British Isles, the stoats of northern Ireland are very different in size and coloration from those in Scotland just across the northern Irish Sea.

Where did these differences come from? One obvious possibility is that islands such as Ireland and mainland Britain were occupied by different colonizing stocks—one larger in body size than the other (Kratochvil 1977). It is true that some physical characters remain stable over many generations, but size is not one of them. Size is too important to be determined merely by ancestry; it is acutely sensitive to contemporary conditions. The mean body size that we observe in a population of weasels is the one that best suits the present environment. To understand how that best size is determined, it is interesting and important to know the history of the colonizing stock, and the size the colonists were when they arrived, but other things also help to determine the outcome.

On the other hand, shape is less critical to survival than size, and local variation in skull shape of weasels may still show discontinuities that can be explained only in terms of different historical origins. Eger's (1990) analyses of skull variation in North American stoats show that pattern well (Chapter 1). In other words, contemporary adaptation and ancestry *both* matter. What we see now is a result of the interaction of contemporary ecological imperatives and historical dispersion.

Stoats in the Alexander Archipelago provide a locally detailed and elegant example of this interaction. Cook et al. (2001) describe three distinct genetic lineages of stoats inhabiting the islands (Chapter 1). Molecular data confirm that these observable differences are not only real and taxonomically significant but also much more locally complex than morphological data can reveal. So the present stoats have not only adapted to local conditions (in this part of the world, stoats are larger in cold, dryer areas on the continent and smaller in moist, warm ones, like the islands), but they also retain traces of their different historical origins (Eger 1990).

The stoats of Ireland have provided endless puzzles. For many years, the only measurements of Irish stoats available were those of Fairley (1971), whose material came from the northern part of the island. The stoats there are intermediate in size between the stoats and the common weasels of mainland Britain, and they were assumed to represent all Irish stoats. One early theory, put forward in a famous paper by Hutchinson (1959), linked the small size of Irish stoats with the absence from Ireland of common weasels.

Hutchinson based his paper on the theoretical idea of character displacement, which proposes that two similar, sympatric species must be different in size to coexist, using resources of different sizes (Chapter 14). This hypothesis predicts that, where one of such a pair of species lives alone, it could shift to an intermediate size suitable for exploiting all resources available. Hutchinson noted that Irish stoats seemed to fulfill these predictions very well. In fact, so also do the very large common weasels in the Mediterranean, especially on islands such as Sardinia that have no stoats (Figure 4.2 and 4.3). The same argument was applied to the very large common weasels of Egypt by Dayan and Tchernov (1988).

Unfortunately, all these examples are invalid. The common weasels in the Mediterranean are large because common weasels consistently become larger toward the south of Europe (Figure 4.3). The large common weasels in Sardinia and Sicily could be merely continuing the trend, and in Egypt they could actually be a different species (Chapter 1).

The stoats of Ireland get more and more interesting as new data appear. Although the stoats in the north are indeed very small, those only 250 km away in the south are about as large as British stoats (Table 4.1; Fairley 1981). Since there is also a southward increase in the proportion of Irish stoats that have the straight-line belly pattern typical of British stoats, Sleeman (1987) suggested that both trends might have been influenced by an unrecorded introduction of British stock into southern Ireland at some time in the past. Lynch (1996) confirmed that small-sized northern Irish stoats are very different in their skull characteristics from the stoats of England or Scotland, and are unlikely to be descended from ancestors that crossed to Ireland from the east on a hypothetical postglacial land bridge. Lynch suggested that the ancestors of the present Irish stoats could have survived the last glaciation on the exposed continental shelf south of Ireland, feeding on cool-climate rodents such as lemmings in the periglacial tundra, but he added that more recent human-assisted colonizations from England cannot be excluded.

Plausible as this idea is, problems remain. For example, we now have to explain why stoats in northern Ireland have not yet (after about 8,000 years) become larger in body size in response to the mild, damper climate of postglacial Ireland. It looks as if some other important factor is involved besides climate and ancestry.

Could the development of small, island stoats be stunted in some way? Those on Terschelling Island, off the coast of the Netherlands, are small on average, and they suffer badly from skrjabingylosis, an unpleasant condition caused by a parasitic nematode in the nasal sinuses (Chapter 11). Van Soest et al. (1972) suggested that this infection could be the cause of the distinctly small size of the Terschelling males. Much less evidence supports this suggestion than contradicts it, even as an explanation specific to Terschelling Island, for reasons discussed further in Chapter 11.

Another possible cause of stunting is that young stoats cannot reach their full potential size if they are not well fed in the first few months of their lives. Sleeman (1987) suggested that the limited choice of prey and short growing season in northern Ireland might restrict the growth of young stoats there. Such an effect can be directly demonstrated in feast-or-famine habitats such as New Zealand beech forests (Powell & King 1997), so there could be something in the idea. Nonetheless, there is no obvious reason why it should be confined to northern Ireland.

In Britain, the northward increase in size of common weasels is matched by a similar trend in body size of field voles (Corbet & Harris 1991). Voles are among the favorite prey of weasels and, in general, the body sizes of hunter and hunted are closely linked (Chapter 6). Unfortunately, we have no information on what aspect of body size of either predator or prey to measure, or any precise data on the size ranges and relative abundance of the prey eaten by weasels on any of the northern-hemisphere islands. We do not know how long we might expect any local adaptation to take or how far it might go. As Levins (1966) pointed out, no theory can be general, accurate, *and* precisely relevant to local conditions. If the size of the weasels on each island is a unique, local compromise, no general theory will explain the whole pattern unless it includes detailed information on the weasels living in all the island habitats, how long they have been there, and where they came from.

On the other hand, we do have some of this information for one group of islands in the South Pacific, and any attempt to construct a general theory about what determines body size in weasels should take account of what is happening there.

ADAPTATION IN BODY SIZE OF STOATS TRANSPORTED TO NEW ZEALAND

The stoats and common weasels that were introduced to New Zealand over the 20 or so years after 1884 were relatively large, because they probably all came from Britain (Chapter 13). They found an environment quite different from the one they left, with a generally milder climate (ranging from warmer than Britain in the north, where there are subtropical rainforests, to colder than Britain in the high mountains, where there are large, permanent glaciers). They also found a prey fauna completely different in size distribution. The newly established pastures where stoats were first released teemed with rabbits, which were familiar prey from England. In the neighboring forests, ship rats and birds were abundant (Chapter 5). But the staple prey of common weasels and stoats almost everywhere else in the world, the various kinds of voles, were completely missing. The only small rodents available were house mice, widely distributed but not nearly abundant enough to compensate for the absence of voles, and not particularly nutritious (Vaudry et al. 1990). Lizards and insects, especially large ground-dwelling forms, were then still common but, generally, the new environment offered more large and fewer small prey than did Britain.

Of course, the comparison is not simple, for two reasons. First, the factors determining the body sizes of male and female stoats are not exactly the same, so the two sexes might not respond in the same way to the same change in conditions. Second, the new environment offered by New Zealand to the immigrant stoats was not the same everywhere. Two main sorts of New Zealand native forest, one dominated by the five species of southern beech (*Nothofagus* sp.) and the other by native conifers (the podocarps) and broad-leaved trees, were quite different from each other and also from anything in the northern hemisphere. The alpine tussock grasslands, open gravel riverbeds, and second-growth scrub elsewhere in New Zealand were also quite unfamiliar. So, the way in which the immigrant stoats responded to the shift in prey size distribution need not be the same in both sexes nor in all habitats. And indeed, it was not.

We can ask two questions about the size of stoats in New Zealand: Have they developed any local differences in size related to habitat? and Have they changed relative to their British ancestors?

The first question is easy to answer in the affirmative. Within New Zealand, the adult males among the 1599 stoats collected in the 1970s from podocarp or mixed forest habitats, mostly at low elevations, were smaller than those from alpine southern beech forests and grasslands, by about 3% in skull length and about 4% in head and body length (King & Moody 1982). The same difference appeared, consistently but less clearly, in adult females and in young of both sexes. It even appeared in samples from the two kinds of forests taken within a short distance of each other; adult male stoats from podocarp forests on the west side of the Main Divide of the Southern Alps were smaller than their neighbors less than 20 km away in the beech forests on the east side, in two quite separate pairs of samples.

As in the Alexander Archipelago, these local variations in size of the New Zealand animals must have arisen since their ancestors arrived on the islands. In both cases, it stretches all credibility to suppose that the pattern could be due to sampling error or to systematic variation in body size of the original colonizing stock. On present data it appears that, within a hundred years (1884 to 1984), stoats in New Zealand developed a range of variation in body size between local populations no less than exists on the whole of continental Europe. Male stoats living in the foothills of the Southern Alps are probably among the largest in the world; in fact, they are near the top of the range of sizes of male long-tailed weasels in North America. Female stoats in New Zealand are larger than any Eurasian females, and near the middle of the range of sizes of female longtails (Table 4.1).

This response to a range of new environments has been extremely rapid, though not unique. Such remarkably rapid shifts of mean body size in colonizing populations have been observed before—for example, in the rats introduced to New Zealand and to other Pacific islands (Yom-Tov et al. 1999). The New Zealand example, however, is one of rather few in the world where substantial shifts in mean body size of stoats are correlated clearly with diet (Figure 4.8), sex, and habitat, and can be precisely dated (the first arrival of stoats in 1884 is well documented; Chapter 13). The key factors controlling body size are not fully understood; meanwhile, even simple phenotypic changes, when consistent, tell us that to a stoat, there is something different about New Zealand. The most obvious difference between New Zealand and Europe is in the frequency distributions of sizes of prey available (King 1991b; Chapter 5).

The second question, on how much stoats in New Zealand have changed size in the last century, is harder to answer. King and Moody (1982) did an extensive survey of the stoats in New Zealand, including the related question of how the contemporary animals have changed compared with their British ancestors. They concluded that male stoats have become larger in the beech forest/grassland habitats of New Zealand, and are unchanged, or possibly smaller, in the podocarp and mixed forests, over the 100 years since colonization. Unfortunately, few specimens of British stoats were available in the late 1970s, most collected in the first half of the twentieth century (before myxomatosis) and preserved in the Natural History Museum in London. King and Moody had to assume that the skulls in this rather modest sample of British stoats were still

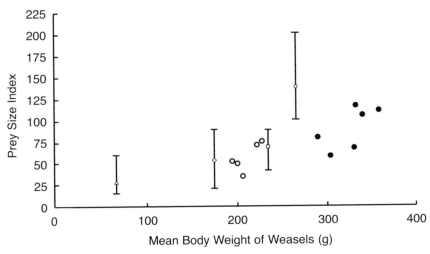

Figure 4.8 Mean body weights of local populations of stoats in New Zealand (filled symbols, males; open symbols, females) plotted against an index of the distribution of prey sizes taken by each population. For comparison, we also plot the means and ranges for the index of prey size for North American stoats (plotted with a small, open triangle), northern European stoats (small, open diamond), North American longtails (small, open square), and British stoats (small, open circle). For diets of stoats and longtails, see Figures 5.1 and 5.2. (Data for New Zealand stoats redrawn from King 1991a: fig 6, and see Figure 5.4.)

much the same size as the skulls of the stoats that were transported to New Zealand in the late nineteenth century.

In the late 1990s, a large, new collection of British stoats and weasels was made by McDonald and Harris (2002). Preliminary (unpublished) measurements of this material have cast some doubt on whether the museum material quoted by King and Moody did in fact represent the stoats of the late 1800s. Tools for statistical analyses are also much more sophisticated now than in the 1970s, so we might get to understand this fascinating story better when extensive samples from both countries can be analyzed with new methods. For example, it would be helpful to combine new information on how early nutrition and intraguild competition can affect body size of stoats (Chapter 14) with DNA analyses and new statistical analyses of traditional morphological measurements. The first DNA data have been provided by Holland et al. (in prep.), who sequenced mitochondrial DNA from 80 stoats from across New Zealand. They showed that five haplotypes (maternal lineages) survive at present, from an unknown number originally imported. The distribution of these haplotypes does not correlate with the variation in stoat body size. Unfortunately, that result tells us nothing about whether variation in body size of stoats is mainly a phenotypic or a genetic response to local conditions, or some combination of both.

Meanwhile, it is still safe to assert that stoats in New Zealand are generally large compared with stoats in other countries, and that stoat body size consistently correlates with prey size (Figure 4.8). Since the climate of New Zealand is generally milder compared with Britain, and the prey available are on average larger rather than smaller, New Zealand stoats find more advantages than disadvantages in remaining, or becoming, large. The effect is most substantial in females and this, too, is just as expected. The cardinal advantage of small size for a female stoat, the ability to hunt rodents into their last refuges, is irrelevant in New Zealand, whereas larger size would help female stoats deal with rats, rabbits, and possums.

The real explanation for the body size of the weasels of Ireland, Terschelling, New Zealand and most other islands is still unknown (Dayan & Simberloff 1998). Our guess is that the average size of the weasels on any island will drift toward whatever gives them, in the local conditions, the best year-round compromise between the advantages of smallness (see Figure 2.8), especially the size distribution of the small mammals available to hunt, versus the upward pull exerted by sexual selection. The point of balance may be determined, in ways we do not yet understand, both by the climate and by the size distribution of available prey. For example, the range of potential prey available on islands is usually smaller than on continents. Weasels colonizing an island may find that familiar, staple prey items are missing and the alternative supplies are different in various ways. On cold islands such as Newfoundland, supporting abundant small prey and no large ones, small weasels would have some advantages and suffer few penalties. On warm islands with fewer small prey than large ones, large weasels would be favored. We hope that somewhere, someone will find a way to put this hypothesis to the test.

5 | Food

THE DIETS OF weasels are relatively well known, because they can be deduced by standard methods. The procedures are straightforward, and the quantities involved are small and more or less painless to the nose. But interpretation of the results is often hazardous, because it is not always easy to deduce useful information from diet analyses.

IDENTIFYING AND INTERPRETING WEASEL DIETS

Scats are dry and inoffensive, and many samples can be obtained from known individuals during a successful live-trapping program. Occasional scats can also be collected in the field, and are useful if they can be correctly identified. By contrast, the stomachs of dead weasels can be collected from trappers, game-keepers, or game wardens, but they give only one sample per individual and are less pleasant to handle than scats.

Teeth and large pieces of bone discarded in dens can often be identified, but dens are hard to find even when inhabited by weasels wearing radio transmitters. Only in alpine and arctic grasslands can weasel dens be found quite easily, because weasels often commandeer the large and conspicuous overwintering nests made by their vole and lemming prey. By systematicly searching immediately after the spring thaw, finding those nests that have been used during the winter by weasels is relatively easy. Weasels leave behind many easily identifiable rodent teeth and small bones, and often the actual number of voles eaten can be calculated (Fitzgerald 1977; Sittler 1995). In more temperate climates where large prey are available, such as rabbits or seabirds, carcasses may show canine marks that sometimes match the distance between the canines of weasels or stoats (Hewson & Healing 1971; Lyver 2000).

Because metabolism in weasels is so rapid, stomachs contain undigested meat only from very recent meals. Further down the intestine, and in all scats, only hairs, feathers, and bone fragments remain. These can be identified from a combination of minute differences in their structure (Day 1966). The extent to which these differences can be used to identify a weasel's most recent meal depends on the characteristics of the potential prey groups and of the habitat of the study area.

The various species of shrews cannot be separated in scat or gut samples from weasels, even though shrews belong to several different genera. Likewise, rabbits and hares can be grouped only as "lagomorphs," because in most samples they cannot be distinguished reliably. The two most common genera of voles, *Clethrionomys* and *Microtus*, are represented throughout the northern hemisphere and are among the most frequently eaten prey of all species of weasels. In Britain, remains of bank voles (*C. glareolus*) in diet samples can be distinguished from those of field voles (*M. agrestis*) (Day 1966), but in the United States, red-backed voles (*C. gapperi*) can seldom be distinguished from meadow voles (*M. pennsylvanicus*) in gut samples (Brown 1952; R.A. Powell, personal observation). Carrion taken from a large carcass, such as predator-killed or road-killed deer or sheep, especially the inner parts without hair, often cannot be detected at all.

The list of potential food items to be identified in a weasel gut is relatively short, since they have evolved as specialist predators of small vertebrates (mammals, birds, and lizards). In their native habitats they eat insects, worms, or vegetable matter (usually only berries) only when extremely hungry. Moreover, since it is fair to assume that weasels never eat hair or feathers except in the course of eating the animal to which they were attached, one can make a rough estimate of the total number of individual prey eaten.

The stomach capacity of most weasels is only about 10 to 20 g, whereas the average weight of a small rodent is about 15 to 30 g. Hence, a weasel cannot eat more than one small rodent at a sitting, so a single stomach, intestine, or scat usually contains the remains of only one item. Conversely, one item can appear in more than one scat, so a group of scats collected at the same time and place has to be treated as a single sample.

The nutritional value of each prey is related to its body size. We can see which items are the most profitable for weasels to hunt and eat by calculating the diet in terms of the weights of the various types of prey eaten rather than their number. The imbalance between large and small items is then corrected, because, for example, seven birds' eggs at 3 g each count the same as two mice at 10 g or one meal of 20 g taken off a dead rabbit. In a sample from which ten birds' eggs, eight mice, and six meals of rabbit were identified, the total weight of prey eaten would be 30 + 18 + 120 = 230 g, of which eggs contributed 13%, mice 35%, and rabbits 52%. That looks quite different from the same data expressed as percentage frequency of occurrence, that is, eggs 42%, mice 33%, and rabbits 25%.

On the other hand, the results of weighting the prey items by body size must be treated with caution, for several reasons. First, the proportion of any one item is, by definition, relative to the total, so the items are not independent. Hence, we cannot say from such figures that weasels from one area rely more on a given type of prey than those from another area. Second, one cannot always discern from their remains whether individual prey were adult or juveniles, which can differ tremendously in size (and in catchability). Third, the results are greatly influenced by how much one assumes a weasel eats from a single large carcass. And fourth, small prey have relatively more bones and hair, which are not equally represented in the remains. Weasels cannot digest hair, so all fur eaten reappears in the scats (Gamberg & Atkinson 1988), although weasels will avoid eating skin and hair if there is plenty of meat. Large bones will also be avoided, and small ones can be partially digested. Therefore, analyses of hair and bones cannot reflect all prey equally. Even so, calculating the proportions of prey apparently eaten is worth doing to show that, in general, small prey are much less profitable (i.e., return less energy for effort) to eat than large ones.

Identifying a weasel's menu is the easy part of the job. Deciding what the figures mean is far more difficult. To begin with, simple lists of prey eaten by weasels from different places will be biased toward the season, the habitat, and the ages and sex of the weasels most often represented. Likewise, samples of different composition cannot be compared with each other directly, because weasels of different sexes or ages, and in different seasons, may eat different things. For example, some foods, such as birds' eggs, are most available in spring and early summer: It would not be valid to compare the proportion of birds' eggs in samples from two areas unless both had been collected in the same season. And, of course, large samples give much more reliable information than small ones. Any analysis has to reach some compromise between splitting, to avoid compounding different effects in one sample, and lumping, to increase the sizes of the samples.

The large and scattered scientific literature on weasels contains many descriptions of what they eat. Figures 5.1 to 5.4 summarize some of them. Much can be learned about the feeding habits of these little carnivores by setting out the available information systematically in this way, but there are limitations. In the original papers cited here, the data were presented in different ways. Some researchers counted the number of mice that were identified, and expressed the total for mice as a percentage of all the food items identified. Others counted the number of samples containing mice and expressed that as a percentage of all the samples examined. Some included empty stomachs in the figure for total samples, while some excluded them.

Wherever possible, we have standardized the data by recalculating all the results as the number of each item counted as a percentage of all items. This is not, in fact, the best way to compare the diets of weasels in different places, because the proportion of each item is not independent. Pie charts calculated in this way, however, do give a quick and vivid impression of the differences in diets of weasels in different environments, and within broad limits these patterns probably do reflect real variations in how weasels make their living in different kinds of places.

FOODS OF STOATS IN NORTH AMERICA AND EURASIA

Far more has been published about the eating habits of stoats than of either of the other species. Ermine fur has long been an important but unreliable item of trade in the far north, and in North America, both the stoat and the long-tailed weasel have been trapped, skinned, and traded for many years. Scientific interest in them goes back at least to the 1920s (Seton 1926). Soviet studies on the biology of stoats date back just as far (Heptner et al. 1967). The Russian literature on stoats, and on their larger fur-bearing relatives such as sables and martens, is extensive, and some of it is now available in English (King 1975b, 1980d).

There are huge differences in mean body size of stoats living at opposite ends of their enormous geographic range (Chapter 4), and in the habitats and prey available to them. Figure 5.1 A and B show the results of a dozen studies that illustrate some ways that these adaptable little animals have responded to such variable living conditions. The smallest, North American stoats may weigh only a quarter as much as the largest stoats of the same sex in the British Isles and New Zealand, and have a totally different way of life.

On the tundra of eastern Greenland, Benoît Sittler collected stoats' scats from lemming nests (Sittler 1995; Gilg et al. 2003). When the lemming population was high, the stoats ate nothing but lemmings (Figure 5.1A, c). When the lemming population crashed, the stoats still hunted predominantly for lemmings but made up the difference with ptarmigan and ptarmigan eggs. Similarly, on the tundra-covered island of Igloolik, at 69°N off the coast of northern Canada, lemmings were at low density in the summer of 1977 when Simms (1978) collected a pile of scats and prey remains from an active stoat den (Figure 5.1A, b). Nevertheless, the stoat was still able to catch lemmings, since they made up more than three quarters of the 142 remains tallied; birds and one insect accounted for the rest. Clearly this stoat killed birds when it could in summer. As Simms pointed out, not much else was available.

The stoats studied by Lisgo (1999) in boreal forests of Alberta and by Northcott (1971) in the boreal forests of the Northwest Territories (Figure 5.1A, a) lived mainly on small rodents, which constituted roughly two thirds of the prey items killed. Males, especially, also killed some red squirrels, which weighed so much more than the small rodents that squirrels accounted for three quarters of the total weight of prey eaten (Lisgo 1999). Nonetheless, analyses of the population dynamics of stoats across northern North America show that fluctuations in their abundance follow those of voles and lemmings rather than squirrels (Johnson et al. 2000b).

Further south, stoats living in the cool temperate farmlands and forests of southern Canada and the northern United States are small in body size (Chapter 4). So, although they have a wide choice of prey, including various species of voles, mice, shrews, lagomorphs, and squirrels as well as birds, stoats in Ontario, Quebec, and New York eat mainly small rodents (Figure 5.1A, d, e, f). In Ontario, meadow voles were almost the sole item on the menu of the stoats studied by Simms (1979b) (Figure 5.1A, d), even though deer mice were also abundant. And in the alpine meadows of the Californian Sierra Nevada, remains of montane voles were practically all there was to show for an entire winter's feasting by the tiny stoats studied by Fitzgerald (1977).

A

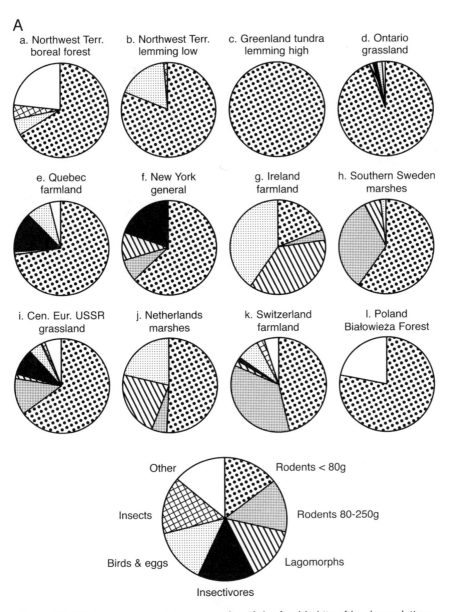

a. Northwest Terr. boreal forest

b. Northwest Terr. lemming low

c. Greenland tundra lemming high

d. Ontario grassland

e. Quebec farmland

f. New York general

g. Ireland farmland

h. Southern Sweden marshes

i. Cen. Eur. USSR grassland

j. Netherlands marshes

k. Switzerland farmland

l. Poland Białowieża Forest

Other

Rodents < 80g

Insects

Rodents 80-250g

Birds & eggs

Lagomorphs

Insectivores

Figure 5.1 (A) Some representative examples of the food habits of local populations of stoats, and a key identifying foods. (a) Northwest Territories (*n* = 77, Northcott 1971); (b) Northwest Territories (*n* = 172, Simms 1978); (c) Greenland (Sittler 1995, Gilg et al. 2003); (d) Ontario (*n* = 305, Simms 1979a); (e) Quebec (*n* = 384, Raymond et al. 1984); (f) New York (Hamilton 1933); (g) Ireland (*n* = 27, Fairley 1971); (h) Southern Sweden (Erlinge 1981); (i) Central European USSR (*n* = 1,055, Aspisov & Popov 1940); (j) the Netherlands (*n* = 61, Brugge 1977); (k) Switzerland (*n* = 690, Debrot & Mermod 1981); (l) Poland (*n* = 58, Jędrzejewska & Jędrzejewski 1998).

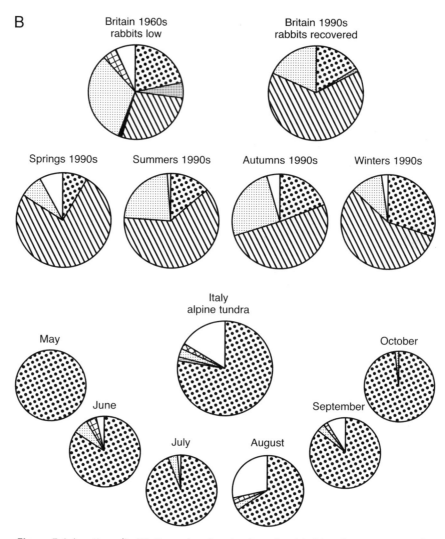

Figure 5.1 (continued). (B) Examples showing how food habits of stoats vary with changing rabbit populations and across a year. Britain, rabbits few in 1960s (*n* = 152, Day 1968); and after recovery of the rabbit population in the 1990s (*n* = 789, Springs *n* = 169, Summers *n* = 148, Autumns *n* = 71, Winters *n* = 80, McDonald et al. 2000); Italy, alpine tundra (*n* = 734, May *n* = 28, June *n* = 26, July *n* = 93, August *n* = 416, September *n* = 94, October *n* = 77, Matinoli et al. 2001). Prey designated as in Figure 5.1A.

The much larger Eurasian stoats still take many small rodents, especially voles of the genus *Microtus*, but also a substantial proportion of bigger prey. A particular favorite is the water vole, *Arvicola terrestris*, the familiar "Ratty" of the *Wind in the Willows*. Both the pioneering studies on the vast Volga-Kama River flood plains of the central European USSR (Figure 5.1A, i), and more recent work in southern Sweden (Figure 5.1A, h) and Switzerland (Figure 5.1A, k) have shown how important water voles are to the feeding economy of stoats and, as a direct consequence, to their population dynamics in those areas (Chapter 10). In contrast, no water voles appeared in a small sample of stoats collected from osier beds and along the riverbanks of the low-lying Netherlands—only the occasional muskrat, an exotic species introduced from North America for fur farming (Figure 5.1A, j). Nor did the stoats in Białowieża forest in Poland prey on water voles, but relied on smaller voles, yellow-necked mice, and amphibians (Figure 5.1A, l).

In the Italian Alps in the summers of 1996–97, small rodents were always the primary summer prey, found in 60% of 734 stoats scats collected from May to October by Martinoli et al. (2001) (Figure 5.1B). But rodents became relatively scarce in July, at the same time as fruits, mostly berries of juniper (*Juniperus cummunis*) and *Vaccinium alignosa*, were becoming more abundant. In August, when fruit was mature and easy to collect, these stoats incorporated fruit as over 30% of their diet. By September and October, fruit was still abundant but becoming dry and less palatable, while rodents were increasing again. Martinoli et al. concluded that the alpine stoats they observed, especially males concentrating on mate searching, minimized their foraging time by harvesting an alternative "prey," the profitable and easily available fruits, during a period of high energy demand in August, and returned to the more time-consuming rodent hunting only when the mating season was over. McDonald et al. (2000) found similar changes in diet in Britain over the seasons (Figure 5.1B).

British stoats are even larger, and small rodents seldom constitute more than a third of their diet even in lean years. Instead, their primary prey is the rabbit—at least, it was until myxomatosis arrived in Britain in late 1953 (Sumption & Flowerdew 1985). The massive population crash of rabbits throughout Britain from 1954 onward was followed by an equally massive drop in numbers of stoats (see Figure 10.2). When the first diet studies of British stoats were done in the 1960s and 1970s, both stoats and rabbits were still scarce. Nevertheless, lagomorphs (still mostly rabbits) comprised about a third of the items eaten by British stoats, and small rodents and birds another third each (Day 1968; Potts & Vickerman 1974) (Figure 5.1B). By the 1990s, British rabbits had largely recovered their numbers, both in the countryside (see Figure 10.2, inset) and on the menu of stoats. The overall diet of the 789 stoats collected in 1995–97 by McDonald et al. (2000) (Figure 5.1B) comprised 65% lagomorphs and 16% small rodents, presumably much as it had been before myxomatosis arrived.

Northern Ireland has no field voles or bank voles, so the only small rodents available are wood mice and house mice. Even though the stoats there are much smaller than those in Britain, they must rely on rabbits, birds, and rats at least as much as their bigger brethren, for lack of anything better (Figure 5.1A, g). In the southwest of Ireland, Irish stoats are about as large as their cousins in Britain (see Table 4.1). Sleeman (1992) collected stoat carcasses from the southern counties, and found that they also depend heavily on rabbits, despite the presence of bank voles in the southwest (introduced by at least 1964) (Fairley 1984).

FOODS OF LONG-TAILED WEASELS IN NORTH AMERICA

Long-tailed weasels range from the southern Canadian snows south through the United States and Mexico and into South America. They encounter a great variety of habitats and climates, and they range in size from much smaller to much larger than stoats in continental Eurasia. It would be a fair expectation that the small ones should concentrate on small rodents and the large ones kill bigger prey when they can, just as stoats do, but the data are rather sparse. In some samples, such as the one from Iowa (Figure 5.2f) and another (not included in Figure 5.2) from Michigan (Quick 1944), the pattern is the typical small weasel type—more than three quarters of the prey are small rodents. In other samples, the diet is more variable and more often includes medium-sized rodents such as ground squirrels, chipmunks, and rats, and also cottontails and shrews (Figure 5.2a, d, e).

Longtails also eat insects and carrion occasionally, as do stoats. Insects are not a staple food for longtails, nor even necessarily a reliable emergency resource, since insects are common only in summer when other foods are abundant. Two longtails were among the scavengers that approached 42 of 64 grouse carcasses set out and monitored by Bumann and Stauffer (2002). An unusual longtail report, but one that illustrates nicely the versatility of these predators, is of one that entered an Indiana barn in broad daylight and climbed into the rafters to a nursery colony of big brown bats. It ignored the farmer and killed three nursing female bats and their five young before the farmer shot it (Mumford 1969).

The range of long-tailed weasels extends into South America, but little is known of the tropical weasels. Longtails in Central and South America are believed also to prey mainly on small mammals, rabbits, birds, and their eggs.

FOODS OF COMMON WEASELS IN BRITAIN AND WESTERN EUROPE

Common weasels have been studied more widely than least weasels, in part because in Western Europe it is the common weasel that lives in the humanized landscapes surrounding cities. University students and professors prefer nearby sites

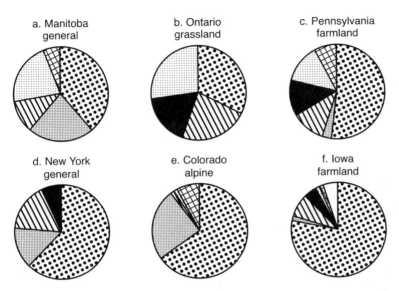

Figure 5.2 Some representative examples of the food habits of local populations of long-tailed weasels. (a) Manitoba (*n* = 200, Gamble & Riewe 1982); (b) Ontario (*n* = 34, Simms 1979b); (c) Pennsylvania (*n* = 112, Glover 1942b); (d) New York (Hamilton 1933); (e) Colorado (*n* = 84, Quick 1951); (f) Iowa, (*n* = 166, Polderboer et al. 1941). Prey designated as in Figure 5.1A.

for field studies, when possible, and have tended to avoid the distant northern climes, with long winters and difficult working conditions, that support least weasels.

The first general study of the foods of common weasels (and stoats) in Britain was that of Day (1968) (Figure 5.3i). Day's work has been widely quoted, and for many years was the only British study to include animals of both species of weasels collected (mostly from gamekeepers) from all over the country. His identification key to the hairs of the small mammals of Britain (Day 1966) opened the door to many later studies. Recently, McDonald et al. (2000) (Figure 5.3j) made a second comparative study of the two species, also from carcasses collected from gamekeepers across Britain. In both studies, small rodents dominated the diet of common weasels (1960s, 56%; 1990s, 68%). Both stoats and common weasels ate rabbits more often in the 1990s than they had done 30 years before (stoats 65% vs. 25%; common weasels 25% vs. 18%), but for different reasons. By the 1990s, rabbits had recovered from myxomatosis, so stoats could return to their specialized niche as rabbit hunters, and common weasels benefited from the greater availability of young rabbits in spring.

Young plantations are an ideal habitat for weasels, because the ground between the trees quickly becomes overgrown with grass, neither mown nor grazed, providing ideal living conditions for the favorite prey of all weasels, voles of the genus *Microtus*. In such places, common weasels eat almost only voles (Lockie 1966). Two of the four habitats observed in southern Sweden by Erlinge (1975) were young

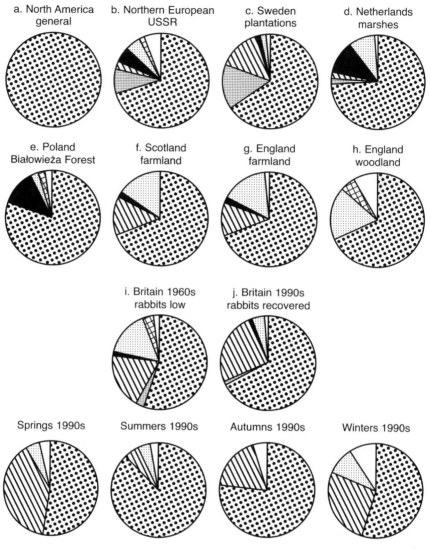

a. North America general

b. Northern European USSR

c. Sweden plantations

d. Netherlands marshes

e. Poland Białowieża Forest

f. Scotland farmland

g. England farmland

h. England woodland

i. Britain 1960s rabbits low

j. Britain 1990s rabbits recovered

Springs 1990s

Summers 1990s

Autumns 1990s

Winters 1990s

Figure 5.3 Some representative examples of the food habits of local populations of least and common weasels. (a) North America (Hall 1951); (b) Northern European USSR, tundra and taiga (Parovschikov 1963); (c) Sweden (Erlinge 1975); (d) the Netherlands (*n* = 161, Brugge 1977); (e) Poland (*n* = 15, Jędrzejewska & Jędrzejewski 1998); (f) Scotland (*n* = 206, Moors 1974); (g) England (*n* = 445, Tapper 1979); (h) England (*n* = 285, King 1980d); (i) Britain 1960s, rabbits few (*n* = 152, Day 1968); (j) Britain 1990s, rabbits recovered (*n* = 458, Springs *n* = 131, Summers *n* = 66, Autumns *n* = 60, Winters *n* = 27; McDonald et al. 2000). Prey designated as in Figure 5.1A.

plantations and in both, voles were the single most important prey (Figure 5.3c). In spring when voles were scarce, male weasels turned to the newly available young rabbits. Scats also contained bank voles, wood mice, water voles, and shrews and, in one of the plantations, also birds and even lizards. Weasels avoided an older spruce plantation and a deciduous alder woodland. The trees at these sites were tall enough to shade out the grass, and few rodents ventured out on to the bare ground.

In Wytham Woods, near Oxford, King (1980b) collected 250 scats from 36 individually marked common weasels caught in wooden box traps. To make sure she got a sample from each captured weasel, she supplied a dead white mouse in each trap. The populations of small rodents in the area were rather low (21 to 39/ha), and the weasels were usually hungry enough to eat the lab mouse even if it became distinctly "ripe." Eating something shoved everything else previously eaten along in the gut, and the usual result was at least one sample of each weasel's last wild-caught meal. White fur in scats showed clearly where wild meals stopped. The Wytham weasels ate mostly bank voles, wood mice, field voles, and birds, more or less in the order of their abundance. Bank voles were by far the most common of the small rodents in the wood, followed by wood mice. Few field voles lived in the wood, but some were killed by the weasels whose home ranges extended to a neighboring young plantation (Figure 5.3h). Wytham had very few rabbits and no rats at the time. There were plenty of shrews and moles, but no shrews and only one mole appeared in the scats.

Interesting and important as it is to know what weasels eat in woodlands of different types, these habitats are in the minority. In Britain, woodland comprises less than a tenth of the total land area, whereas arable and pasture land comprise two thirds of England and Wales, and almost a quarter of Scotland. The majority of British weasels live on farmland. In open country, the small rodents most often eaten by common weasels are field voles, which are at home in any kind of grassland, and wood mice, which often live in hedges and are not afraid to venture out into the farmers' fields, especially before harvest. Bank voles stick to thick cover, so they are less often caught in open country than in woodland.

In general, then, common weasels depend on small rodents and small birds, and where these are scarce, such tiny hunters do not thrive. In every sample shown in Figure 5.3, as well as in other samples not shown, the proportion of small rodents in the common weasels' diet is never less than half of the total number of prey items identified, and is often nearer to three quarters.

FOODS OF LEAST WEASELS IN NORTH AMERICA AND NORTHERN EURASIA

The foods of least weasels have been sampled in the Arctic, at Barrow on the north coast of Alaska, and in the extreme north of European USSR (Figure 5.3a, b). There they live almost entirely on lemmings and voles when these rodents are abundant. In the years when the lemmings and voles disappear, life for least weasels becomes extremely precarious. Alternatives are few and fleeting, except in the short northern summer.

Every year the tundra is occupied by vast flocks of migratory birds that forage and nest on the ground. During summers of lemming crash years, these birds become unwilling providers of eggs and young for hungry least weasels. In the lemming crash year of 1969 at Barrow, the nests of sandpipers and Lapland longspurs suffered heavily. In such straitened circumstances, least weasels also kill less favored alternative prey, such as shrews, and scavenge carrion of large animals they could not kill themselves or leftovers from kills made by large predators (Nasimovich 1949). Averaged over many years, as in the long collection from the Arkhangelsk region in the far north of European Russia, these various subsidiary resources give a false impression of variety in the diet of the northern least weasels (Figure 5.3b). By the following winter, the migratory birds have returned south, the seasonally active mammals have gone into hibernation, and subzero air temperatures freeze carrion solid and prevent all small mammals from venturing above snow for long. Then, least weasels have no choice but to search and search and search for the last few live lemmings or voles under the snow.

In North America, the range of the least weasel extends south of the Canadian border into the northern prairie states and down the Appalachian mountain chain. Here least weasels, still virtually confined to small rodents, share their only resource with many other, more generalist predators. This competition is hard on least weasels, and it means that they are usually scarce except in habitats or at times that are exceptionally good for small rodents (Chapter 10). For example, in the Southern Appalachian Mountains, least weasels are most often captured in orchards, where thick fescue grasses provide superb habitat for meadow and woodland voles (R. A. Powell, unpublished data). In northern Fennoscandia, *Microtus* voles also dominate the diet of least weasels, occasionally supplemented with bank voles and mice (Korpimäki et al. 1991). By contrast, in primary forest in Poland, bank voles and yellow-necked mice dominate the diet of common weasels (Jędrzejwski & Jędrzejwska 1993; Jędrzejwski et al. 1995).

In the end, and despite the differences in their morphology and reproduction, least and common weasels have extremely similar diets and respond in the same way to changes in prey abundance. Common and least weasels are the most specialized of all weasels for preying on small rodents, especially voles (Jędrzejwska & Jędrzejewski 1998).

FOODS OF STOATS AND COMMON WEASELS INTRODUCED INTO NEW ZEALAND

One of the constant themes of this book is the close relationship between weasels generally and the various species of northern hemisphere voles with which they evolved. So close is this relationship that one might predict that no species of weasel could survive where there are no voles. In New Zealand, history has provided a test of this idea.

The stoats and common weasels transported from Britain to New Zealand in the nineteenth century joined a simple community of introduced mammals. The range of body sizes of prey available was different from anywhere in the northern hemisphere: fewer small mammals of mouse size (0 to 50 g), and more of rat (80 to 200 g) and rabbit size (500 to 2,000 g). The only competing predators were feral cats (then still scarce), and the native morepork owls and bush falcons. So, the forests of late nineteenth-century New Zealand were not only well stocked with food for stoats, but also offered them shelter from the harrier hawks and ferrets that preferred more open country.

Not surprisingly, stoats moved into the New Zealand forests at once. Even though their new home contained a radically different array of potential meals, compared with their native land, stoats adjusted quickly and thrived. Common weasels, however, were (and still are) less adept at hunting grown rabbits than stoats, and they found the few available prey of small size (feral house mice, native lizards, wetas) an insufficient substitute for voles. New Zealand has proved inhospitable for common weasels: They are not extinct, but they are scarce and patchy in their distribution compared with stoats.

It would be interesting to know what the first colonizing stoats ate, but that, of course, is now not possible to discover. Almost the only thing we can be sure of is that the Australian brushtail possums, introduced for their fur and now a widespread and important source of carrion, were not on the menu, because they were still only just getting established when the mustelids first arrived. Fortunately, it is just as interesting to learn what contemporary stoats eat, because this gives us unusual insight into how a predator adjusts over time to totally unfamiliar prey.

A survey of the biology of stoats during the 1970s, in which King and Moody (1982) collected 1,599 stoats from all ten national parks then gazetted in New Zealand, showed that the most important single class of food for stoats was birds (all species pooled, found in 43% of the 1,513 guts examined). The remaining categories of food were feral house mice, rabbits (and, occasionally, hares), possums and other carrion, rats (mainly ship rats but possibly some Norway rats or, in one area, a very occasional Polynesian rat), geckos and skinks (small lizards), freshwater crayfish, and insects. In forests, birds and insects between them constituted more than half the items eaten, and rats, mice, and possums were also very important (Figure 5.4). In open tussock grasslands, lizards and lagomorphs substituted for possums and rats. Hedgehogs are very often found dead on roads in New Zealand (King 2005a), but in this survey they were the only animals within the expected range of prey sizes that stoats avoided (found only twice).

Insects, mostly weta (native Orthoptera), were found in an astonishing 41% of the total guts collected in the 1970s. Stoats in New Zealand eat insects all year

a. Podocarp/mixed Forest
b. Podocarp/mixed Forest
c. Podocarp Forest
d. Coastal Habitat

e. Alpine Habitat
f. Alpine/beech Habitat
g. Southern Beech Forest
h. Southern Beech Forest

i. Braided Riverbed Habitat
j. S. Beech Forest mouse high
k. S. Beech Forest mouse low
l. "bush"

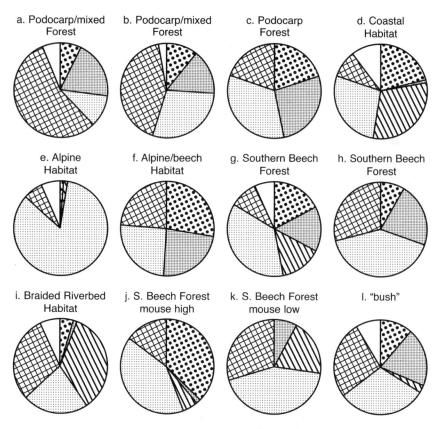

Figure 5.4 Some representative examples of the food habits of local populations of stoats in New Zealand. (a) Podocarp/mixed forest (44 guts, Rickard 1996); (b) podocarp/mixed forest (52 guts, King et al. 1996); (c) podocarp forest (129 guts, King & Moody 1982); (d) coastal habitat (75 guts, Alterio & Moller 1997b); (e) high-elevation shearwater colony (788 scats, Cuthbert et al. 2000); (f) alpine/southern beech forest habitat (93 guts, Smith 2001); (g) southern beech forest, average (75 guts, King & Moody 1982); (h) southern beech forest, mice few (62 guts, Purdey et al. 2004); (i) braided riverbed habitat (196 scats, Dowding & Elliot 2003); (j) southern beech forest during mouse population high (54 guts, Murphy & Dowding 1995); (k) southern beech forest during mouse population low (54 guts, Murphy & Dowding 1995); (l) "bush" habitat (124 guts, Fitzgerald 1964). Prey designated as in Figure 5.1A.

round, much more often than their relatives do in the northern hemisphere (McDonald et al. 2000). This is understandable; even though a weta supplies only a small parcel of food for a stoat, it is still large enough (2 to 4 g) to be worth snapping up in passing, and available all year round. In the cold parts of their enormous ranges, in North America and Russia, stoats and longtails do eat insects in summer, but only occasionally.

Large prey (possums, lagomorphs, and rats, all introduced) clearly supplied half or more of the total food value for stoats. Birds and mice together supplied only a third, while insects supplied only about a tenth, despite their frequent consumption. House mice, the only small mammal prey available to stoats in New Zealand, are obviously no substitute for the voles that are their mainstay in the northern hemisphere.

Since that first national survey, numerous studies have confirmed that birds generally comprise over half the prey captured regularly by stoats in New Zealand (King et al. 1996; Alterio & Moller 1997b; Murphy et al. 1998). A second large national-scale survey of the foods eaten by stoats collected in 1999–2000 is now awaiting analysis (R. A. McDonald and D. Smith, in prep.), and should provide an interesting comparison with the older data.

An intriguing comment on the use by stoats of road-killed carrion was supplied by Murphy and Dowding (1994) from radio tracking work in the Eglinton Valley, one of the study areas that had been surveyed in the 1970s. They tracked several male stoats that actively preferred to hunt along a road, and were often found denning near road-killed possums. One managed to drag a possum (a deadweight of five to ten times its own body mass) to a nearby den.

Stoats are versatile predators, able to switch their attention to whatever prey are most abundant. Prey switching can bring relief from stoat predation for their staple prey (see Figure 13.3), but only if the alternative is nutritious enough. For example, throughout the massive swings in abundance of mice following a heavy seedfall of southern beech, New Zealand stoats still eat as many birds as usual (King 1983b; Murphy & Dowding 1995)—perhaps because, as experiments in enclosures have shown, stoats cannot gain much net energy by hunting house mice (Raymond et al. 1990). Only when house mice reach plague numbers, and then only for a short period and in only one area identified so far, can stoats afford to concentrate totally on mice and give birds a break (Purdey et al. 2004; White & King in press).

Ship rats are also abundant in some forest types, and they also take many birds and their eggs, so to protect birds, ship rats are regularly removed from some areas (Innes et al. 1995). After one successful program to remove ship rats, stoats replaced the rats lost to their diet with birds (Murphy et al. 1998). Stoats have affected populations of many bird species in New Zealand (McDonald & Murphy 2000), and we will return to the problem of how to meet the challenge stoats present to conservation of the vulnerable endemic fauna of the New Zealand archipelago in Chapter 13.

6 | Hunting Behavior

THE HUNTING BEHAVIOR of wild weasels is very difficult to observe. You can go every day to an area you know to be inhabited by weasels, and never see one. Modern tools such as radio telemetry have reduced that problem to some extent; now researchers can find radio-collared weasels in their rest sites, and have a better chance of watching them hunt. Bogumila Jędrzejwska and Wlodzmierz Jędrzejwski (1998) watched wild common weasels hunting in the Białowieża Forest in Poland during early evening. When rodent populations were high, they followed their radio-tagged weasels as they systematically searched the forest floor along logs, under fallen branches, down holes along roots, and up trees into holes and crevices. Without access to that give-away radio signal, few of us can be so lucky.

On the other hand, weasels are so small, as carnivores go, that it is possible to set up fairly natural conditions for them in captivity, where observations are easier to control. By providing weasels with wild-caught prey in a large enclosure furnished with grass and trees, and watching through a window, one can learn much about their hunting and killing behavior. Another tactic that works well in cold climates is to deduce the hunting tactics of weasels by reading tracks and signs in snow. Skilled woodsmen such as Ernest Thompson Seton (1926) could read whole stories from snow tracks. Even chance encounters with wild weasels are important and can give fascinating glimpses of natural behavior, but they have also spawned a whole literature of unlikely stories that are impossible to interpret.

Being a predator is a risky profession. Some types of prey are easy to find but dangerous to attack. For example, rabbits are much bigger than even the big weasels, and in the open, a rabbit is formidable prey for a small hunter. An enraged doe rabbit with young to protect has been seen to chase and kick a stoat halfway across a paddock. Other prey are also well protected in different ways—for example, the spines of a live hedgehog appear to be a completely effective defense. Small predators dealing with prey that are well defended or larger than themselves must choose and approach their targets carefully.

By contrast, other kinds of prey, including most of those sought by weasels, are easy to kill but widely dispersed and well hidden. Most voles offer little physical resistance once found by a weasel, but the weasel runs a high risk of failing to find enough voles each day to fulfill its needs. If that happens often enough, the weasel dies. Death is the ultimate penalty for predators that misjudge the hazards of hunting—not merely the death of their own bodies, for that is inevitable sooner or later, but the worse failure of dying before raising any young.

There are three main kinds of predators, which balance the risks of hunting in different ways. Pack-hunters, such as wolves and African lions, live in groups that depend on, and often live within sight of, large and formidable prey that are more easily found than killed. Ambushers, like mountain lions and leopards, patiently lie in wait for a suitable prey to pass by. Weasels are neither sociable nor patient. They are active, solitary searchers that specialize in exploring every likely place for small and vulnerable prey, especially rodents, that are more easily killed than found. They hunt alone, with restless energy and fierce concentration. They are adaptable and intelligent, and tailor their methods according to their targets and their opportunities.

TARGET SPECIES

Small Mammals

Weasels on the hunt run from one patch of cover to the next, investigating every small hole and occasionally listening and looking around from a vantage point, testing the wind. Their habit of sitting up on their hindquarters is an obvious way of overcoming the disadvantages of having such short legs, and the Germans have coined an expressive phrase describing it: "er macht Mannchen" ("he becomes a little man"). Weasels search for prey using smell, hearing, and sight, probably in that order (Gillingham 1986). They can smell out the exact routes where voles and mice have traveled, and they are alert to the slightest sound. Weasels systematically search the forest floor, snooping along logs and under fallen branches, checking around the trunks of trees and exploring roots to find holes, and entering cavities and crevices up trees. Sometimes a weasel will dig to enlarge a hole, or locate a rodent by sniffing or listening. The weasels watched by Jędrzejwska and Jędrzejwski (1998) explored on average 26 holes per kilometer of foraging.

All weasels tend to intersperse foraging expeditions with periods of rest (Erlinge 1979b; Sandell 1988; Zielinski 2000). They may hunt for less than an hour, or for several hours, then return to a den. In winter, they often sleep most of the night, while in summer, females with growing young forage almost incessantly. All of the weasels can travel up to 2 km in a single trip of a few hours, especially when food is scarce, but in times of plenty they may travel only a few hundred meters on each hunting trip.

Weasels are good swimmers (Sleeman 1989b) and can reach inshore islands at least 1.5 km from a mainland shore. From eyewitness accounts, and from the distribution of stoats on offshore islands, it is quite certain that a stoat can easily reach islands within that distance of the mainland (King & Moors 1979a;

Taylor & Tilley 1984). The swimmer cannot know whether the effort is going to be worthwhile, although sometimes it is lucky. G.C. Phillips watched a stoat swimming "a determinedly straight course" for some 400 m to an island in Baltimore Bay (southwest Ireland) that still had healthy rabbits after the mainland stock had been virtually wiped out by myxomatosis (unpublished observation quoted by King & Moors 1979a). On the other hand, the stoat seen by Morton Boyd (1958) on Eilean Molach, 200 m from the shore of a Scottish loch, would not have stayed long on an islet of less than a tenth of a hectare.

In Finland, stoats move frequently among the thousands of inland islands (Heikkila et al. 1994), which provide patches of high prey density and set the scene for a hide-and-seek drama between stoats and voles. In New Zealand, three radio-collared stoats regularly visited both sides of a fast-flowing river and could only have crossed it by swimming (Murphy & Dowding 1994). In Wyoming, R.A. Powell (unpubl.) has followed tracks in the snow that crossed cold, swiftly flowing streams and small rivers even when the air temperature was far below freezing.

Weasels concentrate their hunting in the habitats where they know, perhaps from experience, that they are most likely to find prey (Nams 1981). The favorite hunting grounds where a resident weasel will spend almost all its time often comprise less than half the area of its home range. In snow, the track of a hunting weasel zig-zags this way and that, missing nothing of interest (Powell 1978a, 1978b). A thick covering of snow protects small rodents from most predators, but not from weasels. Tracks frequently dive under the snow into the hollows at the roots of trees, through the loose snow under protruding branches, and down natural fissures in the snow cover beside logs (Figures 6.1 and 6.2). For example, the tracks of the stoats studied by Edwards et al. (2001) twisted and turned through areas with vegetation low to the ground and where coarse woody debris, saplings, and trees protruded from the snow.

The tracks often pop back up above the snow elsewhere. If a snow crust is buried beneath the surface, the weasel can get no further down, but will move along the covered crust, pushing up the new snow with its back and making characteristic bulges, like those above moles' tunnels. A captive weasel will burrow in leaves or peat the same way.

Weasels forage most widely when prey populations are decreasing (Jędrzejwska & Jędrzejewski 1998; Klemola et al. 1999). Tracks in the snow suggest that large weasels travel longer distances than small weasels, which implies that, *on the average,* the larger (or largest) species living at any site forages more widely than the smaller, and that males forage more widely than females. Yet, the broad overlap in paw sizes, track patterns, and bounding distances among the species and between the sexes makes absolute identification of a track impossible.

Voles make extensive tunnels through the grass and moss under the snow, and weasels are the supreme experts at tunnel hunting. The smaller ones tend to be exactly the right size for getting into the tunnels of the local voles. In

Figure 6.1 Weasel tracks in snow zig-zag back and forth and often dive under the snow. These erratic travel patterns show that weasels inspect everything of interest, but the zig-zags and dips also make movements of weasels unpredictable, thereby making the weasels hard for other predators to catch.

Ontario, Simms (1979a) calculated that small female stoats could get into over 90% of subnivean vole tunnels (22 to 28 mm diameter) and male stoats and female longtails could get into over 70%. The small Canadian stoats whose stomachs were examined by Northcott (1971) had made a speciality of digging fat jumping mice (*Zapus*) out of their winter hibernation nests.

Female common weasels in Scotland can run easily through all but the very smallest vole burrows (averaging 23 mm diameter) and males can get through the largest with a squeeze (Pounds 1981). That most males can squeeze through most vole tunnels is critically important, because pregnant female stoats and weasels (and presumably longtails as well) average the same diameter as adult males (Gliwicz 1988).

The largest weasels are excluded from vole tunnels (Simms' male longtails could get into only 10% of them) but freely enter the wider tunnels of ground

Figure 6.2 The northern weasels are small enough to hunt a vole through its under-snow tunnels and to follow it into its last refuge.

squirrels, chipmunks, rabbits, and water voles. Mole runs (see Figure 8.4) are easily accessible and often used, especially as they usually lead to a large, warm, and comfortable nest (Chapter 11). At dusk one evening, Florine (1942) set two traps at the entrance of a pocket gopher burrow, so that only animals coming from within would be caught. By 9 p.m. one of them held a pocket gopher, and the other a long-tailed weasel. Likewise, in July 1930 at Milner Pass, in the Rocky Mountains National Park, Colorado, Dixon (1931) watched a stoat chasing pikas. It could get into any space between rocks that the pikas could, and just as fast. It followed by scent the winding course taken by the pika with ease and accuracy. Other pikas joined in as the first ones got tired. It was almost as if the pikas were cooperating, forcing the weasel to run a relay race against overwhelming numbers of opponents.

Weasels have no difficulty killing small rodents in confined spaces. Weasels that each of us has kept in captivity could kill voles and house mice in narrow tunnels. The radio-collared weasels observed by Jędrzejwski et al. (1992) hunted during the day, systematically searching underground dens and tunnels of rodents, and cavities 1 to 4 m up old trees. The strictly nocturnal yellow-necked mice were especially vulnerable in cold weather, because then they saved energy by retreating to their nests and falling into short periods of torpor—not a true hibernation, but a sleep deep enough to make them defenseless against an intruding weasel.

On the other hand, if the potential victim is already alert, and especially if it is relatively large, the risk of injury is real, and the weasel does not always come off best. When Durward Allen (1938b) paired long-tailed weasels with cotton-tail rabbits in a cage, the weasel killed the rabbit in half the trials, but the rabbit killed the weasel in the other half. A cage is clearly an artificial place for a weasel and rabbit to meet, yet wild weasels do find and attack rabbits in confined places, and must sometimes be unlucky.

Large Mammals

In theory, the larger a predator is, the larger the prey it can kill—but, also, the larger its food requirements. We would expect small predators to avoid the risk of injury by avoiding large prey. Small prey are often more numerous than large ones, and they take refuge in places more easily searched by small predators than large ones. Weasels are small predators, and small prey certainly are their bread and butter, so to speak. Weasels, however, are also bold and confident out of proportion to their size, and they seem to play by a different set of rules. Their extraordinary courage, strength, and tenacity put them among the few solitary predators able to attack prey larger than themselves (Figure 6.3).

For example, Harestad (1990) witnessed a longtail attempting to attack a Columbian ground squirrel, a stocky rodent weighing 350 to 800 g, considerably more than a longtail (see Table 4.1). The longtail's plan was foiled when two other squirrels mobbed it, and it fled. Such incidents may not happen every day, but eyewitness accounts of weasels attacking large prey are too common to be ignored, and are filled with vivid details describing, for example, the lightning thrusts with the teeth, the legs propped on widely straddled paws, and the tail furiously bristling with excitement.

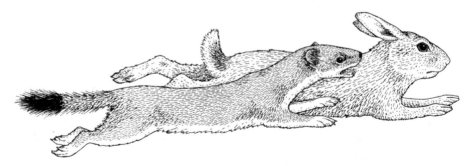

Figure 6.3 Weasels are among the few predators that regularly kill prey larger than themselves without having the advantage of hunting in a pack. The risks from attacking large prey are considerable because large prey can injure or kill a weasel that makes a wrong move. This stoat is showing the typical "bottle-brush tail" reaction to excitement and danger as it makes a grab for the rabbit's neck.

We define "large prey" as those as large or larger than the weasel attacking them. Thus, rabbits, many ground squirrels, and most rats are large prey for all weasels, but even chipmunks and large voles may count as large prey for the smallest least weasels.

Carbone et al. (1999) explored the diets of carnivorous mammals in general, examining especially the amounts of invertebrate prey eaten and the proportion of vertebrate prey as large as or larger than the predator in relation to energy requirements and intake rate. For their size, weasels take an exceptionally large proportion of vertebrate prey, and they eat large prey (relative to their own size) much more often than expected.

Carbone et al.'s model predicted that most predators larger than 20 to 25 kg should eat predominantly large vertebrates, but because invertebrates are easier to catch and can be harvested fast enough to sustain a body smaller than 20 kg, invertebrates should often constitute a large proportion of the diet of predators below that size The weasels stand out starkly among the small predators for having diets like big predators, because they do often kill prey larger than themselves. True, weasels are specialists on small rodents, which are mostly smaller than themselves, but in contrast to other small predators like shrews and hedgehogs, they do not seem able to live on invertebrates.

Why are weasels willing to run the risk of injury that comes with attacking prey larger than themselves? We suggest that this question has two answers. First, large prey provide a lot of food for a successful attacker, and weasels are perpetually hungry. Consequently, large prey may at times be critical for their survival (Powell & Zielinski 1983). The analysis presented by Carbone et al. (1999) suggests that when the density of small rodents falls below some critical level and alternatives are few, weasels must tackle large prey to meet their food requirements. The risk is balanced by the fact that, once a large prey item has been killed and cached, the weasel may be able to eat without hunting for days and thereby avoid exposure to the dangerous world outside its den.

Second, weasels always approach large prey with care. Just as a wolf takes caution when attempting to kill a large deer, which could kill the wolf with one well-placed strike from a hoof, a weasel takes caution when attacking large prey. Once, R.A. Powell (unpubl.) watched a 40-g female least weasel hunt and kill a 60-g female meadow vole. The least weasel approached the vole cautiously under cover. She watched the vole, which appeared unaware of her, from several vantage points before choosing one with perfect access, hiding the weasel until she was very close to the vole yet allowing an escape if necessary. Although the vole was large enough to kill the weasel, when the weasel's attack came it was swift and safe. The vole struggled violently, but only for the briefest time while the weasel held it firmly with her teeth at the back of the head and with all four paws. The weasel made a safe kill of a potentially dangerous prey.

Unfortunately, no one has made extensive, well-documented observations of weasels killing large prey. The abundant anecdotes are difficult to interpret, at best. For example, many well-known stories (some repeated in the first edition of this book) describe stoats killing rabbits and other large birds by "dancing" to distract attention from imminent attack, or stoats mesmerizing rabbits by their behavior or odor. The dancing of weasels and stoats is so well known in Britain that there has been for years a restaurant near Manchester called *The Waltzing Weasel.* "Stoated" rabbits (supposedly mesmerized by stoats), rescued without a mark on them, may recover from their paralysis and totter away, only to sink down and die later. Hewson and Healing (1971) examined carefully several rabbits killed by stoats. They concluded that, as the teeth and jaws of a stoat are somewhat small compared with the well-muscled neck of a rabbit, and the injuries inflicted did not seem to be severe, the rabbits must have "died of fright."

Intriguing though these stories are, there is another, simpler explanation of at least some of these reports, related to diseases. Lagomorphs in North America are periodically ravaged by epizootics of tularemia. During these epizootics, sick, live-trapped animals often die from stress-related causes in the hands of researchers despite every effort to give them humane care. Some rabbits suffering from tularemia or other diseases may be easy prey for weasels, and may die from stress when attacked, or perhaps even when approached.

Opinions are divided on whether "dancing" weasels are merely playing, or deliberately using the "dance" as a hunting technique. In favor of the first interpretation is the fact that these "dances" are not confined to situations offering a potential hunting opportunity. During months of radio tracking stoats in Scotland, Pounds (1981) watched 13 "dances," some of which were performed without any audience at all. In favor of the second interpretation is the fact that weasels in general are intelligent and opportunistic hunters, and if they find themselves surrounded by curious rabbits or birds, for whatever reason, they will certainly take the chance to catch one if they can. If they realize the connection between their behavior and the subsequent kill, they might well learn to "dance" on purpose. A third, completely different explanation is that the "dances" are an involuntary response to the intense irritation caused by parasitic worms lodged inside the skull (Chapter 11), and are quite unrelated to hunting behavior.

Whatever the interpretation of "dancing" or "mesmerizing," one consistent factor is that, when associated with an attack on large prey, these behavior patterns appear to reduce or minimize the risk of injury to the weasel. Such a benefit could eventually reinforce the behavior, whether it was deliberate or not.

Another way to reduce the risk inherent in attacking large prey is for two or more predators to cooperate. Although all weasels are normally solitary, Bullock and Pickering (1982) described in detail an incident involving two common weasels that persisted in attacking an adult brown hare. They suggested that the two weasels were cooperating. More frequent are anecdotal accounts of families of weasels hunting together before the young disperse. But we doubt if these stories are evidence of true cooperation, either. Weasel families do not stay together long, so the time available for learning of cooperative behavior is short. Our own anecdotal observations suggest that real cooperation is limited—rather, the mothers in family groups do the hunting and the youngsters tag along (p. 223).

Birds

Weasels climb trees readily and well (Figure 6.4), even to great heights, visiting the nests of birds and squirrels, running fearlessly along the branches and down again head first (Figure 6.5). While searching along branches, in holes, and in crevices in trees, weasels often find birds' nests and roosting birds (Dunn 1977), or other potential meals. DeVos (1960) once caught a hare in a trap, left it lodged a meter up a small tree, and came back to find a long-tailed weasel dragging it off. He retrieved it and put in into a larger tree, this time 3.5 m up, and stood back to watch. The weasel searched around and eventually found it, climbed up, and got it down again. When the hare was returned to the same place, the weasel found it again within a minute. Every time it climbed up and down with as much skill as any squirrel. The longtail watched by Pearce (1937) climbed spirally, hugging the trunk with its paws and wrapping its sinuous body around it. The weasel appeared to be familiar with this method of climbing, and did not go up more directly even when in great haste (Figure. 6.4).

Birds that feed on the ground in daytime are fair game for stoats, and the development of radio collars has increased the chances that someone will see them doing it. Murphy and Dowding (1995) watched, on two separate occasions, a collared stoat stalking yellow-crowned parakeets and chaffinches feeding on grass seed at ground level.

Song birds and their nests present little potential danger to weasels, but larger birds are different. Weasels do attack them, at least occasionally, and some of these attempts can misfire. A weasel jumping up at the neck of a large bird will often coil up its body and hold on with all four sets of claws, refusing to let go even if carried high into the air. For example, Barrow (1953) saw a waterhen flying overhead with a common weasel clinging to its throat (Figure 6.6). Unfortunately for the weasel, the bird dived into deep water, taking the weasel with it. The same story is sometimes told of predatory birds that have made the mistake of attacking a weasel and failing to grasp it firmly, allowing the weasel to twist in its talons and strike back (Chapter 11)—although in the case of the waterhen observed by Barrow, the weasel appears to have attacked the bird first. Weasels are certainly bold enough to try such an unlikely target if really hungry; one longtail was even observed trying to steal prey from a snowy owl (Boxall 1979). Such fearless behavior "could contribute to the high level of predation on weasels" (Fagerstone 1987)—indeed, it is almost a form of "weasel roulette" (King 1991d).

Figure 6.4 Weasels are nimble climbers, and can easily scale any rough surface. They climb up trees almost as confidently as any squirrel.

Like many predators, weasels sometimes have to tolerate having the tables turned on them by flocks of aggressive birds. For example, Hunter (1969) traced a loud outburst of chattering to a group of sparrows mobbing a common weasel. It darted into a flowerbed, pursued by sparrows dive-bombing it from all directions. Twice it broke cover, and twice it was driven back. It had to work its way under cover about 9 m along the bed, followed all the way by the mob, before it could finally escape. A weasel in such a position can hardly retaliate, since such excited birds would be virtually uncatchable. Mobbing birds usually do a weasel no physical harm, but they alert every other possible quarry in the neighborhood.

Some birds will dive-bomb predators approaching too close to their nests, as did one pair of gulls observed by Hosey and Jaques (1998) defending their egg from a stoat. When one gull struck the stoat with its foot, the stoat bit back, and the gull momentarily lifted it into the air a meter or so before dropping it.

Figure 6.5 Weasels have mobile ankle joints and climb down trees head-first, squirrel fashion.

The stoat, having lost the advantage of surprise, wisely decided to hunt somewhere quieter.

THE KILL

Whether the prey is found by vision, hearing, or scent, the final kill is certainly done by eye and stimulated by movement (Heidt 1972). Voles that respond to the presence of a weasel by running away can be overtaken in a couple of bounds and dispatched with hardly time for a squeak. Some rodents, however, especially deer mice and wood mice, may respond by climbing upward or "freezing,"

Figure 6.6 A weasel that attacks a large bird runs the risk of being carried into the air, since it may not release its grip while its prey is still moving. This common weasel was dropped into a lake by the waterhen.

and both these tactics are often successful (Cushing 1985). A weasel does not seem immediately to recognize as prey a mouse that, though in plain view, is sitting motionless, and in the wild this must sometimes give the mouse a chance to escape. Wood mice react less to the pungent musk of weasels or stoats than do field voles, so are less likely to give themselves away (Stoddart 1976; Gorman 1984).

Even larger, more visible mammals can use this trick so long as they sit still. Murie (1935), perched up a tree, watched a long-tailed weasel trying to catch a snowshoe hare in winter. Both were in their white winter coats, and there was some 15 cm of snow on the ground. The hare came loping along, then criss-crossed its tracks in a small area just in front of Murie's tree and settled down. The weasel followed, tracking the hare's prints through the maze exactly, at times passing quite close to the hare. When it had followed every turn of the trail to within a meter of its end, the hare skipped off. The same game was played twice more before the hare ran off in top gear, and the weasel gave up.

Jędrzejwska and Jędrzejwski (1998) closely watched wild, radio-collared common weasels killing small rodents in Poland. Small rodents in the open were seldom chased farther than a bound or few. When a least weasel entered a ro-dent hole, rodents erupted from all the emergency exits. Only six documented attacks involved a chase of 5 to 20 m, and all of these were on yellow-necked mice. Of the kills observed, a third were down rodent holes or tunnels, especially bank voles' holes. Another 45% were on the ground along fallen logs or in ground vegetation, especially those on voles and yellow-necked mice. Finally, 22% of kills, all of mice, were in tree cavities. In 1990, when rodent populations were high, less than half of all of attacks were successful. Roughly two thirds of the attacks were on single rodents, and one third on rodents in huddles of two to six. Of 19 attacks on huddles of rodents, the weasel killed two only once, providing rodents in groups a 20% lower chance of being killed by a weasel than singles. In summer when rodents are reproducing, 70% of kills may be nestling voles.

Laboratory studies have examined the preferences of weasels for different prey. The only generality that we can deduce from the information reported so far is that, not surprisingly, weasels appear to attack first those prey that are easiest to kill. Derting (1989) found that captive American least weasels caught woodland voles most often, meadow voles least often, and deer mice at an intermediate level. Meadow voles were the largest of the three species presented, and they put up the most aggressive fight. Woodland voles tended to be slow, timid, and easily cornered. Wild woodland voles, however, usually live in extended family groups, in which they can offer each other the double protection of the safety of numbers and of warning behavior (Powell & Fried 1992).

In general, voles are more easily caught than mice (Erlinge 1975). Finnish least weasels, given the choice, preferred bank to field voles (Pekkarinen & Heikkila 1997; Sundell et al. 2003). Males killed bank voles more quickly, and females ate bank voles first, whenever they had the option. Both sexes also ate juvenile bank voles before eating adults. The stoats observed by Raymond et al. (1990) and Vaudry et al. (1990) in a large, complex enclosure were more successful at hunting meadow voles and deer mice than house mice and short-tailed shrews. Male stoats took longer to find meadow voles than did females, especially when the stoats were hungry. Both sexes of stoats got the least energy gain from house mice, and both found the shrews either hard to find or hard to kill.

Such lists of preferences established in artificial conditions are interesting but, in real life, they may not influence a wild weasel's diet except during the short periods when rodents are abundant. When food is scarce, the racing metabolism of a hungry weasel requires it to take whatever it can get.

Where behavior or physiology makes some kinds of prey more vulnerable, weasels might be expected to take advantage. Cushing (1985) suggested that least weasels can perceive the pheromones of deer mice, and prefer to hunt female mice in estrus. Estrous mice emerged from their holes sooner than did diestrous (sexually inactive) mice, and were more likely to flee and less likely to freeze, making them significantly easier to catch. When given the choice in a Y-maze between odors of estrous and diestrous mice, the weasels sought the odors of estrous mice first. Unfortunately, when further Y-maze experiments were done with a larger group of least weasels, each tested only once, and with four age/sex groups of bank voles, Ylönen et al. (2003) confirmed that least weasels used olfactory cues in hunting, but not to distinguish between categories of voles. It is not clear yet whether this difference between mice and voles is real or an artefact of doing such experiments in captivity.

A weasel orientates its killing bite to the back of the head or neck of a rodent, stimulated by movement and guided by visual cues, particularly the position of the eyes and ears of the prey (Heidt 1972). The final strike is made with deadly accuracy. The long upper canines pierce the back of a rodent's skull or the vertebral column, and meet the lower canines entering below the ear or the throat. This neck bite is very characteristic, and death is just about instantaneous. Skinned mice killed by weasels often show no injury besides the two pairs of needle-like punctures in the neck.

If necessary, in the heat of the moment, the weasel may first grab the mouse almost anywhere, often wrapping its long slender body around its victim and using its feet to manipulate it and to gain leverage to transfer its hold to the neck (Figure 6.7). The whole process takes only from a few seconds to less than a minute, though the weasel often keeps its grip until the mouse stops kicking, perhaps shaking it a bit in the meantime. Weasels may attack larger prey by jumping on their backs, to get at the neck from above, or by darting in to reach unprotected parts from below. An animal that backs into a corner or makes cries of threat or fear only increases the weasel's excitement.

Figure 6.7 Weasels can make up in agility what they lack in stature, by wrapping their long bodies around a prey to help contain its struggles. If a weasel's initial bite is not at the head or neck of its prey, it can reorient its bite while holding fast to its prey with all four sets of claws.

Once its prey is dead, a weasel may lick any blood coming from the prey's mouth or wounds before starting to eat, but there is no truth in the old belief that weasels suck the blood of their prey. Weasels do not—in fact, physically cannot—suck blood.

A weasel presented with any of the common small rodent species already dead will eat them equally readily. Nearly always, a weasel begins its meal by eating the brain, often leaving completely the point of the nose and teeth, and then proceeding backward to the rest of the body. The feet, tail, and intestines, especially the stomach, are left until last. If many prey are supplied, a weasel may simply eat the brains and abandon the rest, or eat the brains of all and the body of only one, or even alternate among several carcasses. Strict carnivores, however, often have trouble obtaining enough carbohydrates from their animal diet. Their digestive tracts cannot digest plant material, which are the most ready source of carbohydrates. Consequently, weasels usually eat the small intestines of their prey, and thereby obtain partially digested plant matter with readily available carbohydrates.

CACHING

When several live prey are in sight, a weasel will kill one after another, and even search around for more, until it is exhausted. This is not because it enjoys killing, but because the entire sequence of killing behavior is instinctive and is set off by the sight of moving prey, whether or not the weasel is hungry. A weasel in a chicken coop is psychologically unable to ignore the fluttering of the live ones and settle down to eat one that it has killed. For a weasel, this behavior is completely logical, and has evolved because it is in an individual weasel's best interests to behave that way. In the wild, weasels never find prey as abundant as they are in a chicken coop. Most weasels do not know where the next meal will come from, and must search hard to find it. Consequently, when presented with more than one meal, a weasel will catch as many as it can, while it can, and store the extras for later. People do exactly the same thing when they buy more food at the grocery store than they need for one meal and put the rest in the pantry.

Weasels have a strong tendency to store surplus food. Where prey are very abundant, or unusually vulnerable, carcasses can be accumulated in astonishing numbers. For example, in the days when haystacks were common and always infested with huge numbers of rats and mice, weasel caches were often found inside when the stacks were dismantled for threshing. A typical cache would contain 40 to 50 freshly killed mice with the telltale puncture marks of the weasel's teeth in the neck. Fifteen caches made by common weasels in Russia, found by Parovshchikov (1963), contained an average of 30 carcasses each. Besides various kinds of voles and mice, the weasels had stored water voles, common shrews, moles, frogs, lizards, garden dormice, and goldcrests. One stoat cache found under a rock in Greenland by Sittler (1995) contained the remains of nearly 150 lemmings.

Weasels do, of course, prefer fresh meat, but if none can be had and the need is pressing, they can remember the locations of their caches and return to them. Svendsen (2003) describes a female longtail hunting in an alpine meadow in Colorado. She found two nests of golden-mantled ground squirrels, killed all nine young, and carried them to an old pocket gopher burrow. She visited the cache three times in the next few hours, and added new items to the same store during the following 2 weeks. She was lactating, and had young in another burrow about 175 m away.

In warm climates, caching food may produce only a short-term benefit, or even a liability. Surely, one might argue, it is wasteful if the stored food spoils, and killing multiple prey at once could increase the chances that later hunting expeditions will not be so lucky. But weasels need to eat frequently, although not much at a time, so almost all food is cached for a while. For example, the weasels studied by Jędrzejwska and Jędrzejwski (1998) needed to eat one to two small rodents per day. In summer, when weasels hunted on 90% of days, they captured a mean of two rodents per hunting day. In winter, when they hunted on only 60% of the days, they captured a mean of 3.1 rodents per day. During a frost, when the temperature fell below −5°C, they ventured outside much less often, and captured a mean of only 0.57 rodents per day and made up the difference from their stores. Cached food allowed the weasels to survive the coldest days when they dared not go outside at all.

Even in summer, food does not rot immediately when stored in cool holes in the ground, and a cached prey item, even if somewhat putrid, could make the difference between surviving and starving for a weasel. But in cold climates, where energy demands are high and alternative prey few, the caching habit has real survival value. Oksanen et al. (1985) and Jędrzejwska and Jędrzejwski (1989) suggested that caching should be regarded not as an unusual event but as a positive strategy, at least in winter and in the far north. This is a much more likely explanation: Really damaging behavior should be weeded out in the course of evolution, and after all, weasels evolved in cold climates, and are still completely at home in them.

The killing and caching of prey as a regular and predictable behavior of weasels differs from "surplus killing," as described by Kruuk (1972). Kruuk observed that large numbers of game animals were killed by spotted hyenas during a foggy night. The prey were disoriented and unable to escape from the hyenas, and the hyenas, stimulated by the movement of the prey, continued to kill for as long as opportunity offered. Under normal circumstances, the hyenas might have been able to kill only one or two prey before the rest scattered. Such "surplus killing" may be maladaptive for hyenas hunting large mammals in a hot climate, but for them, such events are very rare and have no evolutionary significance. By contrast, weasels predictably kill and cache as an adaptive strategy.

ENERGY EQUATIONS AND A WEASEL'S CHOICE OF PREY

The food eaten by a hunting animal is determined largely by three things: what there is to hunt, what the hunter can catch, and whether the chase is worthwhile. The best kind of prey for a hunter is one that is easy to find, is easy to kill, provides a much greater return in energy than the hunter spent in getting hold of it, and is always available. These are the reasons that, for weasels, the ideal prey are voles, mice, and lemmings. They make clear runways and burrows leading straight to their nests, marked with scent when occupied; they are not usually dangerous to tackle; they contain a worthwhile amount of meat, minerals, and most of the other things a weasel requires in its diet, conveniently wrapped in a waterproof package; and they may be so abundant that a weasel can supply all its needs for the day in only a few hours. All weasels, therefore, specialize on small rodents, of whatever species are present where they live, because they can obtain the most energy for the least expenditure and risk from making a profession of hunting them. A profitable catch is defined as one that returns more energy to the weasel than it spent in finding and catching it.

Outside their specialization, weasels will eat whatever else is profitable, and this means that they are restricted to a certain range of sizes of prey. The common weasel might receive a good return for the investment of energy it has made in subduing a large animal, but the risks and the size of the investment are high. When small prey are profitable, common/least weasels avoid adult rats and rabbits. These large prey are tackled regularly only by large stoats and longtails. Young rabbits are different, of course, and all weasels feast on young rabbits whenever they can get them.

At the other end of the scale are the items that, like insects and earthworms, are so small that the amount of energy gained from digesting them scarcely exceeds that spent in catching them. Earthworm setae were common in the scats of common weasels at Wytham, where the supply of small rodents was barely adequate (King 1980b), and Osgood (1936) observed a female stoat carrying worms to her young. If a hungry weasel comes across an earthworm lying on the surface of the ground, it might snatch it up in passing, but we doubt that a weasel with a stomach full of vole would bother, nor can we imagine even a hungry weasel actually digging for worms.

In general, the prey killed by real weasels matches well the predictions of the prey that ecological theorists expect would be taken by an optimal predator (Erlinge 1981). Prey species can be ranked according to the net energy gained by the weasel relative to the time and energy required to catch each one. According to theory, a weasel should put its most highly ranked prey type at the top of the list and start by concentrating on that prey type for as long as possible, and then it should add the increasingly less profitable prey one by one until all its requirements are met.

This theoretical idea generates some interesting predictions. One is that a predator should always search for, and attack if found, prey types that are on the list of profitable types, and ignore any prey type if it is not on the list, however abundant it is. If fresh nutritious steak is on the menu, who is going to bother with dry crusts, even if they are piled high? On the other hand, a weasel should be adaptable: An individual of a prey type that is not normally profitable might become so if it is unusually vulnerable. Rabbits would be on the usual list of profitable prey for stoats and longtails, but not for least weasels. But a least weasel finding a dying rabbit that is easy and completely safe to kill could do so. Rabbits in general might not be on its normal profitable list, but "sick or injured rabbits" could be.

Raymond et al. (1990) calculated the profitability of four prey types for stoats: meadow voles, deer mice, house mice, and short-tailed shrews. Meadow voles, the largest of these prey, offered the most gross energy for the stoats in general, but for individual stoats the rank order varied according to their differing abilities to find and handle prey, and their different needs for periods of rest between attempting to find and kill different prey. For male stoats, deer mice ranked at the top of the lists; male stoats in an enclosure could gain 600 kJ/h in energy from hunting deer mice. For female stoats, meadow voles were more profitable than deer mice, because female stoats gained their best return, just under 600 kJ/h, hunting voles. For both, the next ranked prey was almost as profitable, but the equations were influenced by other considerations. For example, male stoats took somewhat longer than did females to find meadow voles, and that additional investment of time lowered their efficiency in hunting meadow voles. Shrews landed squarely at the bottoms of the lists for both males and females.

The main problem about this approach is that specializing on small rodents in general, and on northern voles in particular, commits the weasels as a group to dependence on a very unreliable food resource. The population densities of rodents in the far north fluctuate across a huge range, so that, every few years, weasels can catch far more than they need, while in the intervening years they go hungry, or have to find other food. The least weasels and stoats of the northern forests and tundra have little else to turn to, except perhaps a few berries, or a bit of frozen carrion, when the voles and lemmings disappear (Nasimovich 1949). Larger mammals such as rabbits and hares are absent or too big to tackle; pikas and ground squirrels hibernate, and birds migrate south, for more than half the year.

The consequence of these huge fluctuations in numbers of small rodents, and the variation in seasonal availability of alternatives, is that the lists of profitable prey for weasels change unpredictably from one year to the next. Nonetheless, the distributions of prey killed by weasels do tend to follow the patterns predicted for optimal foragers. When voles or lemmings are abundant, weasels eat little else even when other foods are available (Chapter 5). Conversely, stoats in the Italian Alps ate large amounts of fruit only when this alternative food became (temporarily) more profitable than hunting rodents (Martinoli et al. 2001).

In milder climates, the choice is much better. Birds and rabbits are available all year round, and normally vary less in numbers from year to year than do small rodents. The eggs of birds provide a glut of nutritious and defenseless prey in spring. North America has a wide range of medium-sized rodents and small lagomorphs, including pocket gophers, ground squirrels, chipmunks, and pikas, which are important food for long-tailed weasels. But a weasel's choice is still limited by seasonal and geographical variations: Birds' eggs and young rabbits are not available in winter.

The only common and widespread potential prey that weasels always seem to avoid are the shrews. Previous explanations for this have assumed that shrews must be distasteful. Erlinge et al. (1974) suggested that shrews were avoided because they gave off an offensive smell, which deterred the weasel from making a serious attempt at a kill: "The weasels seemed doubtful at the decisive moment," as they put it. Such a reaction might explain why Raymond et al.'s (1990) and Vaudry et al.'s (1990) stoats searched for and handled shrews ineffectively compared with other prey, and rested more between hunting trips when only shrews were available. Search times, handling times, and resting times were all high for shrews.

Shrews were common in Wytham throughout a 22-month live-trapping study, but King (1980b) found none in any of the 344 weasel scats analyzed, despite looking for them (Chapter 5). Captive weasels will eat dead shrews provided as food if they have to, although only with the greatest reluctance; when given live shrews as prey, they often ignore them. Wild weasels cache shrews with the rest of their kills, but eat them only as a last resort. Shrews tend to accumulate in caches in a higher proportion than the live animals are available (Rubina 1960).

Recent data suggest that taste or smell might not be the right explanation. Shrews were the least profitable prey presented to stoats by Raymond et al. (1990), suggesting that weasels will not bother to eat shrews unless small rodents are extremely scarce. Such conditions might be more often found on small islands and in the far north, and there, where the choice of prey is very limited, weasels do sometimes eat appreciable numbers of shrews. On Tershelling Island (off the coast of the Netherlands) and in Ireland, both places where voles are missing, shrews are eaten much more often than on adjacent mainlands with voles (van Soest et al. 1972; Sleeman 1992). In the far north, only voles, shrews, and weasels are active under the snow all winter, and in vole crash years, hunting shrews may be the weasels' only possible alternative to starvation (see Figures 5.1 and 5.3). The density of shrews there tends to decline at the same time as the voles with which they live (Sonerud 1988).

Whether or not shrews do truly taste bad, weasels do not seem able to gain much energy from them, and that may be reason enough to avoid them wherever other alternative prey are sufficient. In western Finland in spring, small birds rather than shrews are the main alternative prey of least weasels and stoats when voles are low. Experimental removal of least weasels and stoats when voles were decreasing in density had a positive effect on the numbers of small birds compared with nonremoval areas, but none on shrews or game birds (Norrdahl & Korpimäki 2000b).

ACTIVITY: DAY OR NIGHT HUNTING?

Many mammals are active predominantly by day or by night, but weasels have to hunt whenever they are hungry, which is often (Zielinski 2000). This pattern may or may not be detected by studies on captive weasels, which often prefer to stay hidden until nighttime, when human disturbance around their cages has died down. Price (1971) trained least weasels and Kavanu and Ramos (1975) trained longtails to run on activity wheels, and then tested their responses to changes in how much effort was required to earn a reward. Zielinski (1986, 1988) trained both these species and also stoats, so the literature supplies information on all three species. The captive longtails and stoats were both strongly nocturnal, though willing to run by day if necessary. The least weasels were willing to run by night or day, for several sessions a day, consistent with Gillingham's (1984) conclusion that least weasels must eat several meals a day because they cannot eat enough in one meal to supply a whole day's energy needs. A few days of short rations had little effect, but weasels not fed for 24 hours more than doubled their wheel-running activity, much as wild weasels make longer hunting expeditions when voles are very scarce (Chapter 8).

Weasels can see well at any time of day or night, so in the wild their activity is governed not by their visual abilities but by a fine balance of conflicting needs: to find food sufficient to maintain their galloping metabolism, to find mates or feed their young, and at the same time to avoid their two worst enemies, cold weather and larger predators.

Raptors, owls, and foxes are a constant danger, and weasels adjust their foraging schedules in response (Zielinski 1988). For example, common weasels were easier to trap during the day than at night in Wytham Wood, in England, where danger from humans during the day was probably lower than that from tawny owls at night (King 1975c). Morning and evening trap records are only a rough method of estimating activity even when corrected for changing day length, so these data were enough only to show that King's original assumption, that the weasels would be mainly nocturnal, was wrong. Years later, Macdonald et al. (2004) made another study of the Wytham weasels, mainly on arable farmland on a different part of the estate. They confirmed, from data that were vastly more accurate but collected over a much shorter period, that the seven males and three females they radio tracked were exclusively diurnal. In Białowieża Forest, eastern Poland, and in Kielder Forest, northern England, places where common weasels have to dodge the attention of tawny owls, foxes, and many other nocturnal hunters, they are also mainly diurnal (Jędrzejwski et al. 2000; Brandt & Lambin 2005).

Cold winters are a time of serious risk for weasels, and in all northern continental climates weasels absolutely depend on access to a well-insulated den (Chapter 7). This restriction is one of the costs of being a long thin animal (Brown & Lasiewski 1972) in a cold climate, and has more to do with thermoregulation than with hunting strategy. Weasels cannot lay down fat or hibernate (Chapter 2); they can best defend themselves from chilling by staying in their dens and feeding from their cache of stored prey, so that on the coldest or wettest days they can avoid going out at all.

The total activity budget of the Quebec stoats was correlated primarily with ambient temperature, and secondarily with reproductive condition (Robitaille & Baron 1987). Outside the breeding season, there was a strong correlation between air temperature and the periods that these animals were willing to spend outside their dens. The same correlation was observed in wild, radio-collared longtails in Kentucky by DeVan (1982), and in radio-collared common weasels in Poland by Jędrzejwski et al. (2000). The Polish weasels actively hunted for an average of 2 to 6 hours a day, depending on the season, and spent almost all night in their dens. Their hunting trips were always short, most less than 2 hours at a time, but they made many more trips on warm days (>10°C) than cold days (down to –5°C). Variations in temperature were much more important in determining the lengths of hunting trips of the Polish weasels than was the density of rodents, at least over the normal range.

In a milder climate, thermoregulation is less of a problem, so individual activity patterns can be more variable. Stoats radio tracked in temperate southern Sweden divided their lives into a pattern of alternating periods of hunting, eating, and resting around the clock. They were seldom out of their dens for more than 10 to 45 minutes at a time in any season (Erlinge 1979b), totalling 6 to 9 hours per day. In New Zealand, 11 stoats tagged with activity-sensitive transmitters were usually active for at least 40 minutes at a time, totaling an average of about 8 hours a day at all seasons (R. Martin & M. Potter unpubl.). Of the 896 hours of activity data logged during this study, two thirds was recorded during the day, depending on the season, but there was huge variation between the individuals. Some ignored cool temperatures and rain, others avoided them, and this high individual variability meant that, at least in that study area, there were no times or conditions that were better than any others for maximizing trapping success.

One would expect weasels to be less active when the density of prey is very high, because easy hunting allows smaller home ranges and shorter hunting expeditions, leaving more time to rest and stay out of danger in a den—or, in the right season, for interactions with other weasels. In summer in Québec at a time when voles were abundant, the stoats watched by Samson and Raymond (1995) made an average of five expeditions averaging 40 minutes long during the day, and one at night. In total they spent an average of only 5 hours out of their dens each day, perhaps because the study was done after the mating season. By contrast, in one summertime study in northern England, common weasels were *more* active in habitats with very high vole numbers, rather than less (Brandt & Lambin 2005). In Québec, stoats are about the same size as common weasels in northern England, and the summers are warmer, so why the difference? One possible explanation is that common weasels, unlike stoats, can produce a second litter in summers when food is abundant (Chapter 9), and a second breeding cycle demands a lot of extra activity by males searching for mates and by females hunting for their young. Brandt and Lambin's study fits neatly with other evidence of extended breeding by common weasels in vole peak years (McDonald & Harris 2002).

BODY SIZES OF PREDATOR AND PREY

The link between the sizes of a predator and its prey is well demonstrated in "sets" of species with similar hunting habits and tastes but different sizes, such as the three species of *Mustela* in the northern hemisphere. Least and common weasels do usually kill more small rodents, and fewer medium-sized rodents and lagomorphs, than do stoats or longtails living in comparable habitats. Likewise, the stoats of smaller than average body size living in northern lands do concentrate on small rodents much more than do the large British and European stoats, which eat many more rabbits (Chapter 5). But these comparisons can be tricky, since merely finding rabbit hair in a weasel's gut does not prove the weasel killed the rabbit. Lagomorph hair is very distinctive under the microscope and unlikely to be misidentified, but it looks the same in lagomorphs of any age, and the same in those that have been killed by a weasel and in those that have been scavenged on the road. Another problem of interpretation arises because predators living in different places do not always have the same opportunities to choose between large and small prey. One of the reasons that Arctic stoats eat fewer rabbits than British ones is that they meet fewer.

We can avoid that geographic problem by examining the prey choices made by male and female weasels of the same species from the same population. Because males are so much larger than females, but hunt the same prey fauna, we can study the size relationships between predators and prey independently of location.

The latest and most comprehensive analysis found no evidence of any difference in diet between the two sexes of common weasels in Britain, when the analysis of a substantial sample was corrected for age and season (McDonald et al. 2000). Remains of large prey, such as lagomorphs, were more often found in spring, and small rodents in summer, and these seasonal biases can sway simple annual averages. Earlier studies that claimed that male common weasels ate more lagomorphs, and females more small rodents, were not always well enough checked for sampling error. By contrast, McDonald et al. found firm evidence that the diets of stoats were different in the two sexes, and this difference was itself significantly affected by season. Lagomorphs were always more commonly eaten by male than by female stoats, especially in winter, and small rodents more often by females than by males throughout the year. Both sexes were equally likely to take birds' eggs.

In the far north, female stoats are matched to the sizes of vole tunnels better than are males (Simms 1979a); so, females are better at searching for and catching voles (Raymond et al. 1990), and they increase their foraging efficiency for voles more effectively when hungry than do males (Vaudry et al. 1990). It may be that the small northern female stoats are less able to kill a full-grown rabbit than a male, but, more important, they may need to eat fewer lagomorphs because they can make a good enough living on small rodents without the extra effort of tackling large prey. In most places in the northern hemisphere, female stoats can rank lagomorphs so low that they fall off the bottom of the profitability list.

By contrast, female stoats in New Zealand do not have that option. They do eat some small prey (mice and insects) more often than do males, but in many places they also take large prey, lagomorphs and rats, as often as do males (King & Moody 1982; Murphy et al. 1998). How do they do it? Some unknown proportion of rabbits and rats eaten by female stoats everywhere would be young ones taken from nests, which are not too much of a challenge. But for a stoat stalking an adult rat or rabbit, it is a matter of adjusting the equations, balancing the risk of injury if it attacks versus the risk of starvation if it does not. When the second risk looks larger and more certain than the first, an attack is probably worth while.

New Zealand stoats are the prime example supporting the conclusion of Carbone et al. (1999), that carnivores with high energy requirements cannot live on invertebrates, even though they do eat lots of them, especially the females (King 1991b). The traditional small mustelid trump card, their specialization on small rodents, is of no advantage in New Zealand, and the abundant large insects are not an adequate substitute. Stoats there often have few options other than to rely on their relatively large size and their extraordinary boldness and tenacity, and perhaps also to exploit the paralyzing shock to the victim of being attacked by a stoat of any size. For breeding females it is a life-or-death calculation. The fact that stoats have prospered so well in New Zealand shows that they often do beat the odds.

7 | The Impact of Predation by Weasels on Populations of Natural Prey

THE POPULAR PICTURE of the weasels described in Chapter 1 is rather different from the real one. No doubt a vole would agree with the proverbial description of a weasel as "the Nemesis of Nature's little people," but the ecologist has the advantage of not having to see the weasel from the viewpoint of the vole. In a face-to-face encounter, the odds are usually in favor of the weasel, but the weasel has first to make the encounter. In this, the long-term odds are certainly in favor of the vole.

Besides, the meeting of a single vole and a single weasel tells us nothing about the impact of weasels on the numbers of voles, since the fates of populations cannot be gauged from the fates of individuals. Therein lies one of the main reasons why predation by weasels, or any other predator, is such a tricky subject to investigate. The converse questions, of estimating the effect on weasel populations of losses to larger predators and to trappers and gamekeepers, are equally difficult, for the same reasons.

PROBLEMS AND METHODS OF STUDYING PREDATION

Problems

The first and biggest problem has more to do with the characteristics of humans than of predators. Biologists are people, and they like to be sure they're getting interesting and reliable results, preferably in the shortest time possible and without getting wet too often. Field studies on predators, such as weasels, do not meet these requirements, for four reasons.

First, weasels are vastly less abundant than rodents, and their populations are often quite unpredictable. The low density and patchy distribution of weasels mean that a field study on any member of their family is more of a gamble than a project on voles. As graduate students, both of us skated on very thin ice when we began our research. King (1975c) trapped only five common weasels in the first 6 months of twice-daily trap rounds every other week (total 4,032 trap rounds), and none of this substantial effort contributed anything to the final analysis. Powell (1979b) trapped more fishers in his first year, when he had no transmitter collars (the order was delayed 8 months by the manufacturer), than in the following 3 years combined when he had more collars than he needed.

Second, predation is not a fixed process that, like a chemical reaction, reliably reappears whenever and wherever certain conditions are met. The interactions between predators and prey are flexible, and the outcomes variable from year to year and place to place.

Third, one cannot deduce anything about predation from watching the predators on their own: One has to study the prey as well, and this, of course, doubles the work. One must count not only the number of adult prey present, but also the number of young born and when. One must count the numbers added to and subtracted from the prey population by all causes, not only predation. And one must document the local and seasonal variation in all these processes. A proposed study on weasels usually involves too much work for one person, and can be tackled only by a team of at least two: one to study the weasels, the other to study the prey. Additional people who study the habitat and other animals will contribute even more to the study. We were each extraordinarily lucky to work in the same study area as someone else working on important prey species and willing to share data with us (Brander 1971; Flowerdew 1972).

Fourth, all known techniques for counting small mammals, the most important prey of weasels, are imprecise, and errors of estimation will increase at compound interest through the series of calculations needed to estimate predation rates. Moreover, a weasel's estimate of the number of small mammals available will include, for example, the nestlings, transients, and trap-shy individuals that are missed by the human observer, and will exclude those that are unavailable to weasels for some reason but present and counted by the human observer. Finally, all weasels know who their neighbors are, and who else might be hunting the same local stock of small mammals.

It is very important to realize the difference between a study of predation and a study of the food habits of predators. Predation is a matter of rates and relative numbers, and its results can be understood only in terms of whole populations, not only of the predator and of its prey, but also of the predator's competitors and enemies, and of its response to changing conditions. A record of a weasel killing a blackbird is a valid observation of the behavior of that weasel. A list of the prey identified in the stomachs of 1,000 weasels can be a valid estimate of the food habits of the population from which those 1,000 weasels came. But unless the densities of weasels and prey available in each case are known, plus the relative importance of other predators and prey in the same area, neither is a study of predation.

The difference can be illustrated in terms of a familiar analogy: shopping. Clearly, the stock of a certain item of goods in a shop depends not only on the number removed by shoppers but also on its price and replacement rate and on a host of other interactions involving the whole local shopping center. Table 7.1 summarizes some of these interactions, and introduces some useful shorthand terms that cannot be avoided in any discussion of predation.

Table 7.1 The Ecology of Predation Explained in Terms of an Analogy with Shopping

Concept	Predation	Shopping
1. The prey	Voles	Morningmunch breakfast cereal
2. The predators	Weasels	Shoppers
3. The locality	An ecological community	A town
4. The habitat	A particular field	A particular shop
5. Prey spectrum	Total local fauna	Total stock of shop
6. Prey available	Subset of (5) within killing range	Subset of (5) affordable
7. Prey selected today	Subset of (6) according to the opportunities and needs of the day	
8. Searching time	Time to locate vole	Time to find right shop and right shelf
9. Pursuit time and killing power	Ability to catch and kill vole	Ability to find item and pay
10. Prey replacement rate	Reproduction rate of voles plus recruitment rate of population	Manufacturer's production rate plus buying rate of shop
11. Risk factor	Chances of injury to weasel during kill	Chances of overspending
12. Penalty for misjudging (11)	Death	Loss of face and credit
13. Preference for	Most vulnerable of those worthwhile	Best quality of those offered cheap
14. Functional response	Increase in voles taken per weasel with increased density/availability	Increase in items bought per shopper with increased opportunity
15. Numerical response	Increased breeding success of weasels with vole density	Increased number of shoppers with increased opportunity
16. Surplus killing/caching	Killing above requirements and storing surplus when opportunity offers	Stocking up on specials
17. Alternative prey	Shrews	Grapefruit
18. Specialist	Weasel willing to search for voles rather than eat shrews	Shopping around for Morningmunch
19. Generalist	Weasel willing to eat shrews when voles scarce	Person of wide tastes
20. Impact of predation		
Nil	When (14) and (15) much less than (10)	Supply exceeds demand
Controlling	When (14) and (15) exceed (10)	Demand exceeds supply

The widespread conviction that predation must control prey populations is an old one, common among scientists as well as the general public, but it tends to overlook two important points (White 2001). First, despite appearances, few predators are efficient prey-harvesting machines, not even weasels (Chapter 6). Foraging efficiency is very hard to measure in the wild, but in one study of common weasels hunting rodents in forest at high density, fewer than half of all attempted kills were successful (Jędrzejwski *et al.* 1992). Second and consequently, most potential prey animals are not killed by weasels. Far more die for other reasons, especially the very young, without ever meeting a weasel. So, studies of predation impact cannot come to a valid conclusion without working out whether all these other reasons add to or substitute for predation. Not only is that calculation extremely difficult to work out, but also the answer depends on where and when you look, and on the scale of the interactions studied (Powell 2001).

Methods

The problem of studying the impact of predation by weasels on their prey has been tackled in four ways. All are inaccurate and imprecise to some extent, but all provide important information.

The first is an indirect method. If the daily food requirements of individual weasels are known, the researcher can either estimate how much of a known loss could be accounted for by weasels eating prey at that rate, or else compare how much loss could be inflicted by weasels compared with the amount needed to reduce prey of a known density by a given amount. Which of these two estimates is made depends largely on what sort of information there is about the prey. It sounds logical, but there is a snag. This method has to assume that weasels kill a certain, average number of prey per day, but they do not. Weasels are adaptable and intelligent predators that kill as many as they can catch, which is sometimes more and sometimes less than they need, depending entirely on the circumstances (Chapter 6).

The second is the direct method of counting the number of prey alive and then counting the number of them that have been removed by weasels. This method can be sabotaged by spectacular errors of assumption or census. One early study used unrealistically high figures for the density and productivity of least weasels, and produced the calculation that the weasels had eaten over three times more voles than were present (Golley 1960). The opposite error can be equally large if the number of voles removed is estimated from analyses of the weasels' diet, since weasels may kill and cache many more voles than they eat.

The third is the simplest to do and most difficult to interpret: Remove the weasels and watch what happens to the prey populations. The simple underlying assumption is that whatever changes in the numbers of prey might follow must have been caused by the change in weasel numbers. Unless the experiment is carefully controlled, however, it is usually impossible to eliminate any of umpteen other possible explanations. For example, if predation reduces only the production of young, but the density of adults is controlled by something else, removal of weasels and other predators may or may not increase the density of the population. After stoats and other predators were removed from sites in northern and southern Finland, the brood size and survival of young grouse improved (Kauhala et al. 2000), but the benefit to the adult grouse was not so clear.

The fourth is the theoretical approach. Mathematical, graphical, and verbal models of the dynamics of predators and prey can be very powerful aids to thinking, but are helpful and realistic only in direct proportion to the amount of information about real animals they incorporate.

All of these methods have different combinations of advantages and disadvantages, and all have been tried somewhere. The most progress in understanding weasels and their prey comes when different approaches are combined. Many examples of this in the recent literature show that the effects of predation by weasels are not the same in all situations. Weasels that have, for most of the year, almost exclusive access to small rodents in their burrows and under prolonged snow cover in boreal forests and grasslands can have a substantial impact on numbers of their prey. Weasels that are part of a wider community of predators hunting a variety of prey in temperate forests and farmlands typically have much less impact.

WEASELS IN TUNDRA AND BOREAL FORESTS

Weasels were entirely at home in the cold, windswept open spaces of Pleistocene times, and they still occupy the tundra and conifer forests of the far north today. Their speciality is hunting the various species of voles and lemmings that are also well adapted to living in those chilly habitats. In summer, these prolific little rodents make networks of runways through the matted felt of dead grass stems on the surface of the ground, and pull the green stems down through the tangle from below. Voles and lemmings are the favorite prey of all the cold-climate predators: hawks, owls, foxes, martens, and weasels. During summer, these other predators have to hunt voles and lemmings through the curtain of grass, while the small, northern weasels can follow them along their runways and into their nests. In winter, the migratory raptors move south, and deep snow protects the small rodents (at least to some extent) from foxes. Martens can get under the snow, but are too large to use vole and lemming tunnels.

Many northern populations of voles and lemmings display spectacular fluctuations in abundance every 3 to 4 years. In the summers that the rodents are increasing and at high density, raptors flock to kill huge numbers of them, and fledge large broods of hungry young. Predatory mammals reproduce well too, including weasels. The combined functional and numerical responses (defined in Table 7.1) of all local predators increases the toll on small mammals manyfold. But no predators can outbreed voles and lemmings at that stage, so the rodents can more than replace the losses.

The crunch comes in the following winters, when the permanent snow cover is established, the rodents are no longer breeding, and their food is getting scarce.

When the large predators have gone, the weasels carry on alone, searching out the diminishing numbers of rodents with deadly determination. The rodents have few defenses against predators that can follow them right into their last refuges. The weasels' relentless pursuit prevents the remaining rodents from rebuilding their populations, until in their turn, weasel numbers are cut down by starvation. When weasels become scarce, the surviving rodents have a breathing space to build up their numbers again, especially as, while rodents were few, their food supplies were able to recover. Improved food supplies and fewer weasels set the stage for the whole process to begin again. These repeated chains of events used to be described as "cycles," but there is important variation between places and years, so we prefer to use the less elegant but more accurate term "population fluctuations."

The most authoritative predator biologist of the 1930s to 1950s, Paul Errington (1963), did not believe that rodent populations in temperate habitats could be controlled by predation. His opinions were held to apply generally and were not disputed for a long time. In temperate habitats, Errington's judgment was more often right than wrong. But by the 1980s, a strong body of opinion had developed, pioneered by Pearson (1966, 1985) and Fitzgerald (1977) and developed especially by the Scandinavian school of ecologists (Hansson & Henttonen 1985; Henttonen et al. 1987; Sonerud 1988), holding that the northern voles and lemmings are different: Their population fluctuations are caused largely by continued heavy predation during the winters of the decline phase, due mainly to weasels. To test this hypothesis, one must count the numbers of both rodents and weasels in winter, just when fieldwork is most difficult. Fortunately, there is another way.

The nesting behavior of the small northern mammals offers a particularly useful opportunity to estimate predation rates. To survive the winter, each vole or lemming constructs a substantial nest of shredded grass stems under the snow. Weasels use these nests as temporary headquarters for several to many days, and they usually leave visible evidence of their stay in the form of scats and leftovers. Raiding nests can be very profitable for weasels, especially the nests of sociable voles that huddle together for warmth, and the large overwintering nests made by breeding female lemmings. The nests are well protected by snow during the winter, but on the arctic tundra in the first few days after the thaw they are clearly exposed to human view.

As the snow melted on Banks Island, in the Canadian Northwest Territories, Maher (1967) found 153 winter nests of lemmings, of which he reckoned 20% had been occupied by stoats. He believed that the stoats were responsible for the low density of lemmings in 1962–1963. At Barrow, Alaska, in 1968–1969, nearly twice as many lemming nests (35%) were raided by least weasels as on Banks Island (MacLean et al. 1974), and in the following year lemmings were so scarce that the number of nests found dropped from 770 to 0.

In northeast Greenland, numbers of lemming nests counted by Sittler (1995) on his 10-km² study area varied from a high of nearly 3,700 in 1989–1990 to a low of 105 in winter 1991–1992. Stoats occupied the most nests during winter 1990–1991, when large numbers of newly independent young hunters helped to push down a crashing population of collared lemmings. By 1992–1993 the stoat population had disappeared. Where do they go? Presumably, anywhere that there are lemmings. Sittler reports (personal communication) that he has found lemming nests taken over by stoats at 83° 40', in the far north of Greenland less than 500 km from the North Pole.

In the high Sierra Nevada of California, Fitzgerald (1977) worked in alpine meadows buried under 1 to 3 m of snow from late November to mid-April. The small weasels he studied habitually carried voles they had killed back to eat in the safety of their dens, but they usually ignored the front of a vole's skull, including the large incisor teeth. Therefore, Fitzgerald could count the pairs of incisors left by weasels in occupied nests and estimate the numbers of voles killed. He concluded that the stoats and longtails living on the meadows removed up to half the overwintering population of montane voles in the winters 1965–1966 to 1968–1969, and after the winter of heaviest losses the voles were reduced to very low numbers (Table 7.2).

In Canada and Fennoscandia, all the species of small rodents living in one locality generally reach low density at the same time every 3 to 4 years, along with the shrews (Henttonen et al. 1987; Sonerud 1988). Intense searching by hungry weasels willing to kill any small animal they meet in that vast network of subnivean tunnels seems the obvious cause. What else could synchronize the

Table 7.2 Impact of Predation by Stoats on Overwintering Montane Voles in the Station Meadows (14 ha), Sierra Nevada

	1966–1967	1967–1968	1968–1969
Mean no. voles/ha in autumn	25	83	127
No. vole nests/ha	23	65	85
Total no. nests examined	292	783	793
No. stoats resident	3	1	4
% nests occupied by			
stoats	28	5	13
longtails	2	2	4
No. voles killed by			
stoats	159	46	225
longtails	5	13	91
Mean no. voles/ha next spring	10	23	<2
% total losses attributed to weasels	>80	13	54

(From Fitzgerald 1977.)

fluctuations of a whole community of unrelated small rodents with different population dynamics? Therefore, the population fluctuations of weasels and their prey in boreal regions have been studied extensively during the 1990s. Before we discuss the dynamics of these communities, though, we must first look at how weasels interact with rodents in the very different habitats of more temperate climates.

WEASELS IN TEMPERATE FARMLAND AND FOREST

Compared with the far north, most temperate habitats support a much wider variety of prey and a larger community of predators, many of them resident all year round. Snow cover may be short-lived, and weasels must not only share the small rodents with larger predators, but they also must also watch out for themselves (Chapter 11).

Vole populations in many temperate habitats fluctuate annually, increasing during breeding seasons and decreasing between breeding seasons. The swings from low to high and back seldom reach the spectacular amplitudes typical of the far north, but fluctuations in vole populations are often large enough to make a lot of difference to the hunting prospects of individual weasels. In Britain, for example, in the deciduous forest of Wytham Wood near Oxford, the combined density of wood mice and bank voles remained between roughly 10 and 30 rodents/ha throughout the 1960s and 1970s (Southern & Lowe 1982). During those years, tawny owls and weasels were the two most important vertebrate predators in the forest. In 1968–1969, the known resident weasels ate on average about 8% to 10% per month (range 2% to 20%) of each of the populations of bank voles and wood mice (King 1980b). These losses accounted for only a small proportion of the 12% to 55% of voles and mice disappearing each month, and for perhaps 14% of the total production of rodents in the forest (Hayward & Phillipson 1979). Predation by weasels also had little effect on the average survival time of individually marked mice (Flowerdew 1972, pers. comm.). On the other hand, the tawny owls were also taking a regular toll, and at times it was substantially heavier than the weasels took. Tawny owls removed on average 15% to 33% of the mice and voles each 2 months.

These two sets of results from Wytham referred to short periods that did not coincide, but they confirmed that common weasels and tawny owls were the two most important predators of voles and mice in that woodland (Southern & Lowe 1982). Years later, the same conclusion was reached after a much more comprehensive and thorough study in Białowieża Forest in eastern Poland. There, a suite of 23 species of predators (of which the most abundant were tawny owls, common weasels, buzzards, and pine martens, in that order) hunted bank voles and yellow-necked mice all year round. Four species (tawny owls, common weasels, pine martens, and red foxes) accounted for, on average over three successive winters, virtually all (>90%) of the toll on both bank voles and yellow-necked mice (Table 7.3); stoats added another 1% to 2% on each. The total number of rodents removed (28 to 35 voles and 14 to 17 mice per ha over 197 days—about 1 kg of rodent meat per ha) was about the same as the total decrease in rodents over the winter: Voles declined from 35 to 8 per ha between autumn and spring, and mice declined from 24 to 3 per ha (Jędrzejwski & Jędrzejwska 1993).

Yet, all that concentrated predatory power could not prevent the Polish rodent populations from irrupting when the oak, hornbeam, and maple trees produced a heavy crop of tree seeds. In New Zealand, Blackwell et al. (2001, 2003) and Ruscoe et al. (2003) likewise concluded, from independent data sets, that predation by stoats is not likely to prevent an irruption of feral house mice after a heavy seedfall in southern beech forests, even though predation may have a significant effect on mouse populations *between* irruptions (Choquenot & Ruscoe 2000).

By contrast, in southern Sweden the local populations of common voles and wood mice vary through the year, but hardly at all from one year to the next (Erlinge et al. 1983, 1984). On a 4-km² study area of meadows and grazed pasture near Lund, common voles ranged only from 8 to 10 per ha in May and June to about 50 per ha in August and September. Erlinge and his colleagues set out to document the effects of the entire community of predators on the small mammals. They estimated that, in 1975 and 1976, the number of rodents produced roughly equaled the number eaten each year by the generalist predators (fox, feral cat, badger, buzzard, and tawny owl) plus the number eaten by the predators that specialize on rodents (stoat, kestrel, and long-eared owl). Even

Table 7.3 Impact of Predators on Bank Voles and Yellow-Necked Mice in Białowieża National Park

	Number of rodents removed from 1 ha in 197 days in autumn and winter											
	1986–1987				*1987–1988*				*1988–1989*			
	n		*%*		*n*		*%*		*n*		*%*	
Species	*Voles*	*Mice*	*Voles*	*Mice*	*Voles*	*Mice*	*Voles*	*Mice*	*Voles*	*Mice*	*Voles*	*Mice*
Tawny owl	16.3	11.7	58	70	19.3	11.1	56	71	22.3	9.8	65	70
Common weasel	4.1	2.3	15	14	7.3	2.2	21	14	5.1	2.0	15	14
Pine marten	3.6	1.6	13	10	5.1	1.5	14	10	4.9	1.6	14	11
Red fox	2.6	0.6	9	4	1.4	0.4	4	3	0.7	0.2	2	1
Stoat	0.5	0.1	2	<1	0.5	0.1	1	<1	0.5	0.1	1	1

(From Jędrzejwski & Jędrzejwski 1993.)

though rodents constituted only about 15% of their diets, the generalist predators accounted for over 75% of the rodent losses (Table 7.4). The stoats' share in the total was under 10%.

The key conclusion from Erlinge's wide-ranging study, the first to give us a clear picture of predation processes at the *community* level, is that community-level processes are important. The data allowed the losses due to the combined force of predators to be set against the production of all the small mammals together. The losses to the rodents were especially heavy in spring, when other favored prey, such as young rabbits, were not available. The voles started their breeding season surrounded by persistent predators poised to snap up the young as soon as they emerged from their nests. This heavy predation delayed and reduced the rodents' recovery from the winter nonbreeding period.

Predation was also particularly heavy in autumn, because of the rapid functional response of the generalist predators to the large increase in numbers of rodents through the breeding season. The net result was that rodents could never escape the attention of predators: The generalists were always ready and waiting, and the specialists joined in when they could. Stoats in this situation were almost as much at risk as their prey, because, when the population of rabbits declined, the stoats were forced into severe competition with the generalist predators for the remaining rodents (Chapter 10).

Populations of voles on meadows and arable land are capable of larger annual fluctuations than those of woodland species, but predators can still account for most new rodents produced. The total, annual production of field voles in England is estimated to be between 677,000 and 982,000 (Dyczkowski & Yalden 1998). The total annual consumption by all predators combined is estimated to be roughly 980,000 voles, of which two specialist predators on voles, common weasels and European kestrels, plus two generalist predators, red foxes and feral cats, account for 85%. Predators of 10 additional species divide the other 15% between them. Common weasels are estimated to kill some 22% of the voles lost to predators, and stoats 4%. During most years, the total production of young voles is roughly matched by total predation, explaining the relatively small changes in the field vole populations from year to year in these habitats.

Table 7.4 Annual Production and Mortality of Field Voles and Wood Mice on 4,000 ha of Marshy Meadows and Pasture in Southern Sweden

		No. eaten per year by predators			
	No. produced	*Total*	*Generalists*	*Specialists*	*% of total eaten by stoats*
Voles	171,400	156,865	120,700	36,165	9%
Mice	20,100	21,546	17,180	4,366	7%

(Erlinge *et al.* 1983.)

These ball-park figures necessarily conceal some important local and seasonal variations. For example, in one study in Kielder Forest, common weasels in overgrown clearcuts were seldom numerous enough to account for more than 5% of the variation in vole survival, reaching 20% only for a few months in summer when the weasels were most abundant (Graham 2002).

Goszczynski (1977) described how a team of Polish workers, interested in the causes of the population fluctuations of the common vole that periodically damage agriculture, attempted to account for all predation processes over 3 km² of mixed farmland. They examined the diets of martens, foxes, badgers, feral cats, and four species of hawks and owls as well as of common weasels. The study ran for 3 years, and covered a complete fluctuation in numbers of voles, from a low in late 1970 through the peak in 1971 (>330 voles per ha) to the next low in 1973. The proportion of voles removed by all the predators combined was high to start with, when the voles were scarce, and about three quarters of the total mortality of the voles at that time was due to predation. But by the time the voles had reached their peak, they were so abundant that predation could account for very few of their numbers and less than half of their mortality, even though by that time all the predators were living almost entirely on the voles, and some of them, including the common weasels, had also increased in numbers.

On the other hand, as the vole numbers declined, the predators were still numerous, so the ratio of numbers ceased to favor the voles. The increasingly desperate predators searched out almost every single vole that was left, reducing the population to about 1 vole per ha. At the height of the slaughter, predation removed an estimated 31% of the energy available to predators in the vole population. The common weasels' share was calculated at 11%, which put them in third place after foxes (37%) and feral cats (29%) in the number of common voles eaten. Unfortunately, the team's density estimate of one weasel per 3.9 to 4.5 km² was probably an underestimate (see Table 10.2). Our guess is that the real contribution made by common weasels to the toll was much higher, especially when the vole population was decreasing.

On North Farm, a game estate in Sussex, England, Tapper (1979) followed the changes in the numbers of field voles from 1971 to 1976. The 2.5-km² study area of gently rolling chalk downland was mostly divided into huge arable or pasture fields with patches of woodland and rough grass. Field voles and common weasels both avoided the open fields, which were frequently rolled, mowed, or heavily grazed, so both lived together in the few undisturbed areas and could be censused there easily. The voles were declining at the beginning of the study, fell to very low numbers (about 20 voles per ha) in 1973, shot up to around 300 voles per ha in 1974–1975, and declined again in 1976. The numbers of weasels caught followed the numbers of voles, but lagged behind by about 9 months (see Figure 10.7). The weasels ate about three to four times more voles per head in the years when voles were most numerous (54% of their diet in 1975 compared with 16% in 1973 when voles were few). Weasels also doubled their own numbers, thereby increasing the number of voles they could remove almost 10-fold.

Wood mice are usually much less common than voles, and also weasels find them more difficult to kill (Chapter 6), so predators hunting mixed populations of voles and wood mice tend to take fewer mice than voles (Table 7.3). The mixture of the two types of prey has consequences for the weasels. On a 150 ha area of farmland in France, the weasels observed by Delattre (1984) removed a substantial proportion (16% of total rodent biomass per month) of a mixed population of rodents when it was dominated by voles, but when mice predominated, both the weasels and their impact declined (Table 7.5).

Severe predatory impacts by weasels in temperate countries are seen only on the rare occasions when a population of rodents is confined to a limited area. In the days before combine harvesting, field crops were stacked into ricks in late summer, and dismantled only in autumn or winter when farm workers had time to do the threshing. A large, well-built grain-stack was a least weasel's idea of paradise: warm, dry, safe from large predators, and overflowing with thousands of rodents. (Modern hay-barns are good, too, but without the concentrated supplies of grain stored in the old grain-ricks, they are not quite of the same class.)

Farm workers welcomed weasels to a rick, with good reason. Over the winter of 1948–1949, least weasels occupied fully 90% of the ricks examined in the Miknov district, near Moscow, by Rubina (1960). In nearly all occupied ricks, there were fewer rodents than in the few ricks not visited by weasels. A more recent example is the study of the overwintering habits of radio-tagged meadow voles in New York (Madison 1984) that was wrecked by a stoat who entered the enclosure. The stoat killed more than half the tagged voles and took over a nest, lining it with fur and carrying back to it the voles it had killed, complete with their transmitters.

Predation by weasels in temperate countries affects not only small mammals. Wytham Wood in England has been the site of more than 50 years of research on the population dynamics of tits (Paridae). These small birds nest in

Table 7.5 Impact of Common Weasels on a Mixed Population of Common Voles, Field Voles, Bank Voles, Wood Mice, and Yellow-Necked Mice on 150 ha of Farmland in France

	Spring 1978	Spring 1979	Spring 1980
Combined density/ha	7.2	6.2	6.8
Proportion mice/voles	41/59	81/19	80/20
Number of weasels			
Male	9	5	2
Female (pregnant)	6 (all)	2 (1)	0
% biomass rodents taken by weasels	16	10	3

(From Delattre 1984.)

natural holes in trees, but also readily use the 1,000 artificial wooden nest boxes provided by researchers. Annual records of nesting success include data on which nests are destroyed by predators, most often by common weasels (Figure 7.1).

For the first 10 years of monitoring, until 1957, weasels raided only 0% to 8% of the nest boxes per season. Then, suddenly, up to half of the boxes were raided in each of the next few years. Dunn (1977) analyzed the detailed records available and showed that year-to-year variations in two factors alone were sufficient to regulate the population density of the tits: clutch size, which was determined largely by food supplies, and hatching success, which was strongly influenced by weasel predation. In turn, the extent of predation in any one year was affected mainly by whether small rodents in the wood were at low density during the nesting season (see Figure 13.3). The weasels sometimes also managed to catch the female bird on the nest (each nest box has only one entrance, and only females brood the eggs), and this might explain why female tits live slightly shorter lives than males.

Figure 7.1 A common weasel in the act of raiding a nest box of a great tit and removing a chick. (Redrawn from a photograph by C. M. Perrins, published by Dunn 1977.)

One might ask why the effect of weasel predation on the tit population was apparently greater than on rodents in the same habitat. There are several possible answers: The nest boxes were conspicuous and easily found; the nests were available only over a short season; lost clutches were rarely replaced more than once; on average there were fewer nests than rodents per weasel (nests about 2 per ha; rodents about 21 per ha); and weasels were by far the most significant predator on tit nests, whereas tawny owls were probably more significant than weasels as predators of rodents.

Since 1976 the problem for the tits (and for the researchers working on them) has been resolved by the installation of concrete, weasel-proof nestboxes (McCleery et al. 1996). After that, nest predation dropped from 30% to less than 5%, the mean number of fledgelings per nest rose from 3.4 to 6.3, and the chances of adults living to the ripe old age (for a small bird) of 5 years was greatly increased. The weasels no doubt continue to raid the unprotected nests of other birds, as they do all over the world (p. 307).

RESPONSES OF VOLES TO WEASELS

The vole that can avoid becoming weasel food for longest has the best chance of leaving surviving offspring. A vole that makes a bad move, and ends up moving through a weasel gut rather than through its own tunnel system, has less or no chance of contributing to the next generation. In the relentless calculations of evolution, natural selection favors the family prospects of voles that maintain constant vigilance for predators. (We refer here to voles because much of the research on this subject has been done on voles; other small mammals may behave similarly.)

Voles appear to respond to the presence of weasels and other predators on at least two different levels. One is the level of immediate, deliberate behavior: Stop, look, listen, and "lay low" (Jędrzejwski et al. 1992, 1993; Borowski 1998). These responses clearly benefit the individual vole that escapes a weasel for one more day. The second is the level of long-term, unconscious effects. Perhaps cautious behavior and stress induced by the prolonged presence of danger may inhibit the feeding rate and reproductive success of voles (Ylönen 1989). Both these responses have costs. The vole that lives but fails to reproduce contributes no more genes into future generations than does the vole that fails to avoid being eaten by a weasel.

When a vole detects a weasel entering its runway or nest, it skedaddles (Jędrzejwski et al. 1992). Flushed voles will run in the open if necessary, but they attempt to escape into thick vegetation or under dead leaves as soon as possible. Presumably, their best bet is to dive for a known hole for shelter, though they will climb twigs or sturdy, herbaceous vegetation if closely pursued. On the other hand, that strategy can quickly give away its position to a weasel that spots movements more quickly than motionless shapes. So, if it finds a relatively safe place, a vole (or, especially, a mouse) will freeze. Because the eyes of weasels are especially alert to detect movements, a small rodent sitting stock still but in plain view has a chance of melting into the background for long enough to avoid notice.

Voles can recognize and distinguish the odors of different predators and take appropriate action. For example, bank voles have a variety of antipredator responses to common weasels and stoats, but fewer reactions to foxes or polecats (Jędrzejwski et al. 1993). This difference must surely be because in some habitats bank voles contribute more than half the diet of common weasels and stoats (Pekkarinen & Heikkila 1997), but less than 10% of the diet of foxes and polecats (Jędrzejwski et al. 1993). A vole is safe from a fox in a narrow tunnel, and in the enclosure experiments reported by Jędrzejwski et al., voles fled into tubes simulating underground burrows when presented with the odor of a fox. But a weasel can follow a vole into a tunnel, so voles presented with the odor of a weasel avoided the tubes and reduced their overall activity for as long as a day or more (Jędrzejwski et al. 1993; Bolbroe et al. 2000). These antipredator reactions certainly seem to be innate and of long standing, since Orkney voles react to the odor of foxes even though these voles have lived in isolation from foxes since Neolithic times (Calder & Gorman 1991).

On the other hand, although avoiding a weasel is a clear benefit to a vole, a flight reaction that drives a vole into areas that it does not know well may constitute an even greater danger than does a nearby weasel. In addition, prolonged reduction in activity also has a potentially serious cost, in that it must reduce foraging rate and food intake. Over the long term, natural selection favors a compromise between different kinds of risk. Borowski (2000) showed that weasel odor affects the behavior of voles only over short distances and for brief periods, and does not cause a vole to leave its normal home range.

If voles react to the mere odor of a weasel and to its actual presence in similar ways, the impact of weasels on the population dynamics of voles might not be confined to killing those that they can catch. Weasels might also have effects on voles they do not catch, for example, by suppressing their reproductive activity. This hypothesis has had a good run lately, but the jury is still out.

Laboratory experiments indicated that female voles placed in close proximity to a weasel or its odor could slow their estrous cycles, reduce pregnancy rates, increase their aggression toward males when approached even when in estrous, and slow their growth (Ronkainen & Ylönen 1994; Ylönen & Ronkainen 1994; Koskela et al. 1996). Population models suggest that reproductive delays could benefit females in fluctuating populations (Ylönen 1994; Kaitala et al. 1997). But reproductive delays have not been observed by all researchers (Norrdahl & Korpimäki 2000a) nor in all vole species, and the experimental designs of the published studies have been severely criticized (Hansson 1995; Lambin et al. 1995). More realistic enclosure experiments have failed to support the reproductive suppression hypothesis (Wolff & Davis 1997; Mappes et al. 1998).

We do not, however, conclude that weasels have no effect on the reproductive output of their prey. On the contrary, we can see two possible ways in which they might: (1) When food is scarce for voles, for example, during a population decrease, the presence of a resident weasel may cause some voles to delay breeding, because if they move about less to avoid meeting a weasel, they will also reduce their chances of finding a mate, or enough food to support a pregnancy (Korpimäki et al. 1994; Norrdahl & Korpimäki 1995a); (2) if weasels preferred to kill large female voles, those capable of reproduction, disproportionately to their numbers (Klemola et al. 1997; Norrdahl & Korpimäki 1998), that would also decrease the reproductive output of a vole population, yet Y-maze experiments showed that weasels had no clear olfactory preference for any particular category of voles (Ylönen et al. 2003).

The reproduction suppression hypothesis is not quite dead yet, since the hunting behavior of wild weasels is only partially observable in captivity, but we suggest that it is usually a small effect compared with that of direct predation. Death in a weasel's jaws is surely a less subtle but far more effective end to the reproductive prospects of any vole.

COMMUNITY DYNAMICS: PREDATORS AND PREY TOGETHER

Clearly, weasels and other predators can kill significant numbers of small rodents, and they do increase their kill rate with rodent numbers. The consequence for the community depends on a combination of two rather different processes, which operate at the level of the individual (the functional response) and at the level of the population (the numerical response) (Table 7.1). Weasels do not hunt more often when rodents are abundant (rather the opposite), but they can make more kills per hunting expedition. To demonstrate this, Sundell et al. (2000) held least weasels and sibling voles in 0.5-ha enclosures. The numbers of voles killed by least weasels increased quickly with increasing vole densities, but only over the range from 4 to 16 voles per ha. The weasels' interest in hunting leveled off at around three voles a day (10 voles caught per 72 hours), even when vole density was increased to 100 voles per ha. In other words, as rodent numbers increase, an *individual* weasel can increase its kill rate (by a functional response), but only up to the limit set by its willingness to continue investing energy in hunting.

When vole numbers increase beyond that limit, the kill rate can increase further only if *populations* of weasels increase their own numbers, that is, by making a numerical response. When food is abundant, this numerical response of weasels can be impressive. For example, in the first half of 1990 in Białowieża when the vole population was high, Jędrzejwski et al. (1995) watched the density of a population of common weasels grow from a low of about 19 individuals per

10 km² at the end of winter (when they were not yet breeding) to 102 per 10 km² less than 3 months later.

All species of weasels are capable of making functional and numerical responses, although the results are not always the same. In some places, populations of rodents periodically escape the effects of predation and reach very high numbers, at least for a while, before they crash; in other places, vole populations tend to fluctuate annually over a more modest range. Two somewhat different but not mutually exclusive hypotheses have been proposed to explain this difference.

The Predation Hypothesis

The Scandinavian school of ecologists has developed the predation hypothesis, which postulates that, in a stable community in temperate climates, many generalist predators combined with some vole specialists can keep vole populations from fluctuating widely. Erlinge et al. (1983, 1984) concluded that voles in southern Sweden start their breeding season surrounded by persistent predators poised to snap up the young as soon as they emerge from their nests. Such predation delays and reduces the recovery of the rodent population from the winter nonbreeding period. Predation is also heavy in autumn, because of the rapid functional response of the generalist predators to the high production of rodents through the breeding season. The net result is that rodents can never escape the attention of predators. The generalists are always there, ready and waiting. In addition, the complementary habitat preferences of different predators reduce the abilities of voles to escape predation altogether (Korpimäki et al. 1996).

A single specialist predatory species cannot exert such a comprehensive effect on multiple prey populations. So, the other side of the predation hypothesis states that predation by the northern rodent specialists, the stoats and least weasels, is actually a necessary prerequisite for vole populations to fluctuate. In northern habitats, winter snow cover is prolonged and least weasels and the small northern races of stoats are the main or only predator on voles for much of the year (Wilson et al. 1999; Norrdahl & Korpimäki 2000a). At the very least, the predation hypothesis asserts that weasel predation in the far north, particularly on *Microtus* species, synchronizes the cycles of all small rodent species across very large geographic regions, including islands (Henttonen et al. 1987; Heikkila et al. 1994) (Figure 7.2).

Extensive evidence from northern Europe and Greenland confirms the expected correlation between heavy predation by weasels and population crashes of voles and lemmings. Extended predation can then keep rodent populations to low levels for several seasons, allowing the vegetation to recover and, thereby, ensuring future increases in rodent numbers (Korpimäki et al. 1991; Norrdahl 1995; Norrdahl & Korpimäki 1995b; Sittler 1995; Klemola et al. 1997; Korpimäki

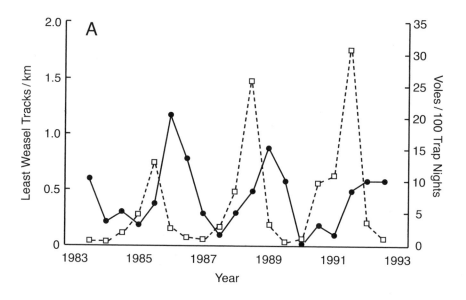

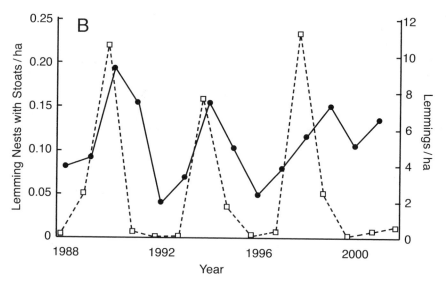

Figure 7.2 The linked population fluctuations of weasels and small northern rodents are well illustrated by two examples involving different species and different field techniques. (A) *Microtus* and least weasels sampled by snow tracking. (Redrawn from Norrdahl 1995.) (B) Collared lemmings and stoats sampled by spring censuses of lemming nests. (Redrawn from Gilg et al. 2003.) In both examples, solid circles and lines represent the weasel population and open squares and dashed lines represent the prey population.

& Norrdahl 1998; Gilg et al. 2003). It may even be able to explain the so-called "Chitty effect" (the observation that voles tend to be larger in body size during the peak of their numbers than in the decline and low phases). This would be easily explained, say Sundell and Norrdahl (2002), if the only voles to survive the population crashes were the smallest ones able to take refuge in burrows too small for the smallest weasels to enter.

Predation has always been among the possible explanations for vole cycles, right back to the early work of Charles Elton (1942), but it was not widely accepted at first. Reservations about the role of predation seemed reasonable at the time, partly because the influence of Paul Errington (p. 142) lingered well into the 1970s, partly because the proponents of the idea assumed that all vole populations are cyclic when many obviously are not, and partly because the crucially different effects of specialist and generalist predators were not distinguished by anybody. Great confusion was created in the literature by field experiments testing different questions with contradictory results (Norrdahl 1995). The confusion could have been expected, however, since the effects of predation, parasites/diseases, and malnutrition on small mammal populations are interactive (Korpimäki et al. 2004).

In recent years, the Scandinavian school has developed the predation hypothesis into a set of clearly defined propositions. They have done a lot of field experiments and population modeling that support the hypothesis as an explanation of how rodent numbers change under given conditions (Oksanen et al. 2000; Hanski et al. 2001; Korpimäki et al. 2002; Ekerholm et al. 2004). In its present form the hypothesis even explains some previously puzzling exceptions. For example, Hanski et al. (2001) developed a model that predicts variation with time in the behavior of rodent populations. Field research has confirmed that rodent numbers have fluctuated less regularly in parts of northern Fennoscandia since the early 1980s (Hanski & Henttonen 1996; Hörnfeldt 2004).

The predation hypothesis also predicts that a change in the population dynamics of voles should correlate with a change in the community of predators. To test this prediction, Korpimäki and his colleagues (2002) organized a large-scale, comprehensive experiment comparing the densities of field voles on four pairs of large (2 to 3 km^2) study areas. Each pair included predator-removal and nontreatment areas, and was followed for 3 years. The densities of the vole populations free of predation grew twice as fast as the nontreatment populations (subject to predation) in the increase phase, and doubled their autumn densities in the peak phase (Figure 7.3).

A model based on these results, that included seasonal effects and limited density dependence for the prey, predicted the type of population dynamics that is actually observed in the weasel-vole community in northern Fennoscandia. Korpimaki and his team could induce a change from multiyear cycles to seasonal fluctuations in the simulated vole populations in their model simply by reducing the proportion of specialist predators (especially least weasels) relative to

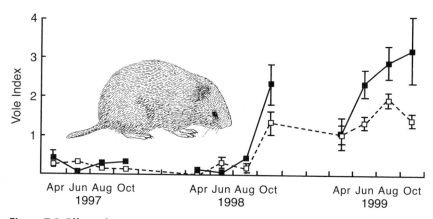

Figure 7.3 Effect of removing weasels on the dynamics of northern voles. Solid squares and lines represent vole populations where predators were removed; open squares and dashed line represent vole populations subject to predation. In this experiment, vole populations lacking predators reached significantly higher densities than those subject to predation. (Redrawn from Korpimäki et al. 2002.)

generalist predators. And in turn, the relative importance of predation by weasels was closely correlated with the duration of winter snow cover (see Figure 1.5), which itself is only loosely correlated with latitude (Strann et al. 2002).

Alternative Hypotheses

On the other hand, there is also evidence that argues against the predation hypothesis, or at least against assuming it is broadly valid outside Scandinavia. In a coniferous plantation in northern England, not only were generalist predators unable to prevent unfenced field vole populations from fluctuating (Lambin et al. 2000), but also specialist predators (common weasels) were neither necessary nor sufficient to drive the fluctuations (Graham 2002). And some populations of northern rodents do not behave as predicted by the predation hypothesis, for example, the population of lemmings in Canada studied by Krebs and his coworkers (1995; Reid et al. 1997), and those on the coast of northern Norway with a mild climate and no permanent snow cover (Strann et al. 2002).

Perhaps, then, the apparent north-to-south variation in rodent dynamics is not due to, or not due only to, the distribution of different kinds of predators, but involves other processes that have not previously been considered. Jędrzejwski and Jędrzejewska (1996) and their coworkers suggested that the missing piece in the puzzle is the thickness and productivity of vegetation on the ground— which is, in turn, affected by the length of the winter snow cover. Voles and lemmings live on or under the ground, and their numbers must be determined at least as much by food supplies as by predation.

The vital measures of food supply most relevant to voles are not closely correlated with latitude, because the production of ground vegetation under the thick canopy of a temperate forest is as low as in the tundra, whereas it is high in open grasslands of the temperate zone. In contrast to the Scandinavian school, the Polish team found no significant differences between the impacts of generalist versus specialist predators on rodents. All predators have their greatest impact on rodent populations during the period when numbers are already decreasing, in part because at that time the individual rodents removed are not being replaced.

According to Jędrzejwski and Jędrzejewska (1996) and Jędrzejewska and Jędrzejwski (1998), even when predation accounts for much of the observed rodent mortality, it does not determine whether the rodent populations will fluctuate substantially across several years (as in "cycles") or merely vary seasonally. Instead, the supply of winter food seems to be the most important precondition permitting rodent populations to increase beyond one season, usually by winter breeding. The simple distinction between the strongly fluctuating northern rodent populations and the more stable southern populations merely represents the two most obvious ends of a continuum governed by an interaction between winter food supplies and predation.

Where the standing crop of ground vegetation is always poor, as in temperate forests, a sudden massive crop of seeds dropped in autumn and lying on the ground all winter certainly stimulates populations of forest rodents and their predators. In Białowieża Forest, a contingent of generalist predators and high densities of weasels account for almost the entire winter mortality of the resident bank vole and yellow-necked mouse populations, but cannot prevent their numbers from fluctuating wildly between years in response to variations in tree seedfall (Jędrzejwski et al. 1995). The population irruptions in forests are short-lived, ranging from 8 to 29 rodents per ha, and never reach the extreme densities attained by voles in temperate farmland and steppe (143 to 490 per ha) (Jędrzejwski & Jędrzejewska 1996). The Poles concluded that the main prerequisite for the northern vole cycles is not predation by specialist vole hunters such as weasels, but rather a mean standing crop of ground vegetation of over 4,000 kg dry weight per ha in summer.

Bank voles are socially intolerant and, where they are the dominant vole species, their populations can be locally stable, possibly controlled more by social behavior than by weasels. Elsewhere, regular fluctuations are imposed on bank voles by interactions with other rodents, for one of two reasons (Oksanen et al. 2000): (1) weasels switch to bank voles as alternative prey when *Microtus* voles are scarce, or (2) masting events in deciduous forests supply enough surplus food to disrupt their normal spacing behavior.

The question of whether predators limit prey or vice versa is a simple question to which there is no simple answer. Communities with weasels are good examples of ecological assemblages with several trophic levels (plants, herbivores, primary predators, and secondary predators), whose components respond on different scales to the same environmental changes (Powell 2001). Weasels typically inhabit ecological systems in which food supplies for their main prey vary from one year to the next, as do the foods of voles or mice through a population fluctuation.

These systems consistently generate two kinds of interactions (Ostfeld & Keesing 2000). The bottom-up effect (herbivores reducing plant matter) is accompanied by top-down cascades (predators reducing prey numbers, and thereby releasing plant abundance). Predation usually reduces herbivore populations below the level they could maintain without predators, yet year-to-year changes in productivity of plants will affect year-to-year changes in herbivore populations more than the parallel year-to-year changes in predator efficiency. Obviously, food affects productivity, and predation affects survival, so the density of any given population at a given time is the integrated result of both processes. Consequently, as Jędrzejwski and Jędrzejewska's (1996) analysis demonstrates, most communities show complex changes through time that depend at least as much on the vagaries of weather (and its effect on plant production) as on predation.

Whether predation by weasels is or is not central to the population dynamics of rodents, the instability of rodent numbers can lead to dangers for the weasels themselves. Erlinge (1983) found that when the population of rabbits decreased, the stoats had to compete with the generalist predators for rodents. When weasels of any species find themselves in that situation, they are simultaneously both hunters and hunted (Powell 1973, 1982; Korpimäki & Norrdahl 1989a, 1989b).

Working out the consequences of this dilemma at the level of the ecological communities formed by weasels and their prey is not easy, as we noted at the beginning of this chapter. Even manipulative field experiments, which are usually regarded as the most authoritative type of evidence, almost always fail to give definitive answers to questions about the behavior of wild populations (Raffaelli & Moller 2000). We need to concentrate on improving our field techniques, and on using models more effectively to create hypotheses to test in the field, so that we can continue to improve our understanding of predator–prey interactions.

WEASELS AND OTHER PREDATORS COMPARED

The 3- to 4-year interactions between weasels and voles have a close parallel, played out over a longer term, in the 10-year cycle of snowshoe hares in Canada. The hare cycle involves a regular fluctuation in abundance of vegetation, snowshoe hares, and lynxes and other predators (Stenseth 1995; Krebs et al. 2001). When snowshoe hares are abundant, predator populations increase, especially the populations of lynxes, which specialize on snowshoe hares. When hare populations begin to decrease because of food shortage, pressure by abundant predators on the remaining hares is intense. Hare populations are reduced and kept at low levels until predators starve. During that period, plants recover and can support high hare populations again. And so the cycle continues.

By contrast, predation by weasels on small rodents, and by lynxes on hares, differs in many ways from predation by wolves on deer. Some ideas developed from studies of larger carnivores simply cannot be applied to weasels. For example, like all large predators, wolves prey as much as possible on prey that are easy to catch. For wolves, the easiest targets tend to be old, sick, disabled, or temporarily disadvantaged deer, because for wolves the risk of injury in attack is very real. Therefore, over the long term the wolves tend to cull unfit deer from a population. But the average mouse is no match for most weasels, so there is less need for weasels to select between individual prey. As Pearson (1985) put it, weasels do not wait until the meadows are overflowing with insecure or maladjusted voles; they can kill almost any vole with ease, and search them out even when vole populations are low.

It is true that a small female least weasel approaching a large, female vole must be cautious (Chapter 6), and so is any weasel attacking a well-grown rabbit. Nonetheless, on average, weasels can kill voles more easily than wolves can kill deer, and that makes a great difference to the economics of hunting by large versus small vertebrate predators. The difference between easy targets for weasels (such as nestlings and newly independent rodents, lagomorphs, and birds) and less easy ones is not obvious, but natural selection needs only a small differential to produce a real effect over many generations. To the extent weasels do select some prey rather than others when they can, they constitute an agent of natural selection favoring bright, alert prey just as wolves do, and it would be interesting to examine this question in the field.

Other ideas on predation developed from studies of invertebrates are not applicable either. Invertebrate predators and their prey both produce large numbers of young with only a very short time between generations. The rate of increase of populations of invertebrate predators is much closer to that of their prey than is that of any of the weasels to theirs (Chapter 12).

Clearly, the old idea of the "balance of nature," of predators and prey living in a dynamic balance, is irrelevant to weasels. Weasel populations are seldom stable, and no balance can be established, let alone maintained. Not only is the concept of any balance in nature difficult to define, but most ecologists now agree that no such balance exists. Perhaps better than any other example, ecological communities containing weasels illustrate that natural populations of predators and prey can fluctuate wildly. Those fluctuations are normal, natural, and just fine.

8 | Adjustable Living Spaces

ROMANTIC POETS WHO speak of the freedom of a wild animal generally mean freedom from captivity in a cage. There are other sorts of captivity, and some are worse (or, at least, sooner fatal) than life in a cage. Few wild animals are free of the daily task of searching for food, and few are free to wander wherever they like. Most animals have a home range on which they live, and to wander onto another's home range may invite trouble, because it might be seen as trespassing. Small animals such as weasels are never free from the need to keep alert for danger constantly, both from visible hazards such as raptors and from invisible ones such as loss of body heat in winter.

A weasel may be safe from all these dangers in a den or temporary resting site, but it cannot stay there indefinitely. Sooner or later, it must venture out to hunt and, in season, to search for a mate. The hungrier a weasel is, the more urgent is the search for food; on the other hand, the colder the weather is, the more energy can be conserved by staying in a nest. Least weasels and stoats living under the arctic snows have to balance the opposite necessities of spending time outside hunting, and inside keeping warm.

In general, weasels tend to conduct their hunting and social affairs in the least possible space and time; for them, sleep is a positive defense against the twin enemies of larger predators and, more important, cold (Buckingham 1979). Weasels range no farther and hunt no longer than they have to, and as soon as their needs are fulfilled they return to a den. Naturally, that takes longer if prey are scarce. Records of weasel home ranges and activity observed at different times and in different places therefore show enormous variation.

METHODS AND PROBLEMS OF ESTIMATING HOME RANGES

Burt (1943: 351) defined an animal's home range as

> that area traversed by an individual in its normal activities of food gathering, mating, and caring for young. Occasional sallies outside the area, perhaps exploratory in nature, should not be considered part of the home range.

This definition is useful and has lasted well because it tells us what a home range is and what its biological basis is. But to estimate animals' home ranges in these terms, one must be able to follow the movements of known individuals, and that is not always easy.

Collecting the Data

Individual common weasels can be identified by their unique belly patterns (see Figure 1.12), but only when they are in the hand. The sight of a weasel in the wild is never close enough to see a belly pattern, and certainly not close enough to read an eartag number. Weasels have few regular habits that might offer an observer a sure observation "ambush," and their dens are hard to find in most habitats except with the help of a trained dog.

Snow tracking (Chapter 10) can give a lot of information without disturbing the animals, but it is not possible in summer, or at all in mild climates. More important, one cannot tell which weasel made a given set of tracks. If tracks in snow are found in the same area day after day, it is a fair bet that they were made by the same animal. But since successive residents on the same ground tend to use the same dens and runways (Musgrove 1951), one would prefer the certainty of identification offered by an eartag.

The traditional way to observe weasels year-round in temperate habitats is indirectly, by catching them in live traps, marking them, and then releasing them in the hope of collecting a series of new location records. Unfortunately, live trapping weasels is often unrewarding, at any rate in certain places and in some years. To begin with, while one may fairly assume that where a weasel has been caught it must at least be present, the reverse is certainly not true.

Modeling studies using the frequency distribution of live-trapping data (so many individuals caught once, so many different ones caught twice, so many three times, etc.) can be used to infer the presence of individuals in the population that are *never* caught, and they could be a substantial proportion of the local population. In one New Zealand study of live-capture data for stoats, the probability of being captured for the first time was only about 17% per day (King et al. 2003a).

These data also showed that even the weasels that have been captured once are less likely to be caught again, and explains why some quickly learn to become downright trap-shy. They confirm the stories told by trappers such as Cahn (1936), who had a battle of wits with a stoat that learned, from a single experience, to avoid a further 32 carefully sited traps. Murphy and Dowding (1995) set a ring of live traps around a den known to contain a litter of young stoats, of which up to four at a time were seen playing around the den entrance, but none was ever caught.

Worse yet, different categories of weasels vary a great deal in the way they react to traps. Resident individuals, which are often older and confident of winning any encounter with an intruder, tend to be bolder and less shy of traps than the nonresidents, which are usually younger, insecure, and continually on the defensive. In the New Zealand data the probability of recapture for adult males was higher than for adult females, and both were higher than for young of either sex (King et al. 2003a).

Again, these calculations confirm the long experience of field biologists that females are more difficult to catch than males. Many studies of capture-mark-recapture records of weasels report that collecting a series of location records for females is especially difficult (Lockie 1966; Erlinge 1974; King 1975c). On the other hand, common weasels caught in live traps set for rodents are much more often females than males (King 1975a). The difference is not due to any attribute of the traps, but to the fact that traps set for small rodents are always laid out closer together than kill traps set for weasels by gamekeepers. Additional reasons for this bias are that females travel less each day and have smaller home ranges that enclose a smaller number of traps laid out at any given density (Buskirk & Lindstedt 1989).

What proportion of a local population of stoats is caught and marked in a live-trapping survey? Attempts to answer this question are usually frustrated, because, aside from any difference between individuals in reactions to traps, field methods almost always provide more opportunities for some stoats to be captured than others. For example, King and McMillan (1982) reported a simple experiment using regular live trapping to monitor a large number of stoats marked with eartags. Of one group of 21 stoats, tagged on or before January 15 and known to be present in the same area on or after January 25, nine were not caught on any of the 7 days on which the traps were set between January 15 and 25. In other words, a full third of the 21 stoats known to be alive were not recaptured in a whole week's trapping. Since the traps were set in a transect, which could not sample all home ranges equally, it may be that some individuals were simply not close to a trap during those days.

These data referred to a very short period during a period of high numbers of mice, when the probability of trapping a stoat may be lower than normal (Alterio et al. 1999; King & White 2004), but its general conclusion was later confirmed over a longer period and using completely different technology. Dilks and Lawrence (2000) monitored the responses of stoats to bait stations using miniature video cameras placed inside the tunnels. Over a period of 5 weeks of continuous observation, they filmed 45 occasions on which a stoat approached the entrance, but on 8 of these (18%) it did not enter.

Clearly, one has a better chance of catching a resident animal by setting several traps in its home range. Therefore, traps set on a transect line must not be set farther apart than the width of the average home range. Early field experiments in southern New Zealand showed that close-set traps (100 m between sites) catch a higher proportion of the locally resident stoats, but they also cost much more time and effort to operate (King 1980a). Intensive radio-tracking work has confirmed that the widest trap spacing for stoats, that minimizes effort and still puts at least one trap in every home range, is less than 1 km between sites in grassland (Moller & Alterio 1999) and 900 m in podocarp forest (Miller et al. 2001). For smaller stoats or weasels elsewhere in the world, these spacings must be smaller still.

Closer spacings increase both the work and the number of traps per home range, which in turn increase the probability of putting a trap in the right place. To get five traps in every female's range, spacing must be cut down to about 250 m (Moller & Alterio 1999). Yet, some traps constantly catch many more animals than others, although it is hard to define the exact characteristics of the most successful trap sites. Experienced trappers get an eye for good sites, but, whether by judgment or by luck, the way the traps are set out decisively influences the number, sex ratio (see Table 13.2), and proportion of the total population captured. Spacing, baiting, number of days set, and so on are all important.

A live-trapping study will, therefore, not sample the local population very accurately. While the male residents are being trapped day after day and the female residents occasionally, many nonresidents will pass through the area unseen and uncounted. As a further complication, the probability of trapping all classes of weasels may also change with the density of prey (Alterio et al. 1999; King & White 2004). It is hard for the field biologist to balance the size of the study area and the density of traps so as to sample both sexes adequately without dying of boredom from tramping around too many empty traps. Moreover, trapping is not an ideal method of working out the home ranges of any animals, because a resident held in a trap is unable to continue its normal life until it is released, usually many hours later.

So what are the alternatives? The ideal is to observe identifiable live weasels in the field by nonintrusive tracking, and in recent years various methods have appeared or are being developed. Some depend on trapping to start with to equip each weasel with an identifying mark; other methods make use of existing individual characteristics.

Natural Marks

The wanderings of a weasel with a unique footprint could be followed with minimal interference using a large number of tracking tunnels (Jones et al. 2004). For decades, biologists have recorded animal tracks in simple tracking tunnels. Smoked plates or sooted paper (now conveniently made with spray-on "sight black") (E. Rogers and D. Tiller, unpublished) or ink and paper (a pad of "ink" and two papers sprayed with a chemical that reacts with the "ink" to produce an indelible blue dye, or even simple food coloring), have all been used successfully (Mayer 1957; King & Edgar 1977; Zielinski 1995). Weasels readily enter well-set tunnels, especially if they are baited, and this method could give many records for each marked animal in a short time.

The problem is how to give each weasel a unique footprint. Field biologists in the United States have long marked small rodents by surgically removing one or two toes (Powell & Proulx 2003), but extension of this method to the larger toes of weasels, even under anesthetic, probably would not and should not be permitted by any Animal Ethics Committee. Instead, it may be possible to take advantage of computer technology to recognize natural differences between footprints.

For example, Herzog (2003) found that the footpads of fishers are unique, much as human fingerprints are. Herzog digitized the patterns on imprints of the pads left on tracking papers by captive and free-ranging fishers, analyzed them, and showed that individual fishers can be identified from their tracks alone. The tracks of weasels are smaller, and obtaining very clear and detailed tracks from them is often difficult, but the idea is intriguing. The two-component dye technique described by King and Edgar (1977) was adapted for field use from the method used by police to take finely detailed fingerprints, and could in principle produce animal prints clear enough so long as a suitable glossy white paper is used.

Another new method of locating animals with even greater promise is to identify individuals from their DNA. In species that regularly use conspicuous latrines, samples of DNA can be collected from feces (Kohn et al. 1999), and that could work for weasels in some habitats, such as the far north where winter dens are easy to find. In other habitats, it would be simpler to collect weasel DNA from "hair traps" (Woods et al. 1999). A network of tunnels, each with a miniature curry comb or a piece of Velcro inside, can grab loose hairs from animals passing through. DNA extracted from the roots and follicles of hairs collected in them can identify every individual, and the practical application of this method is well advanced in New Zealand (Gleeson et al. 2003). One temporary deterrent is that DNA analysis is presently rather expensive, and hair-collecting devices have to be closed after the first sample and cleared manually, but advances in technology are sure to bring the labor and processing costs down rapidly.

Artificial Marks

The main drawback of tracking naturally marked animals is that the weasels will be recorded only where the investigator has set tracking tunnels or hair traps. In addition, almost nothing can be found out about each animal other than its unique track pattern or its DNA profile. What we need is information about each individual, and a means of following it wherever it chooses to go.

Radiotelemetry is now a very sophisticated technology and is the method of choice for many studies. It provides a more accurate picture of the movements and activities of individual animals than can trapping or tracking tunnels or hair traps, and allows researchers to collect extensive data. Telemetry has not been applied much to the study of weasels, however, for several reasons.

First, it is still important to be able to trap the animals to fit them with radiocollars in the first place, and to retrap them whenever the transmitter is lost or the battery runs down. In animal transmitter packages, the batteries use more space and weigh more than all other components combined. A weasel's small size limits how big a transmitter package it can carry and, hence, how long the package will transmit. Weasels are often uncooperative about returning to have their transmitters serviced, which is a problem because battery life can only be short for small transmitters. At present, even the most efficient systems can provide no more than 6 months of battery life (Gehring & Swihart 2000), although that is a distinct improvement over the 30 days expected at best from early transmitters for weasels (Erlinge 1977b).

Second, designing a collar that will stay on a weasel's muscular neck is difficult. A collar that is too tight risks abrading hair or skin, while a collar that is too loose will surely slip off over the weasel's head. Weasels tend to shed their collars quickly, most within a couple of weeks (Murphy & Dowding 1994; Hellstedt & Kallio 2005); on the other hand, a well-fitted collar can remain in place too long. One male common weasel was recaptured, still carrying its useless burden, more than a year after the 21-day battery had expired (Delattre et al. 1985). Implanted transmitters can avoid the collar problem, but they are too large for weasels and have a much smaller transmission range than collars.

Third, the range of any transmitter is usually short (from 50 to a few hundred meters in thick vegetation) compared with a weasel's daily movements. Range can be increased by transmitting stronger signals or using a vehicle-mounted antenna, which can locate a 6-g collar carried by a male longtail up to a mile away in open country (Gehring & Swihart 2000), but the cost is a shorter battery life or a slower pulse rate for the transmitter beeps, making the weasel harder to find. Because all weasels make such good use of cover, even radio-tagged weasels are seldom seen (Sleeman 1990).

Fourth, there is the worry that a collar or implant will limit a weasel's normal activity by affecting the size of hole it can explore. Erlinge (1977b) reported that his stoats disliked their collars and scratched at them at first. The weight (6 to 10 g) and bulk of a transmitter must be maddening for such a lithe animal. In time they seem to accept them, though it is not clear whether they can hunt in all the usual, confined spaces and behave completely normally. Some studies using telemetry have assessed in captivity whether transmitters appeared to affect the weasels, and affirmed that they did not (e.g., Gehring & Swihart 2000). Others have come to the opposite conclusion, even with much lighter (2.5 g) collars (Delattre et al. 1985), so the question has not yet been answered for certain.

Ultimately, any study aiming to obtain adequate data on known, individual weasels must start by catching them in live traps, marking them, releasing them, and then collecting a series of new location records, either through successive captures, relocations using radiotelemetry, or a combination of methods. Smarter live trapping is therefore the key to further progress.

Improving the design of live traps would be a great step forward, especially if they could be equipped with alarms to call for attention when sprung. Finding good live-trapping sites also takes some skill and a lot of luck. Traps must be set in places weasels are likely to visit (in or under thick vegetation, in the roots of old trees, on fallen logs, alongside streams, under stacks of wood or heaps of stones, in stone walls and hedgerows), and one soon gets an eye for suitable sites (Figure 8.1).

Traps must also be protected from direct sun, excessive rain, potential flooding, and other dangers of the environment. Wooden traps offer the best protection, which is why we both prefer to use them, even though they are heavy and inconvenient to handle. Weasels are highly strung animals, and they really *need* the darkness, the privacy, and the insulation afforded by a wooden trap provided with adequate bedding. To avoid catching mice and voles, which are much more numerous than weasels and as fond of exploring holes, the mechanism can be weighted so that only animals heavier than, say, 30 g can be caught. This is worth doing because unselective wooden traps will often be blocked against weasels and damaged by gnawing. A trap suitable for stoats, incorporating a separate, insulated nest box that can double as an anesthetizing chamber, can be hand made from published plans (King & Edgar 1977).

By contrast, a night spent in a metal or wire-mesh trap (Belant 1992) can be a cruel experience for a weasel. The sizes and placement of treadles on ready-made, welded wire traps leave too little space for adequate bedding to keep weasels warm during their protracted stay, and it is difficult to protect open-mesh traps from drafts, rain, or excess heat. In addition, captured weasels bite and pull at the wire, often breaking their teeth and skinning their noses. Solid metal traps such as the Sherman can become ovens or freezers, sometimes even both on the same day, unless a detachable wooden nest box is added. We strongly advise against using any kind of metal trap for weasels.

In some habitats, placing traps where weasels must find them is so easy (e.g., in stone walls through open country and along ditches and hedgerows on traditional farms, or along streams or dry stream beds: R. A. Powell unpubl.) that weasels can be trapped without bait, although baited traps tend to catch more (King & Edgar 1977). In forested study sites, attractive trap sites must be carefully constructed and well baited to entice the local weasels to their doors. In some countries, traps can be baited with a live mouse, which both lures the weasel in and provides it with a fresh meal while it is waiting to be released (Erlinge 1974; Lawrence 1999), but providing for the needs of the mouse while it lives adds a lot of extra work.

If traps are left locked open before or between trapping sessions, the weasels get used to going in and out, and sometimes use them as temporary dens or larders. The residents can be counted as soon as the traps are set, and some shyer ones may be caught (at least once) that would otherwise never have ventured into any trap.

Figure 8.1 Three representative trap sites and sets for weasels. Placing a trap in a covered place and making the entrance look like that of a tunnel entices weasels to snoop around and inside the trap.

We consider (others differ) that live-trapped weasels are best handled under general anesthesia, preferably using oral (inhaled) anesthetics. The use of anesthetic ether, pioneered by Lockie and Day (1964), has now given way to more modern drugs such as halothane or ketamine, always after instruction from a vet and under supervision or permit. Present anesthetics are relatively safe for the animal and easy to use; they minimize fright to the animal and the risk of bitten fingers; and they simplify data collection and recording, outfitting with a transmitter, and inserting eartags (Figure 8.2). Tiny PIT (passive integrated transponder) tags implanted under the skin are even better than eartags, because they are less likely to be lost.

Captured animals recover quickly and, if handled carefully, seem not to find the experience traumatic. At least some residents can be recaptured often (King 1975c; King & Edgar 1977; King & McMillan 1982; Gehring & Swihart 2000; Purdey et al. 2004). On the other hand, recovery from anesthesia is never guaranteed; released animals do sometimes die when they get wet or cold; and keeping animals in captivity to recover can affect the social structure of the population. To overcome this problem, Gehring and Swihart (2000) developed a most commendable technique that allowed their long-tailed weasels to recover in a wooden nest box (rationed with a couple of mice) in the field. They opened the nest box remotely after a couple of hours to allow the weasel to leave at will, and removed the box later.

Estimating Home Ranges

The simplest method of presenting the results of a trapping or telemetry study is to map all the positive records collected for each individual and draw a line around the outermost ones. The home range is defined as the area of the minimum convex polygon that includes all the locations known to have been visited. This method is simple and has a long history of use (Hayne 1949), but it often also includes large areas that are never visited, so grossly oversimplifies, and can actually muddy, the real picture (Powell 2000).

Minimum convex polygons are strongly influenced by extreme locations, those that are at the edges of where an animal travels, so missing an extreme point or including an exploratory trip can make a huge difference to estimates of an animal's home range. But in fact, almost all the studies of weasel home ranges that we review here have presented their results in terms of minimum convex polygons, because the simplest method is often the only one whose scrutiny the field data can bear.

A more important problem with minimum convex polygons is that they fail to use all of the hard-earned data from the interior of a home range, and they concentrate on the question of the *size* of the area an animal covers at the cost of more interesting questions about how it *uses* that area. For example, a weasel's

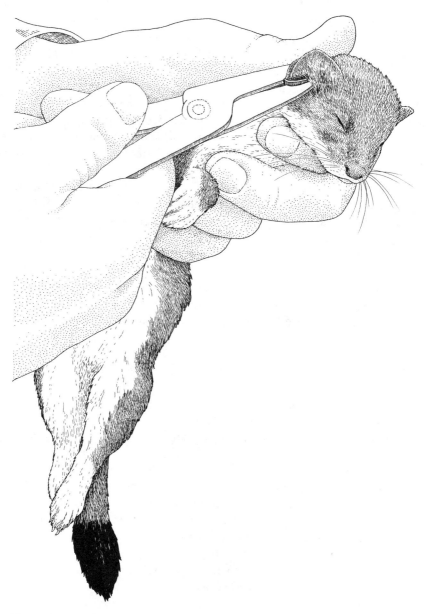

Figure 8.2 Attaching a tag onto the ear of an anesthetized stoat. (Redrawn from a photograph by C. D. McMillan.)

home range should really be described in all three dimensions of space plus the fourth dimension of time. Weasels of all species regularly explore the airspace both above the ground, by climbing trees, and underground, by running through burrows, and they use some parts of their home ranges much more often than other parts.

The best way to understand this complex pattern of use is to build a probability distribution of how a resident animal uses space—or, better yet (when someone figures out how to make one), a distribution showing how important the different parts of a home range are to its owner. For weasels, the question of how each uses the inside of its home range for foraging and avoiding larger predators is at least as important as the question of how it maintains its boundaries and interactions with other weasels outside.

If enough data are available, which requires successful radiotelemetry on a large number of individuals, new techniques for analyzing home ranges can begin to answer questions about how weasels move about on the ground most familiar to them. The simplest of these new methods is to divide the study area into small cells by drawing grid lines on a map and to count the number of times each animal was located within each cell. The result looks like a series of pillars rising from the map, and the height of each pillar shows the number of locations for that animal in each cell, thereby estimating how each animal divides its time across its home range (Powell 2000).

The next step is to use a fixed kernel estimator (the best method of analyzing home range data available at present) to convert the separate pillars into an undulating smooth surface (a utility distribution) whose highest spots are above the places the animal uses the most. It is easy to step from that to a probability distribution, showing the probability that the animal will be found in any given part of its home range. For example, in a nesting colony of burrowing seabirds in New Zealand, around 80% of the radio fixes from the stoats observed by Cuthbert and Sommer (2002) were located among seabird burrows, even though burrowed ground occupied on average only about 20% of each stoat's home range.

Utility distributions are demanding of data but extremely handy for analyzing home ranges. For example, superimposing an animal's utility distribution over a habitat map allows one to estimate the probability that the animal will be found in each habitat. Superimposing the utility distributions for two neighboring animals allows one to estimate how important the overlap is to them. If the overlap is in areas little used by each, it is probably of little importance to the animals, no matter how large its area. Cuthbert and Sommer's stoats tolerated a high degree of overlap in the 80% of their ranges outside the seabird colonies where they spent the least amount of time. If the overlap is in areas used much by both, the overlap is surely important, even if small. Unfortunately, the data requirements of kernel estimators are hard to meet, and few studies of weasels have attempted to use them.

Terminology: Home Range or Territory?

Burt's (1943) definitions of these two similar words are simple, but have proved useful for decades: A home range is the area over which an animal travels in the normal course of its activities, and a territory is a home range (or part of a home range) that a resident animal defends against other members of its own species, or to which it has priority use. The word "territory" is appropriate for an exclusive, defended area, whereas "home range" is a better word when there is considerable overlap and tolerance between neighbors.

Weasels seem to display all variations from one extreme to the other. Most resident weasels of both sexes have a home range, but not all home ranges include defended territories, so the word "home range" is often more useful to describe what weasels do than the more restricted definition of a territory. Moreover, it is not always possible to say to what degree each resident's ground is defended, and there is nearly always some degree of overlap between neighbors —indeed, there has to be a common zone where each can deposit scent marks for the other to find.

Once a resident animal has established its home range, it patrols its ground more or less regularly, setting and renewing scent marks in the course of each hunting expedition, notifying all other weasels that it has priority of use of the area. The home ranges of the two sexes usually overlap, and the cores of their home ranges are defended as territories against members of the opposite sex for most of the year. This pattern of intrasexual territories is quite typical of mustelids in general (Powell 1979a). At least in favorable times and places, males defend their ground against other males, and females against other females, while both tolerate (or avoid) members of the opposite sex. At other times, overlap is extensive, or weasels simply abandon their home ranges and move on. The more recent data reviewed here and elsewhere (e.g., Johnson et al. 2000a) have generally supported Powell's description.

The amount of effort a resident makes to evict intruders depends on whether the resources contained in its area are defensible. A den is a point location and can be defended quite easily, at least while its owner is in residence. Food resources are widely scattered, and usually cost too much energy to defend inside an exclusive area, especially as the chances are high that the owner cannot detect all intruders anyway. Males will, therefore, maintain strictly exclusive territories only where prey are concentrated, especially in winter when males are not interested in large-scale searching for females. In spring the same males may completely change their behavior.

Females seem to stick to familiar ground all the time, and their problem is not, or not only, to evict other females but to watch out for the males that use the same ground. Females can be just as intolerant of each other as are males (Figure 8.7), but they meet neighboring females less often than they meet the male whose home range overlaps with their own. On the other hand, spacing patterns are not a general characteristic of any species, including weasels, because the spacing of individuals depends on the local conditions, especially the distribution of food, which can change from one month to the next (Powell 1987, 1989, 1994; Clutton-Brock 1989).

HOME RANGE IN COMMON WEASELS

Carron Valley, Scotland

The first live-trapping study of the home ranges and social organization of weasels has become a classic and is often cited. It was done in the Carron Valley, near Stirling in Scotland, by Lockie (1966), in a young pine plantation overgrown with thick grass full of field voles. On his 32-ha study area (surrounded by about 800 ha of similar plantations), Lockie found 10 male common weasels, each jealously guarding a plot of 1 to 5 ha (Table 8.1; Figure 8.3). Transient males (at least 20) passed through all year round, especially in late summer and again in early spring. These were usually caught only once. Occasionally one would settle for a while, but its movements were much restricted, often to only one trap. The three known resident females lived on much smaller areas and each was caught repeatedly in only one trap. Unlike the males, they had no contact with each other. Six transient females were caught once each.

When Lockie began his study in November 1960, the weasels' territories were already established. They remained stable until November 1961. Then, for reasons that have never been explained, and at a time when weasels are normally settling into steady home ranges for the winter, the system broke down even though the voles were still very abundant (almost 300 per ha). Some residents died, others disappeared, and those that were left seemed to lose contact with each other. Over the next 2 years, to the end of 1963, enough common weasels entered the area to populate it, as it had been in 1961, 10 times over, but never more than two lived there at one time, and the system of contiguous, defended territories was never reestablished. By May 1962, the numbers of field voles were down to 44 per ha.

Wytham, England

Intrigued by Lockie's results, King (1975c) began to work on the common weasels of Wytham Wood near Oxford, England, in early 1968. Wytham is a deciduous woodland, quite different from Lockie's study area, and the only common small rodents within its boundaries were wood mice and bank voles, whose combined density was always very much lower than that of the teeming field voles of the Carron Valley. In the first 6 months of trapping over the 27-ha study

Table 8.1 Some Representative Estimates of Weasel Home Ranges[1]

	Country, years	Habitat	Sex	Area (ha)	Reference
Common weasel	Scotland, 1960–1963	Young plantation	M	1–5	(Lockie 1966)
			F	<1	
	England[2], 1968–1970	Deciduous woodland	M	7–15	(King 1975c)
			F	1–4	
	England, 1991–1992	Farmland	M	21–192	(Macdonald et al.
			F	4–29	2004)
	Scotland[3], 1971–1973	Farmland	M	9–16 (W), 10–25 (Su)	(Moors 1974)
			F	c.7	
	Scotland, 1977–1979	Farmland	M	2.4	(Pounds 1981)
			F	1.2	
	Poland, 1990–1991	Deciduous forest, rodents high	M	24	(Jędrzejwski et al. 1995)
		Deciduous forest, rodents low	M	167	(Jędrzejwski et al. 1995)
Least weasel	Iowa, 1925–1957	Farmland	M+F	4–10	(Polder 1968)
	Finland, 1952–1958	Mixed	M	0.6–3.0	(Nyholm 1959b)
			F	0.2–2.1	
Stoat	Scotland, 1977–1979	Farmland	M	254	(Pounds 1981)
			F	114	
	Sweden, 1973–1982	Pasture and marshes	M	8–13 (W)	(Erlinge 1977b)
			F	2–7	
	Ontario, 1973–1975	Mixed	M	20–25	(Simms 1979b)
			F	10–15	
	Switzerland, 1977–1980	Alpine	M	8–40	(Debrot &
			F	2–7	Mermod 1983)
	Finland, 1952–1958	Mixed	M	29–40	(Nyholm 1959b)
			F	4–17	
	Finland, 1998–1999	Subarctic birch forest, tundra	M	121–207 (Su)	(Hellstedt &
			F	35–66 (Su)	Henttonen, in press)
	Russia, 1970–1971	Meadows, scrub forest	M+F	11–69	(Vaisfeld 1972)
			M+F	120–124	
	Alberta, 1996	Mixedwood boreal forest	M	123–205	(Lisgo 1999)
			F	66–95	
	New Zealand, 1990–1991	Nothofagus forest, mice abundant	M	93 (Su/A)	(Murphy &
			F	69 (Su/A)	Dowding 1995)
	New Zealand, 1991–1992	Nothofagus forest, mice scarce	M	206 (Su/A)	(Murphy &
			F	124 (Su/A)	Dowding 1995)
	New Zealand, 1996	Nothofagus forest, mice scarce	M	223 (Sp)	(Alterio 1998)
			F	94 (Sp)	
	New Zealand, 1997–1998	Podocarp forest	M	256 (Sp)	(Miller et al. 2001)
			M	145 (A)	
			F	44–123	
	New Zealand, 1992, 1995	Rough grassland	M	66–215	(Moller & Alterio
			F	32–135	1999)

Table 8.1 (Continued)

	Country, years	Habitat	Sex	Area (ha)	Reference
	New Zealand, 2001–2002	Braided riverbed	M	313 (Sp)	(Dowding & Elliott 2003)
			M	185 (Au)	
			F	127 (Sp)	
			F	116 (Au)	
Longtail	Michigan, 1940	Mixed	M+F	32–160	(Quick 1944)
	Colorado, 1941–1946	Mixed	M+F	80–120	(Quick 1951)
	Kentucky, 1970–1975	Farmland	M	10–24	(DeVan 1982)
	Indiana, 1985	Mixed	F	41	C. Vispo unpubl.
	Indiana, 1998–2000	Mixed—kernel	M	180	(Gehring &
		method	F	52	Swihart 2004)
		Mixed—polygon	M	237	(Gehring &
		method	F	39	Swihart 2004)

1. For comparison, most of these areas were calculated by the minimum convex polygon method, although some authors also presented estimates derived from the same data using other methods (see text). Data refer to year-round ranges unless specified as (Sp) spring, (Su) summer, (A) autumn, or (W) winter.

2. This study and the following one were both done in the same area, but in different habitats.

3. This study and the following one were done in the same area, but Moors allowed a "corridor" along fencelines of 40 m, whereas Pounds allowed 10 m.

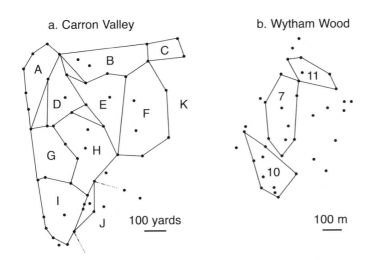

Figure 8.3 The patterns of winter home ranges of male common weasels. a. At high density in a young plantation in Scotland, April–October 1961 (Lockie 1966). b. At low density in deciduous woodland in England, January–February 1969 (King 1975a). Dots represent trap sites. Both sets of data plotted by the simple but debatable convex polygon method of joining the outermost traps visited by each animal. Kernel home range estimators (see text) provide more accurate estimates of home ranges but demand more data than we have found published in map form.

area, only five weasels appeared in the traps, one of them three times. But help came just in time.

While King, in some depression, took a break in August 1968, the Wytham gamekeeper borrowed some of her traps and caught four weasels in a week (and let them go, unmarked). So, he asked, what was the problem? He had applied the old gamekeeper's trick of tipping the guts of a rabbit into a plastic bag, stirring the mess around with a stick, and wiping the fragrant aroma on the entrances of the traps. King wasted no further time before applying the same method, and by the time the fieldwork part of the study ended, in June 1970, 36 common weasels had been caught a total of 348 times.

Only four males lived in the woods at any one time, each occupying at least 7 to 15 ha. Of the four, only one or two lived entirely in the woods, and the rest had parts of their ranges outside. Adjacent to the woods on one side was a young plantation, full of tall, unkempt grass and field voles. Some resident weasels living on the edge of the woods extended their ranges on that side, and their scats showed that they caught a lot of field voles in the plantation. Conversely, weasels resident in the plantation and in adjacent fields never visited the woods.

On average the resident males held their ranges for only about 7 months. When a woodland resident died, his home range was either shared between the nearest neighbors or a new weasel came in from outside. The resident weasels were well aware of their neighbors and adjusted their behavior accordingly. Excursions onto a neighbor's ground were carefully timed to coincide with the owner's absence. The four females observed always had much smaller home ranges than did males, never more than 1 to 4 ha (Figure 8.3).

Since common weasels leave their family groups when they reach independence at about 3 months of age, and most of them lived only a year or so, a different set of residents occupied the woods each year. Surprisingly, the weasels did not live any longer in the protection of Wytham (a reserve) than on game estates (King 1980c), perhaps because the density of small rodents in the wood was marginal (21 to 39 per ha).

When the density of small rodents fell still lower, in 1977–1980, the resident weasels simply disappeared. Those were the years an unlucky doctoral student, Hayward (1983), had chosen to renew research on weasels in Wytham Wood, in the same places and using the same techniques King had used. During the 12 months after August 1978, despite intense effort, Hayward caught no residents at all and only three nonresidents, once each. He came to the wry conclusion that even established weasel populations are liable to local extinction when prey resources collapse. This episode illustrates well that, contrary to popular belief, small predators do not live off the fat of the land; for most of the time, life is more chancy for them than for their prey.

The common weasels in Wytham often traveled along mole runs (Figure 8.4). King watched one, just released from a trap, darting straight down a not very obvious mole hole with every sign of confidence and familiarity. That might have

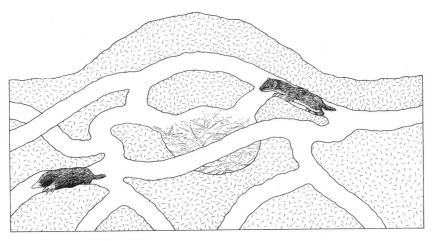

Figure 8.4 Common weasels seldom eat moles, but they frequently use mole runs and borrow the large and comfortable nests made by moles.

been a case of "any port in a storm," but, in fact, more convincing evidence showed that the Wytham weasels regularly ran along mole's runways and borrowed their nests. Among the fleas collected from them were two species specific to moles and rare on all other small mammals (see Table 11.7). The Wytham weasels very seldom ate moles, so they could have picked up those fleas only in mole nests and runways. Gamekeepers know well that common weasels use mole runs to get into rearing pens for game birds, and weasels are sometimes caught in traps set for moles.

With the development of radiotelemetry, the home ranges of the common weasels in Wytham could be worked out in much finer detail. Macdonald et al. (2004) recorded the ranges of four adult males (mean 113 ha) and two non-breeding females (28 ha), which lived in the network of linear habitats criss-crossing the farmland adjacent to the woods, but none moved more than 5 m from the cover provided by hedges and ditches. Successively tracked males occupied overlapping ranges, confirming that every vacated home was quickly taken over by others. On average, these weasels spent nearly half their time sleeping, or at least resting—all except a breeding female, which spent two thirds of the time she was tracked actively moving about, mostly in the area near the nest that probably sheltered her litter.

Aberdeen, Scotland, with Live Traps

In 1970, Moors (1974) began research on common weasels in Scotland near Aberdeen, using some of the same live traps that had proved successful with weasels in Wytham Wood. He set them in the stone walls and fence lines around the pasture and arable fields of Miekle Tarty Farm. His weasels also lived mainly in the stone walls and surrounding long grass, and seldom ventured into the open fields. They could move rapidly from one end of their home ranges to the other, often covering more than 1 km in an hour. Counting only the area the weasels used, a 40-m wide corridor along each wall or fence line, Moors' male weasels used 9 to 16 ha in winter and about 10 to 25 ha in summer. The few females caught had smaller ranges, averaging only about 7 ha (Table 8.1).

Summer and winter home ranges were different because the weasels apparently reorganized their home ranges every spring. In both winters of Moors' study, the males' home ranges were stable, and each animal stayed within his own area. In spring of both years, however, this settled pattern was totally disrupted. By April, none of the known males remained on his winter range; many had disappeared, and the rest were wandering widely. The social system settled down again in late summer, when a new crop of young weasels appeared and contended for home ranges with the surviving adults.

Sweden

In the early 1970s, Erlinge (1974) began research on common weasels in a mixed-age conifer-deciduous alder/oak woodland in southern Sweden. The study area was covered with a rich field layer of grass in all but the oldest spruce woodland, where the trees had grown thick enough to shade out the grass, and had stone walls to provide the weasel's favorite refuges. Field voles were common in the grass among young spruce trees, and so were wood mice and bank voles in the alder/oak woodland, but few small rodents, if any, lived in the oldest plantation.

As expected, the resident common weasels lived where most small mammals lived. When the study began, in winter, three males and two females resided in the grassiest areas. Nonresidents visited the alder woods occasionally, but no weasel was ever caught among the oldest spruces. The residents seemed content to remain on their chosen home ranges, although there were large areas of similar habitat around each. Four young and three adult males visited but did not settle.

In spring, the resident males began to range farther, covering larger areas and seldom staying in one place for long. By the following autumn, a stable new system was reestablished, including five resident males, of which only one had been present when the study began. As before, the residents settled in the grassy areas and ignored the closed-canopy spruce woodland. No breeding females, and few nonresident males, were caught in the second autumn. Erlinge did not report the sizes of the males' ranges, but mentioned that the females' movements, even in spring, were "less extensive . . . one female was trapped 29 times in March and April over an area of about 1.5 ha."

Aberdeen, with Radiotelemetry

In 1977, Pounds (1981) began his doctoral studies at the same field station where Moors had struggled to make sense of weasel biology only a few years before. Pounds had the advantage of having many basic problems already solved; the techniques of live trapping and of maintaining weasels in captivity were by then almost routine, and he also had a fair idea of the distribution and diet of the local weasels from Moors' work. But he was one of the first to apply radiotelemetry to weasels, so he was able to follow individual animals around their ranges and observe their movements and interactions with each other at a level of detail that previous workers could only dream about.

In the winters of 1977–1978 and 1978–1979, Pounds fitted seven male and two female common weasels with radios and tracked each for up to 28 days. As expected, they all used the stone walls, field margins, and rough grassland along fence lines around the fields, avoiding the open fields. Each had two to five den sites that it used frequently and three to ten occasionally used resting places scattered along its regular routes. Each had several favorite hunting areas and would stay in one for a while and then set off on an apparently purposeful straight-line excursion to another one. The weasels could be active at any time, but the longer trips were undertaken more often by day than by night. Some hunting areas were exclusive, although ownership could transfer, and others were shared, although not at the same time. Pounds watched one male hunting in an open turnip field during the harvest, running along the rows, pouncing on the small rodents disturbed by the workers.

The total areas calculated for the winter home ranges of these common weasels were very large. Those of the seven males averaged 34 ha, and those of the two females, 38 and 12 ha. But the "exploitable ranges" (i.e., excluding the generally unused open fields) were 2.4 ha for males and 1.2 ha for females. The difference between the figures given by Moors and by Pounds for weasels observed in the same area only a few years apart (Table 8.1) is due to the difficulty of deciding how large a boundary strip to allow along the field edges. Consistent with all previous studies using only the clumsier method of live trapping, Pounds confirmed that females occupied smaller areas than males, that weasels traveled surprising distances in short times, and that they avoided directly meeting each other but were quick to replace each other in possession of favorable ground.

Poland

Wlodek Jędrzejwski and Bogumila Jędrzejwska began to work on carnivores in the Białowieża Forest in Poland in the 1970s, and in the mid-1980s they extended their work to the entire community of vertebrate predators and prey (Jędrzejwska & Jędrzejewski 1998). Białowieża Forest is an exceptional example of European temperate deciduous and mixed forest, and a third of its area remains in old growth forest of natural origin that has never been logged. The Forest is large, spatially integrated with other extensive forest tracts in Poland and Belarus, and sparsely populated by people (two to three people per km²). Jędrzejwska and Jędrzejwski snowtracked common weasels each winter starting in 1985, and radio tracked 12 weasels in 1990–1991. Bank voles and yellow-necked mice were common in the forest, and root voles were common in marshlands. Densities of rodents were generally moderate (25 to 75 rodents per ha) but irrupted following the synchronous seeding of oaks, hornbeams, and maples, peaking in 1990 at more than 300 rodents per ha and then crashing.

The weasels were active predominantly during daylight, with a peak around noon (Jędrzejwski et al. 2000). On average they left their nests for a total of just under 4 hours per day, depending on the season. In winter, weasels went out once or twice each day for 1 to 2 hours at a time, while in summer they usually went out three to four times, and some days as many as seven times. They were, as one might expect, most active during the mating season.

When rodents were abundant on the forest floor, the ranges of five radio tracked male common weasels ranged from 11 to 37 ha, but after the rodents disappeared, the two different males radio tracked after the crash in rodent densities covered 117 and 216 ha (Jędrzejwski et al. 1995; Jędrzejwska & Jędrzejewski 1998). These are amazing areas for a small, short-legged animal to cover, but snow tracking confirmed them. During the peak, a rough estimate of weasels' ranges could be obtained from plotting tracks along transects through the Forest and making some limited assumptions about distances between tracks of different weasels. The assumptions were tested as far as possible with the telemetry data. Snow tracking yielded an average home range diameter (both sexes) of about 400 m, making a circular area of 13 ha during the rodent peak, and an average diameter of 1000 m, or 75 ha, after the crash. As predicted by theory (Powell 1987, 1989, 1994), when rodents were extremely abundant, male weasels tolerated extensive overlap of home ranges and shared their ground with transients.

Conclusions on Common Weasel Home Ranges

These various independent studies document important consistencies among common weasels, which in their turn illustrate important differences among populations and recurring problems with research methods. When and where prey are abundant, common weasels reach good numbers ("good" here means only "sufficient to support a study," because weasels can never be described as "abundant"), they have small home ranges, and they may even allow partial home range overlap with neighbors. Where prey are scarce, weasels are few, they do not establish home ranges, and transients do not linger. Therefore, differences in prey abundance are predictably correlated with differences in numbers, home range sizes, and defensive behavior of common weasels.

Every study done so far has reported that males' ranges were larger than females', often double or more in size. Males use more space than females, in part because they have larger bodies. At first inspection, males' home ranges appear disproportionately larger than their larger food requirements would require, compared with females. Many other predatory mammals show this same pattern, for example, martens and black bears (Powell 1994; Powell et al. 1997). Recently, however, Yamaguchi and Macdonald (2003) showed that decreased availability of prey caused by home range overlap requires male minks to have extra large home ranges; that is, the home ranges of males are not really disproportionately large because they have less food than previously thought. Ultimately, the reason for the sexual differences in home range sizes is probably related to a complex interaction of food requirements, the importance of different prey for members of the two sexes, and the males' need to know where females are located before the breeding season starts (Chapter 14).

Estimates of areas differ also because different methods have been used to estimate home ranges. The advance of technology has brought benefits, but at a price—it has permitted live trapping, which by definition interferes with what needs to be observed (the movements of animals around their ranges), to be replaced with radio-location data, which is inevitably short term. Moors (1974) and Pounds (1981) studied weasels in the same habitat on the same study area but with trapping and radiotelemetry, respectively, and calculated very different estimates of home range sizes. Both studies demonstrated that weasels avoided agricultural fields and concentrated their activities in the thick vegetation along walls and hedgerows. But the boundaries and areas of ground attributed to each individual changed with the two authors' different assumptions on how far weasels will venture from the cover of stonewalls and hedgerows.

Likewise, King (1975c) and Macdonald et al. (2004) also studied weasels in the same area, one with trapping and the other with radiotelemetry, but they concentrated on different habitats and produced very different estimates of home range areas. Inevitably, King's trapping records in the woodland, ranging from three to 109 recaptures per resident animal (totaling 348 recaptures over 27 months) tell us less about the details of weasel movements than Macdonald et al.'s radiotelemetry data from the farmland covering from 10 to 177 hours per animal (totaling more than 5,000 fixes over 103 days), but the same basic method of quantifying weasel home ranges, using minimum convex polygons, was used in both studies. The different limitations of the available field methods still determine what we can learn about these elusive little critters.

HOME RANGE IN LEAST WEASELS

The few studies of the activities and home ranges of wild least weasels have all been done by snow tracking. For example, over a 13-day period from December

20, 1939, Polderboer (1942) recorded weasel tracks on his farm in Iowa. He found the trails and resting sites of what appeared to be four least weasels, and calculated that each occupied an area of less than 1 ha. The weasels lived in the fencerows and slept in nests of grass and corn husks, originally made by mice, entered through burrows of 2.5 cm in diameter or less.

Polderboer confirmed the identity of the animals making the tracks by trapping each one at a den in a steel trap. None of the dens was lined with fur or had a latrine nearby, which suggests that the weasels were not long-term residents and that their real home ranges were undoubtedly larger than 1 ha. Nevertheless, this figure entered the literature at a time when little else was known about the ecology of least weasels, and when plugged uncritically into calculations of the extent of predation pressure exerted by an average population density of least weasels, produced comic results (Chapter 7). A later estimate of the ranges of least weasels on winter-ploughed fields was 4 to 10 ha (Polder 1968) (Table 8.1). Their dens were concealed under the furrows, and they pursued mice and voles through the small spaces between the turf or stubble sods.

In Finland in the winters of 1952–1958, Nyholm (1959a) systematically observed the tracks left by least weasels in snow-covered fields and copses, and along the banks of rivers and lakes. He reckoned that least weasels ranged over less than 3 ha, although he remarked that it was difficult to determine the ranges exactly because the weasels moved about so much under the snow. Naturally, much depends on how hard it is to find food. In the Kola Peninsula, in the far north of European USSR, voles and least weasels were numerous in the summer of 1938. The voles crashed to very low numbers in the following autumn, and least weasels tracked during the winter of 1938–1939 were hunting over extended home ranges of up to 10 ha (Nasimovich 1949).

Klemola et al. (1999) tracked least weasels and stoats during winter in Finland for 6 years through the 1980s. At the same time, they monitored the abundances of field, sibling, and bank voles. As expected, least weasels traveled farther (males 1,395 m, females 1,015 m per trip) when voles were scarce than when they were abundant (660 m, 110 m, respectively). Weasels also traveled farther in farmland than in forested areas, but exactly what this means is not clear. One possible explanation is that weasels actively preferred to forage for rodents in farmland, but an alternative is that the researchers could more easily find and follow their tracks in open country than under trees.

HOME RANGE IN STOATS

Finland and Russia

Probably more fieldwork has been done on stoats than on any of the other small mustelids. The pioneering studies of their home ranges were all done by snow tracking. In Finland, Nyholm (1959a) followed one stoat around its home range, which was divided into several separate hunting areas, visited in turn. Each area had at least one den or refuge, usually in a barn or a pile of logs, or in the banks of ditches. He tracked many other stoats as well, and calculated the home ranges of 63 of them (Table 8.1). They usually traveled about 500 m in a night; one ran 1.8 km within its own territory, and another, presumably a nonresident, ran almost 6 km.

In Russia, Vaisfeld (1972) mapped the winter ranges of stoats in the provinces of Arkangel and Kirov, in the very far north of European USSR. In the vast flat flood plains of the great northern rivers, the stoats' trails were concentrated around patches of scrub. Every home range (varying from 21 to 69 ha) had at least some scrub cover; no stoats lived out in the open meadows. In some areas, the scrub had been cleared so as to extend the meadows, and there were windrows of bulldozed scrub piled up for burning. These windrows provided even better cover and food for small mammals than the natural scrub, so were greatly favored by the stoats. At the beginning of the winter of 1970–1971, 18 stoats lived on 80 ha of this habitat (i.e., averaging one per 4 ha). Within a few months, 15 of them had been caught by hunters. The three that remained lived in one corner, on home ranges of 11, 13 and 17 ha—not so as to avoid the hunters, but because human disturbance (the piling up of the scrub) had so increased the prey resources of that area.

Sweden

Sam Erlinge and Mikael Sandell (Erlinge 1977b, 1977c, 1979b, Erlinge & Sandell 1986; Sandell 1986, 1988, 1989) observed the home ranges of stoats in southern Sweden, first by live trapping only and later also using radiotelemetry. Their main study area was a 40-km² expanse of pasture, wet meadows, and marshes, crossed by streams and stone walls. The stoats' home ranges were concentrated in the most favorable spots, in the marshes and along the stone walls where small mammals were most abundant. The four main groups of residents were separated from each other by unoccupied open fields, and stoats from one group seldom visited another except in spring. Each group included several adult males and females plus a larger number of young of both sexes. Track surveys during snowy periods confirmed that the trap captures reflected the real distribution of the animals.

Erlinge's stoats established home ranges in late summer, as the year's crop of young entered the population and sought to establish home ranges for themselves for the winter. Males and females lived separately, with little contact. An adult female sometimes used part of an adult male's home range, but avoided him whenever possible, while young of both sexes kept well clear of his area altogether. Females had little to do with each other, and their home ranges were generally well spread out. They did forage throughout their home ranges, and on one occasion two happened to be at a common boundary at the same time. They came no closer than 60 m before moving away in opposite directions. Females generally spent a lot of time in deep rodent tunnels.

Males were in at least indirect contact with their neighbors of the same sex more often than were females. Two males whose ranges overlapped spent a lot of time in the boundary zone. There were never any signs of fighting or chasing or other direct confrontations, but the boundary was obviously set by social contacts between the two males. Adult males whose home ranges included or overlapped those of one or more females confidently moved about wherever they pleased, since they were always dominant in any encounter. When a male visited a female's den when she was not receptive to visitors, she hissed defensively or screeched, and he retreated.

All the resident stoats hunted over some parts of their ranges more than others, usually in short bursts of 10 to 45 minutes separated by longer periods (3 to 5 hours) of rest. The stoats obviously concentrated on the places where rodents were most abundant. Each visited the core of its home range almost every day, but other parts every few days. Some areas clearly known to the stoats were ignored altogether. The stoats covered more ground, and spent more time hunting, when rodents were scarce. They tended to be nocturnal in winter and diurnal in summer, and this change was connected with a pronounced seasonal reorganization of stoat society.

In spring, the settled system of winter ranges gradually broke down, just as in Lockie's common weasels. Some males set off on long excursions (up to 5 or 6 km), others stayed at home but moved about far more actively than before, and still others disappeared (Sandell 1986). The difference in behavior was correlated with body size and, presumably, social dominance. Large, assertive males ("roamers") wandered widely in spring, whereas small, low-ranked males ("stayers") stayed put. Erlinge and Sandell (1986) suggested that the reason for this seasonal change in behavior is that the decisive resource for males in early summer, receptive females, is more widely dispersed and less predictable and defensible than the decisive resource during the winter, usually food.

Dominant males can probably be sure of gaining access to any females they can find (Erlinge 1977a), so they can score the most matings by roaming in search of females across a large, but not *too* large, area. Seaman (1993) and Powell et al. (1997) found that wandering, dominant male mammals can keep good track of only a limited number of females. If a male tries to locate too many females, he cannot stay in close touch with each of them, and misses potential matings. The receptive period of the female stoats is short, and if a roamer mistimes his visits to a female, a low-ranked stayer already on the spot might take advantage. A low-ranking stayer cannot dispute the possession of a female with a dominant roamer if it comes to a fight, but by staying close to one female he could be in the right place at the right moment.

Mating tactics in general are too complex to be explained by single-factor models (Sandell & Liberg 1992), but this one does seem to account for changes in activity of stoats across the seasons, the pronounced differences in activity patterns of males and females in spring, and the strong seasonal swings in the sex ratio of stoats caught in traps.

Canada

The spacing and activity patterns of the small stoats living in southern Quebec, Canada, were observed by Robitaille and Raymond (1995), by live trapping stoats on farmland, predominantly hayfields (ca. 40%), cornfields (ca. 25%), pasture (ca. 15%), and woodlots (ca. 10%), from 1978 through 1980. The number of resident males was smaller than the number of transients each year, but the reverse was true for females, and more females overwintered than males. Nonetheless, residency times for females (mean = 157 days) exceeded those for males (82 days), with huge variation in both. One juvenile male lived on the study site for only 28 days, while one female was there for at least 443 days (almost 1 year and 3 months).

During the breeding season, males traveled farther between captures than females, but when the search for females was over, the travel distances for males decreased. By contrast, the period after breeding was the time that travel distances increased for females, as they took up the task of hunting for a whole family. Home ranges of 11 live-trapped males averaged larger (20 ha) than those of 12 females (5 ha) and home ranges overlapped only between the sexes.

The rapid improvement in radio-tracking technology through the 1990s has produced many new data on home ranges (Table 8.1), although stoats are still difficult to work on and samples are sometimes small. Samson and Raymond (1995) radio tracked six female and five male stoats in summer 1988, following a male and a female across 35.3 ha and 15.6 ha, respectively. The stoats moved at speeds from 0.5 to over 23 m per minute, and traveled on average just under 500 m daily. As might be expected, they used the available habitats in ways that both favored good hunting and also ensured protection from larger predators. For example, an adult female preferred to use brush piles within a clearcut, while a juvenile male used the edges of a stream and many trails within a conifer plantation. Prey densities were high during the study, and the very small stoats of Quebec are more effective rodent hunters than their larger European cousins (Chapter 5) (Samson & Raymond 1995).

Lisgo (1999) studied stoats in the mixed boreal forest of east-central Alberta, in an area including black spruce, aspen, larch, and recently logged clearcuts piled with slash and overgrown with weeds. During most seasons except deep winter, the average home ranges of four radio-collared male stoats was 150 ha, and of four females, 80 ha, but they used the habitats differently. Females appeared to prefer the logged areas and avoided the other dominant habitats, while males avoided the logged areas and aspen and preferred the less extensive black spruce, larch, and birch and scrub habitats. Unfortunately, Lisgo did not estimate the amount of time the stoats spent in each habitat, only the relative proportions of different habitats in each home range area, which does not tell us how important each habitat type was to the stoats. Slash piles in logged areas supported high populations of small mammals, however, and the female stoats with radio collars often hunted there, consistent with Lisgo's analyses of habitat preferences.

New Zealand

For weasels, home range area clearly does not depend only on body size of the hunter, but also on the type and abundance of the prey. Such a generalization can often be tested by watching how animals react to new situations. The introduction of stoats to New Zealand provides just such an opportunity. The home ranges of stoats have been studied in several different ecological communities there, supporting prey resources completely unlike anything in the northern hemisphere. The story that emerges is, in fact, mostly nothing new, and is important because of that.

Murphy and Dowding (1994, 1995) radio tracked stoats in a southern beech forest during two summers when feral house mouse populations boomed and then busted. When mice were abundant in 1990–1991, three males had mean ranges of 93 ha, and four females, 69 ha, and the stoats were often active during the day. In the same area in the next summer, when mice were scarce, the numbers were 204 ha for four males and 124 ha for five females. In another southern beech forest when mice were scarce, the home ranges of four males averaged 223 ha, and of seven females, 94 ha (Alterio 1998). The home ranges of females were evenly spaced and overlapped slightly at a few extreme data points. The same was largely true of males, except that the home range of one male did overlap the home ranges of the others. In yet another habitat, a coastal grassland with grazed and ungrazed portions interspersed with dense scrub, the home ranges of six male stoats averaged 133 ha and of two females 83 ha. In this area, stoats preferred the ungrazed patches intermixed with coarse, woody vegetation (Alterio et al. 1998; Moller & Alterio 1999), where they hunted both by day and by night (Alterio & Moller 1997a).

In two podocarp forests in south Westland, New Zealand, a habitat completely different from either beech forest or coastal scrub, Miller et al. (2001) fitted 27 stoats with radio transmitters and tracked them from July 1997 to May 1998. They collected extensive data on 19 animals, and analyzed them by an unusually sophisticated series of methods—two for home ranges (minimum convex polygons and restricted edge polygons) and a different one for core areas (hierarchical cluster analysis). The mean size of the male home ranges across all seasons calculated by the minimum convex polygon method, most useful for comparing with other studies, was 210 ha, and for females, 89 ha. Males extended their ranges during the breeding season (256 ha) compared with the nonbreeding seasons (149 ha), but females did not. Home ranges overlapped within and between sexes in all seasons, most extensively between male and female ranges.

These comparisons suggest that the large stoats in New Zealand generally do have larger home ranges, closer to those recorded in Scotland by Pounds (1981) (where a single male used 254 ha in 10 days, and three females averaged 114 ha) than those of the smaller stoats of Quebec farmland (20 to 40 ha in males). However, these comparisons must be made with caution, because differences in prey density are critical. New Zealand forests of both types have no voles, and when feral house mice are scarce, that is, in most years, these forests may offer only marginal habitat for stoats (Murphy & Dowding 1994). In years of heavy seedfall in New Zealand beech forests, high densities of mice are quickly followed by high densities of stoats, which live on smaller home ranges (Murphy & Dowding 1995) and are nonterritorial (Alterio 1998). Even during a postseedfall mouse plague, the large male stoats of New Zealand could not survive for long on a Quebec-sized home range of only 20 ha.

Conclusions on Stoat Home Ranges

Stoats can range over enormous areas, which are very hard to document. One might comment that the huge home ranges of the New Zealand and Scottish males might be possible only because these stoats are relatively large, except that the small males studied by Lisgo (1999) in Alberta maintained home ranges averaging 150 ha. At the other extreme, the small stoats of Ontario and Switzerland have home ranges nearer in size to those of common weasels (Table 8.1).

By comparing all these stories, from repeated studies done both in their natural homelands and in New Zealand, we can make a few secure generalizations about spacing behavior of stoats.

1. The home ranges of males are larger than those of females, and members of each sex mostly do not use the same ground at the same time; both differences are likely to be correlated with the different hunting strategies of the two sexes, which in turn is linked (but not directly) to the difference in their average body sizes (Chapter 4) and diets (Chapter 5).
2. In both sexes, home range sizes are determined by an *interaction* between body size and prey density; stoats of all sizes need larger home ranges when prey are scarce.
3. Stoats are perpetually hungry and may need to hunt at any time of the day or night. Theory predicts that the home ranges of predators

should be large and have the least overlap when prey are scarce, but still be able to support a healthy predator population (Carpenter & MacMillen 1976; Powell 1987; Powell 1994). But because home ranges are not a fixed character in stoats, some exceptions to these generalizations (Alterio 1998) are always to be expected.

HOME RANGE IN LONG-TAILED WEASELS

As for stoats, the earliest data on longtail home ranges came from snow tracking. Polderboer et al. (1941) found the trails and dens of four long-tailed weasels on an Iowa farm. They reckoned that, at that time, each longtail seldom traveled more than 100 m in any direction from its primary den, so their trails rarely crossed. Each had access to as many as five or six food caches within this distance. Polderboer dug out the dens and found layers of rodent fur, skins and skulls, and heaps of scats—all the signs of an established resident weasel. But he could not make accurate estimates of the longtails' home ranges, and the assumption of a radius of movements of only 100 m seems much too low, given what we know about longtails now.

On 260 ha of farmland near Ann Arbor, Michigan, Quick (1944) mapped 52 trails made by four longtails in early 1940. He deduced that each weasel had a primary den and hunted within about 300 to 600 m of it. The average length of the 52 trails mapped was 2 km, ranging from 20 m to 5.5 km. The longtails ventured out even when the temperature was very cold (including once when it got down to $-20°C$), but not every night; they would sometimes stay in their dens for days. Quick calculated the areas of their home ranges assuming they were roughly circular and centered on the primary den (Table 8.1).

The ranges overlapped, but the four rarely crossed trails on the same day. When they did meet, each weasel took care to leave its mark. Once, two trails met at a post on a fence corner; each weasel deposited a scat, and then went its separate way. A month later, the trails met again at the same post, and then ran along the fence together for about 20 m. Unfortunately, snow tracking does not show whether the animals that made crossed trails actually met, nor what sex they were (but see pp. 162, 239, 286).

When Quick (1951) snow tracked long-tailed weasels in Colorado, a similar story emerged (Table 8.1). The male longtails that Glover (1942b, 1943) snow tracked in Pennsylvania went on excursions from their dens averaging 215 m in a single night ($n = 11$, range 18 to 773 m) and females averaged 105 m ($n = 10$, range 6 to 433 m). Svendsen (2003) made a passing reference to unpublished data suggesting that the average home range size in longtails is 12 to 16 ha.

The first attempt to radio track longtails was made by DeVan (1982) after he trapped seven male longtails in northern Kentucky (Table 8.1). One was trapped on two occasions, 18 months apart; this male, both times, plus one other male, was also radio tracked. The weasels hunted through the brushy overgrown vegetation along creeks and in patches of woodland, and seldom crossed open fields. Each had at least one well-hidden den.

One of DeVan's radio-tracked males was observed over 2 weeks in January–February 1975. He would come out of his den, on the bank of a dry creek, about 2 hours after sundown, and hunt up and down the creek for about 75 m each way, often revisiting the den for 2 to 10 minutes at a time during his active period of about 5 hours. The other radio-tracked male was followed over 3 months starting in November 1974. He once left his den and traveled to another den 850 m distant, robbing a baited trap on the way. Thus well supplied, he then stayed in the second den for the whole of the next day. In winter, this was a good strategy to avoid exposure to the cold.

Weasels can sometimes have the last laugh. One collared male longtail was tracked by DeVan to a burrow in an overgrown field, where he apparently stayed without moving for 3 days. DeVan grew suspicious, and finally dug out the burrow. In a grassy vole nest at the blind end he found a few drops of blood, some vole fur, and the transmitter. The weasel had gone.

DeVan caught only males, whereas Conrad Vispo (unpublished) followed a 133-g female for almost 2 months in October and November 1985 (a total of 348 tracking hours), in a nature reserve in Indiana (Table 8.1). She, too, was fairly strictly nocturnal. She nearly always came out at about sunset, around 5 p.m., and remained active until about 9 p.m. when she returned to the den and rested until about 11 p.m. After that she was active on and off for the rest of the night, especially during the last couple of hours before dawn. Unusually short excursions were often due to the weather. A heavy shower, or a passing cold front bringing a sudden drop in temperature, would drive her back to the shelter of one of her dens. During the 2 months she was located at seven different dens, all below ground and most in remnant patches of oak woodland.

The most extensive work on longtails was recently completed in rural Indiana by Gehring and Swihart (2004), who followed seven males and four females using radiotelemetry. The 200-km² study area was agricultural (76% in agricultural fields, predominantly corn and soybeans), with patches of forest (11%), grassland (4%), and wetlands (2%). From a total of 555 radio locations, the sizes of the longtails' home ranges were estimated by making utility distributions with an adaptive kernel estimator (Seaman & Powell 1996), and an individual home range was defined as the smallest area containing 95% of the utility distribution. From these home ranges, Gerhring and Swihart analyzed how the longtails used space and how they used the available habitats, but with the same problem that Lisgo had, of failing to weight the animals' use of habitats by the time they spent in them.

The home ranges of four adult male longtails averaged 180 ha, and those of four adult females 52 ha. Three juvenile males stuck to much smaller ranges (22 ha). All these home range estimates came out smaller when calculated by the minimum convex polygon method (137, 39, and 17 ha, respectively). In general, the longtails' home ranges included the various habitat types available roughly in proportion to their area available. Within their home ranges, weasels preferred the forest patches, fencerows, and ditches, and used open fields and grasslands less. These preferences were, not surprisingly, correlated with the distribution of prey. Fencerows supported the greatest abundance and biomass of small mammals and rabbits, followed by forest patches. Forests and fencerows also had the highest counts of rabbit pellets. The longtails with the largest home ranges were those that lived where the total biomass of prey was lowest, just as do stoats. In the breeding season the males seemed to abandon their settled winter homes and range widely in search of females, just as do stoats.

Gehring and Swihart (2003) were especially interested in the responses of native predators in agricultural landscapes to the fragmentation of their habitat caused by human activities. They hypothesized, and we agree, that small mammals, even species as agile as weasels, view patches and strips of good habitat within a landscape as havens of safety in a hostile sea. The longtails used fencerows and ditches as corridors between forest patches, because they offered good hunting and overhead cover to avoid avian predators. In that respect at least, habitat fragmentation might be viewed as a positive *advantage* to longtails, because such corridors become a network of connections between larger habitat patches that channel hunting weasels and their prey together. On the other hand, habitat fragmentation clearly reduces the total amount of important habitats for weasels and, therefore, reduces weasel populations. As has long been known in Britain, where the total length of hedgerows has been halved since 1945 (Robinson & Sutherland 2002), weasels really need these corridor systems and would be disadvantaged by any further homogenization of the landscape in the interests of agricultural efficiency (McDonald & Birks 2003).

SCENT COMMUNICATION

Weasels are solitary animals for most of the year, but that does not mean that they are totally nonsociable. Weasels must be well aware of their neighbors, even if only to avoid them. They keep in touch with each other by a well-developed system of scent communication—an efficient mechanism for small animals living on large home ranges covered with thick vegetation. Even under snow, the conditions are good for scent communication—cool, dark, humid, still, and quiet—and odor signals last well, although they cannot be advertised over long distances by a breeze.

Hidden under the tail of a weasel is a pair of large, muscular sacs, in which is stored a substantial quantity (up to 100 mL in male stoats) of musk, a thick, rather oily, yellowish fluid with a powerful and unpleasant (to most humans) smell. The musk is produced in modified skin glands, which are grouped together at one end of the sac and empty into it. Musk is a complex substance, in which the important components are lipophilic compounds of low molecular weight, several containing sulphur (Brinck et al. 1983). When the musk is exposed to the air, it is metabolized by bacteria into various carboxylic acids, perhaps in different combinations according to which bacteria are present.

Voluntary muscles control both the openings of the glands and the walls of the sac, so a weasel is able to expel musk at will. Normally, only a little is produced at a time, but a severely frightened weasel is able to evacuate the entire contents of the anal sacs at once—the famous "stink bomb" well known to careless trappers. The effective defense system of skunks may well have evolved, eventually accompanied by warning coloration and behavior, from such a beginning.

All mustelids have a system of anal glands, constructed slightly differently in each genus. The anatomy of the scent glands gets progressively more complex in a series from *Meles* through *Lutra* and *Martes* to *Mustela*, which seems to be the most advanced of all (Stubbe 1970, 1972), apart from the skunks.

Weasels also have smaller glands in the skin of the body, especially along the belly and flanks, and on the cheeks. Weasels regularly mark their home ranges by depositing scent from these glands in strategic places. This behavior is hard to see in the wild, but can be observed in very tame captive animals. For example, Drabble (1973) described how a hand-reared female stoat marked her territory: "she pressed herself flat and anointed the surface she claimed by rubbing it with her belly."

Like most carnivores, weasels also make use of two other strongly scented substances, scats and urine. For example, in the course of traveling around his home range, a typical male will seldom pass a place where he has previously found or deposited a scat without visiting it again, carefully sniffing, and then turning and depositing a new one—often with an expression of fierce concentration. Weasels seem to take delight in taunting researchers by leaving proof of a visit on the top of an undisturbed live trap (Figure 8.5). Scats may be found singly along trails, sometimes in different stages of weathering, suggesting regular marking of familiar routes. They are obviously powerfully attractive to any weasel using the same path, because they make good bait in live traps (Rust 1968; R.A. Powell, personal observation).

Scent-marking behavior is well known but not well understood. At the very least, scent marks convey information on social and reproductive status, and probably also individual identity. The various forms of marking behavior have been closely observed in captive Swedish stoats by Sam Erlinge and his coworkers (Erlinge et al. 1982).

A stoat performing anal drag presses his anal area to the ground, with his tail raised, and wriggles forward, pulling himself along with his forelegs. Females do it, too, and also young as soon as they begin to move about outside their

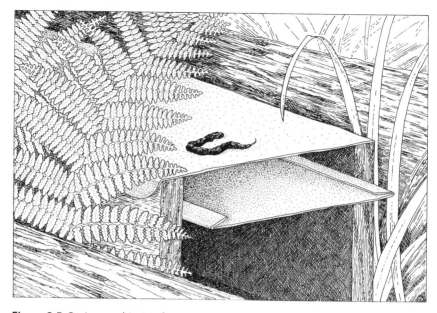

Figure 8.5 Scats are objects of great interest and usefulness to weasels, and are carefully placed in strategic positions around a resident's home range—sometimes including the top of a live trap that had failed to catch the weasel.

nest. The idea is for a weasel to permeate the whole of its home range with its own scent, to mark new objects encountered, and to cover over the marks made by others. Stoats will do it when they are alone, and at any time of day.

Body rubbing, by contrast, is used as a threat signal, especially by a dominant of either sex during an aggressive encounter with a subordinate. The dominant vigorously stretches himself out along logs or stones, scraping the scent from his cheeks and sides against them (Figure 8.6). He does the same thing when he deposits a kill in a cache or when he takes over a den formerly occupied by another stoat. Captive stoats finding cloths impregnated with human scent (unwashed t-shirts) in their cages reacted with energetic body rubbing (Winder 2003). The message is definitely a belligerent one, rather than the mere labeling of property.

It seems, from watching the reactions of one weasel to marks made by another and the way that weasels use these cues to space themselves out with minimum open conflict, that the scents of each are not only individually distinguishable, but also very informative. For example, a subordinate stoat faced with the scents left by a body-rubbing dominant reacts with obvious fear. It shows various signs of uneasiness and anxiety; it gives the little trilling call that imitates the cries made by young and that would mollify the aggressive reactions of the dominant if it were still within earshot, and it will search around for a way to escape.

Figure 8.6 Resident weasels deliberately rub their body scent on prominent objects around their home ranges as a sign of ownership, which is recognized and understood by intruders.

By contrast, a dominant stoat will show no obvious reaction to the marks, apart from marking over them. Dominant individuals also set marks more often than subordinates. They simply appear more self-confident.

The world of a weasel is full of meaningful scents, which are just as informative in total darkness as in full sunshine, and whose messages last longer than those of the fleeting visual images on which we rely. Our noses are useless to read them but, fortunately, the chemical composition of the secretions produced in the scent-marking glands of weasels can be displayed by gas chromatography. If scent marks really do convey important information between individual weasels in the wild, as field and behavioral observations imply, the chromatograms should show subtle differences between individuals, consistent over time in the same individual—and indeed, they do (Brinck et al. 1983).

The component molecules separate out according to their weights, and the patterns produced show distinct differences from one animal to the next. Better still, the differences reappear in successive samples from the same animal. The origin of the differences is uncertain. Perhaps the molecules of one animal's musk are structurally different from those of any other, or perhaps each animal has a unique combination of bacteria producing a different set of metabolites of musk (Gorman 1976). Either way, it seems likely that, to a weasel, scents are as unique as faces are to us. Like faces, scent marks give one weasel a lot of useful information about another—not only its sex, identity, social status, and breeding condition, but also the probable outcome of a confrontation.

The advantage of scent marks is that they persist for some time, so that, for example, a resident male can give out information to potential intruders in many places at once. An intruding male, on the other hand, has the information on which to decide whether to risk an encounter with his unseen neighbor. He knows how long ago his neighbor was last here. Weasels can fight ferociously (Figure 8.7), and are well able to injure each other, so it is an advantage to a stranger to assess the likely outcome while escape is still possible. In fact, avoidance and retreat are much more common than all-out offense in deciding the local dominance

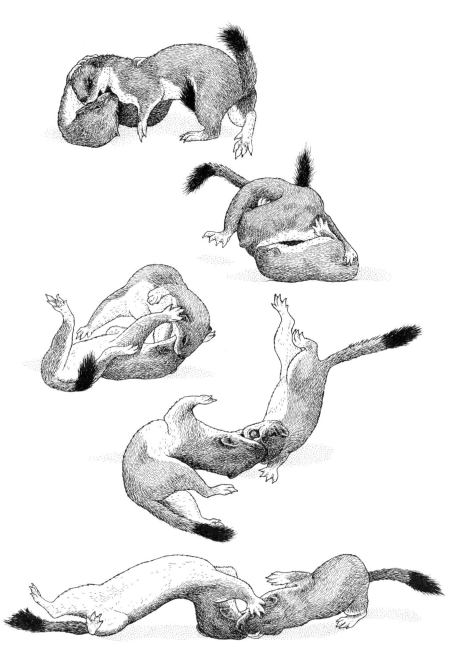

Figure 8.7 Stoats normally avoid each other but will fight vigorously when necessary. Here, two females fight, possibly over territory boundaries. (Redrawn from a series of photographs by Derry Argue.)

hierarchy and distribution of individuals. But running battles do happen, especially in spring. Vernon-Betts (1967) gave a particularly vivid account:

> During the last week in April I was driving down a lane when I saw two [common] weasels fighting on the grass verge. I stopped the car within a few feet and watched them for several minutes fighting with the concentrated ferocity of a couple of bull terriers. I released the brake and free-wheeled after them as they conducted a running fight for more than 100 yards down the road. . . . Both animals went onto the verge opposite a . . . gate in the hedge. Through the gate I saw a hunting cat with ears pricked advancing on tiptoe; clearly it could hear the squeaks of the weasels but had not yet seen them. Suddenly the pursuer saw the cat and made a run for the hedge. . . . The pursued ran for the gate and apparently straight into the jaws of the cat. . . . When I reached the gate, only the weasel was to be seen, white belly up, paws in the air, a bright bead of blood behind one ear and apparently dead. I picked it up. . . . It was an undoubted male, and the same size as its antagonist. I began to carry it back to the car. As I was passing through the gate I noticed its belly was pumping up and down; after a further two steps it gave a violent wriggle so that I dropped it. It hit the ground running and darted into the hedge where it immediately burrowed into a pile of dead leaves. . . . Half an hour later I returned the same way, and, though I stirred the leaves with a stick, there was no sign of it.

This extraordinary incident illustrates not only a particularly vigorous dispute between males in the breeding season, but also the so-called "sham-dead" trick of weasels in response to immediate mortal danger. It may be, as some believe, a deliberate defensive ploy. Others interpret it as the effect of violent exertion on a brain already under pressure from parasitic worms (see Figure 11.8).

When weasels live at high density, such encounters between a resident and a neighbor or an intruder are more likely. For example, Lockie (1966) had already inferred, from the distribution of trapping records, that the resident common weasels he was observing worked to keep others off their "own" ground. Then he actually saw it happening:

> I . . . once [saw] a territory holder escort a transient from its territory. . . . Both animals suddenly appeared running towards me, the chaser shrieking now and then. They paid no attention to me and passed close by. At the known boundary of the territory the owner broke off and returned into his territory where he was shortly after trapped and examined. The other animal kept running and disappeared from view a quarter of a mile down the track. I was unable to catch and examine

the chased weasel, but since none of the known residents was missing I presumed it to be a transient. On another occasion I trapped a transient stoat which squealed as it came out of the anesthetic. . . . Immediately the presumed owner of the territory appeared racing towards me apparently to see what was happening. . . .

Lockie recognized that such incidents were rare, and it is certainly in the weasels' interests to avoid them by use of more subtle means of communication. In 852 hours of radio tracking common weasels, Macdonald et al. (2004) heard three and witnessed only one fight between a collared and an unknown male; collared (resident) males simply avoided each other.

Likewise, Sandell (1988) witnessed only one encounter between stoats, even though he was tracking several males in the same area at the same time. He was waiting for a collared male to come out of a den, when an unmarked male approached from the other direction and entered the same den. Soon after, the unmarked male rushed out with the radio-collared male at his heels, and the chase continued for several hundred meters. Such encounters are rare, Sandell concluded, because resident males normally use scent marks to locate and avoid each other.

In a detailed study of scent marking by wolves, Peters and Mech (1975) documented patterns of behavior that are easier to observe in these larger animals and help to explain how weasels use scent marking to maintain their spacing patterns. Wolves scent mark everywhere within their territories, especially whenever and wherever they meet the marks of other wolves. A dominant wolf would even sneak 100 to 200 m into the neighboring wolves' territory to leave a scent mark, then retreat. Consequently, even though wolves spend relatively little time at their territory boundaries, the boundaries get a disproportionate share of scent marks. In addition, the boundaries tended to be fuzzy, from constant pushing from each side.

The density of scent marks allows wolves to know where they are with respect to territory boundaries, and it also lets nonresident wolves know the safest places to travel: along boundaries where scent marks are dense but wolves are not. Weasels may well do something similar. When a weasel notes that his neighbor is not renewing scent marks, he knows that something is amiss and the neighbor could well be out of the way. Transient weasels, moving along a territory boundary, would also be able to note that scents on one side are old and not renewed, and so identify a possible new home.

Game theory, a branch of mathematics originally developed to understand gambling by humans, was quickly applied to animal behavior and has provided some fascinating insights into why scent marking works (Maynard Smith 1979). Game theory helps to analyze the costs and benefits of developing rules for territorial behavior, such as (1) chase your neighbor out of your territory when he invades (but don't chase fast enough to catch him); (2) run from your neighbor when he finds you on his territory (do not fight on his ground if you can help it); and (3) scent mark on top of his marks. If weasels follow these rules, they minimize their chances of getting hurt in a fight while maintaining their territories. More important, (4) when conditions change and no territory has enough prey to support a single weasel, move out.

And, indeed, weasels do follow these rules. Neighbors do not fight when they meet at territory boundaries, but a resident will chase an intruder (who always runs). When vole populations crash, weasels no longer attempt to maintain territories, and the stable spacing system breaks down.

THE HOW AND WHY OF HOME RANGES

How are we to understand the great variability in the spacing behavior of weasels? It helps to remember the basic economics: A weasel will spend the time and effort required to defend its home range only if it can thereby guarantee priority access to a resource in short supply, such as a population of voles, for itself or to protect its investment in a litter (Carpenter & MacMillen 1976; Powell 2000). If the voles become more abundant, a weasel can economize on energy by reducing the size of its home range, but only to a certain extent, so the size of the area over which a resident weasel hunts depends on the density of the local small rodents, down to a minimum of a few hectares.

If the habitat is patchy, with good spots and bad spots for voles (which is always the case), at some level of abundance the good patches together provide more voles than the resident weasel needs. To delete one or some of the patches from its home range, however, could leave the owner of such a territory short in the future, so it is likely to keep all patches and allow home range overlap with neighbors. This often happens in autumn, when voles are at their seasonal peak and transient weasels can settle into rich areas overlapping the home ranges of residents. At the other extreme, if the voles become so scarce that no home range can support enough of them to feed even one weasel, then the whole system breaks down and the resident weasels move on to seek better hunting elsewhere.

The home range sizes of weasels do in fact show variation of just the sort predicted. For example, the male common weasels studied by Lockie (1966) when field voles were very abundant (110 to 540 per ha), did well on home ranges of only 1 to 5 ha. Those at Wytham, where wood mice and bank voles were much scarcer (together only 21 to 39 per ha), had to hunt over much larger areas and were still perpetually hungry (King 1975c).

If prey density is the most important consideration deciding home range size, the limitations of the weasel's own body is probably the next one. Large weasels can cover more ground than small ones, but the smallest weasels are better able to exploit small rodents, so can make a living on a small area where a large weasel, excluded from tunnels, would starve. The home range estimates compiled in Table 8.1 are influenced by a host of different variables, but there is still a broad correlation between the size of the weasel and the area it occupies. Every study shows that the ranges of females are smaller than those of the males of the same species, measured at the same time and place. The ranges of common and least weasels are clearly smaller than those of stoats and longtails; the smallest local races of stoats (e.g., in Ontario) have the smallest ranges of that group.

When the home ranges of a male and female weasel overlap, the hunting activities of each reduce the prey available to the other. Each catches voles and, probably more important, alerts prey while hunting, making the prey wary and harder to catch (Chapter 7). Female weasels simply tolerate the loss of prey to males, except perhaps in a small, core area worth defending, but have smaller appetites and greater searching efficiency instead. Males, however, benefit from overlapping ranges that provide information on where females live, which is especially useful in the breeding season (Powell 1994). Field data confirm that home ranges often overlap, particularly between sexes (Murphy & Dowding 1994, 1995; Alterio 1998), but many individuals have separate core ranges that do not overlap.

Why do weasels bother to establish a home range and try, to varying degrees, to keep others out of it? The most critical reason is that local knowledge is valuable, for several reasons. First, the weasel that knows its home range knows where rodents are to be found and where they are not, and knows when it can return to each good hunting area. Weasels are good dispersers and are fully capable, when the local rodent population crashes, of wandering across a landscape in search of new habitats dense with voles and unclaimed by other weasels. That is one of the characters of small, short-lived "weed" species like rodents and their specialist predators. But since rodents are not evenly spread throughout all parts of any habitat, and the dangers of wandering on unfamiliar ground are high, a weasel will obviously stick to a good area when it finds one.

Second, weasels are always vulnerable to larger predators, and the best defense against attack is to have an intimate knowledge of one's own ground, the position of every refuge, and the safest, quickest way to get to it. A weasel who knows where the good escape holes are has a better chance of surviving if spied by a hawk (Chapter 11).

Third, energy conservation is all-important to weasels, which means that there is a high premium on minimizing the energy spent in hunting (Chapter 2). A weasel that knows exactly where to go to find a meal is more likely to be able to meet its needs and to get back into its warm nest in the least possible time. Once a weasel has established a home range, keeping others out allows it to keep track of the best times to return to favorite vole-rich patches.

Obeying rules of trespass and scent marking to warn others that it is willing to challenge intruders allow a weasel to maintain its ground with minimal cost. One might say that, through evolution, weasels have become experts in economics, game theory, and information theory by daily experience in the School of Hard Knocks. All these are important to weasels, because their day-to-day survival depends on finding ways to meet the huge costs of living in a small, thin body.

9 | Reproduction

REPRODUCTION AND SURVIVAL between reproductive seasons are the two most important things in life for any animal. The details of the reproductive machine—the anatomy of the organs, the whole complicated process of the production and care of the young, and the extent to which this process can be adjusted to the prospects of success—are matters of intense importance to individuals, and are always under the unrelenting scrutiny of natural selection. This chapter gives a simple description of the machinery. Variations in its performance, which profoundly affect the population dynamics of weasels, will be taken up in Chapter 10, and the curious matter of delayed implantation (embryonic diapause), one of the greatest puzzles of mustelid reproduction, will be discussed in Chapter 14.

THE REPRODUCTIVE ANATOMY OF WEASELS

The testes of males are simple oval sacs within the furry scrotum. The coiled tube of the vas deferens leaves the epididymis, at the distal end of the testis, and ascends back into the body cavity again. The penis is stiffened by the baculum, a small rod-shaped bone attached to the pelvis by muscles at one end, which acts as a rigid support during copulation. The urethra fits within a groove on the underside of the baculum, shown in Figure 9.1. Normally the whole apparatus is hidden inside the body, and in living animals the baculum can only be felt through the skin, like a matchstick lying under the midline between the small tuft of hair at the orifice and a point just forward of the testes. But when the baculum can be dissected out, as from carcasses, it becomes one of the most informative items of a weasel's anatomy.

The baculum is useful to biologists for two reasons. One is that the shape and size of the baculum are reliable characteristics of each species (Figure 9.1), so that an incomplete skeleton of a male can often be identified from the baculum alone (Baryshnikov et al. 2003). The bacula of all species are roughly the same in general design: Each has a more or less straight shaft, with a curve at the distal end and a knob at the proximal end. But the baculum in *nivalis* is relatively short and thick, with the distal curve formed into a distinct hook; the bacula of *erminea* and *frenata* are longer and more slender, and their distal curves are more gentle.

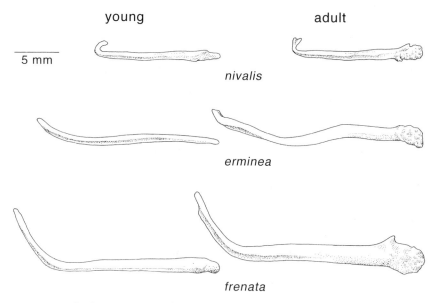

Figure 9.1 The bacula of male weasels are diagnostic of both age and species. (Redrawn from Burt 1960.)

The second reason is that, within each species, the baculum indicates age. Wright (1950) showed, by a series of laboratory experiments, that the development of the baculum is controlled by androgens. The proximal knob is characteristic only of adults; juveniles have a thin shaft hardly broader at the end than along its length. In males castrated as juveniles, the knob did not develop at all. The essential role of hormones in this development was proven by the classic reverse experiment: Castrated juvenile males treated with implants of testosterone propionate developed nearly normal knobs. In intact males, the knob first develops at puberty and continues to grow in size, and therefore in weight, for several years, and probably throughout life (Figure 9.2). Presumably, the strengthening of the bone stimulated each breeding season by testosterone has a cumulative effect.

In male stoats and longtails, it is possible to define the minimum baculum weight marking sexual maturity, although the actual threshold figure depends on general body weight. For example, in the small male stoats of northern Ireland (average body weight 233 g), the minimum baculum weight of adults was 30 mg (Fairley 1971); in the larger males from the Netherlands (284 g), it was 32 mg (van Soest & van Bree 1970); and in the even larger males from New Zealand (324 g), it was 38 mg (King & Moody 1982). Similarly, in the small subspecies of longtails the bacula are smaller than in the larger subspecies (Wright 1947). In *erminea* and *frenata* the weight of the baculum clearly distinguishes the young males, but in *nivalis* the transition from the juvenile form

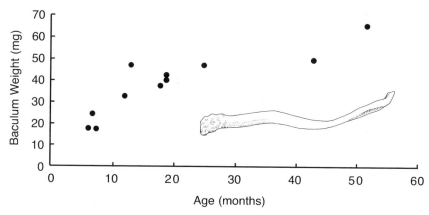

Figure 9.2 The weight of the baculum has long been a useful method of distinguishing subadult males (independent but not yet fully grown or sexually mature) from adults. These bacula of New Zealand stoats, however, all of known or part-known age, show that baculum weight increases for several years, perhaps throughout life. (Redrawn from Grue and King 1984.)

and weight to those of the adult is smooth, allowing no separate age categories based on baculum weight (Table 9.1).

In females, the uterus is a simple tube with two branches (usually called horns) joined at the base and pressed against the dorsal side of the body cavity. The ovaries are quite conspicuous, round and flattened and rather yellowish, lodged in the free ends of the two uterine horns. The ova develop in follicles just under the surface of each ovary, and when they are ripe, the follicles burst and the ova are released to pass down the fallopian tubes to the uterus. The uterus enlarges somewhat before estrus, but this phase is short-lived and seldom observed except in unmated females in captivity. Estrus is best detected externally, from the swollen, moist, doughnut-shaped vulva. A preserved uterus betrays no sign of whether it is carrying young, or has done so before, until the embryos become visible as evenly spaced swellings about 3 weeks before full term.

Table 9.1 Baculum Weight in Relation to Age and Species

	nivalis	*erminea*	*frenata*
Young	5–21 mg	10–30 mg	14–29 mg
Adults	15–59 mg	50–89 mg	53–101 mg
Locality	Britain	Britain	Montana
Reference	(Hill 1939)	(Deanesly 1935)	(Wright 1947)

The mammary glands are set toward the rear of the long abdomen, and are invisible except during lactation. The nipples are tiny pimples hidden under the fur, both in juveniles and in adults until shortly before a litter is born. The nipples of adult females that have recently suckled young remain elongated for some months, but those of adults that have lost their young or failed to rear them remain practically invisible. Female stoats and longtails usually have four or five pairs of nipples, and common and least weasels have three or four pairs. Only those nipples that are being suckled remain active, so fewer than the total possible number will be visible on a female with a small litter.

The most useful feature of the reproductive anatomy of females, at least for researchers interested in weasel reproduction, is the corpus luteum. Each ovum released at ovulation leaves a space behind, which is quickly filled with a dense mass of hormone-producing cells. These cells are slightly yellowish in color and are clearly visible to the naked eye, hence the name (corpus luteum is Latin for "yellow body"). One corpus luteum forms for each ovum released. The function of the corpora lutea is to produce progesterone, the hormone needed to maintain the pregnancy. The beauty and usefulness of the corpora lutea (plural) lie in the two facts that ovulation in weasels has to be induced by copulation and that each corpus luteum marks the site of an ovum that has been released. Hence, counting corpora lutea is a convenient way to estimate which animals have mated, plus their total potential fecundity (King 1981a), and these are very important data for population studies (Chapter 10).

MATING BEHAVIOR IN ADULTS

Mating is a very vigorous affair in all weasels. It has to be, because the stimulus of copulation is needed before ova can be released. All attempts to stimulate ovulation by injection of gonadotropins, the hormones that usually have this effect in other animals, have failed (Rowlands 1972; Gulamhusein & Thawley 1974), and the ovaries of unmated females have no corpora lutea.

Females are subordinate to males for most of the year, and normally avoid them, but a female is well able to reject unwanted suitors with displays of ferocious aggression. Only when she reaches full estrus will she accept a male's advances. He approaches cautiously, rubbing himself on the ground with twisting movements and making excited trilling calls. If she answers in kind, the brief courtship begins. They sniff at each other, trilling incessantly, and follow each other around. If she is fully receptive, or if it is not the first encounter between the pair, she may leap playfully around him, whereupon he immediately grabs her by the scruff of the neck (Figure 9.3). First encounters require somewhat more lively negotiation before getting to this stage, but the result is the same.

Figure 9.3 Stoats mating. The male holds the female firmly by the neck with a bite that may be locked, avoiding both release of the female and injury to the female. Ovulation in weasels is induced by copulation, so a female must be well stimulated. The baculum enables a male to maintain the extended and vigorous intromission required. Copulation is prolonged and may last a quarter hour to several hours.

He drags her about, never letting go of her neck with his teeth (DonCarlos et al. 1986). Indeed, in the domestic ferret his bite becomes locked on her neck, which means that he cannot bite her too hard (Ewer 1973). She may break loose from the leg grasp, but not from the neck hold. She remains limp and passive while being carried, in a condition called *tragschlaffe* ("carry-sleep") by German observers. Young weasels behave in the same way when being carried from one den to another by their mother.

The male then grasps the female round the chest with his front legs, and arches his supple back to make pelvic thrusts, usually while both are lying on their sides. The baculum ensures that she is well stimulated, and that intromissions can be energetic, prolonged (usually about 15 minutes, but they can last several hours, with alternate periods of thrusting and resting), and frequent (up to several times per hour). Afterward the partners may rest, together or separately, and may repeat the procedure over the next 2 or 3 days, even though a single mating with intromission is usually enough to stimulate the release of luteinizing hormone (LH), which causes ovulation (Murphy 1989).

A female has no noble ideas about loyalty to one partner, however, and may accept other males she meets during her short receptive period. Mustelid sperm survive for much longer than those of most mammals, so the result can be litters with more than one father. Holland and Gleeson (2005) observed the genotype profiles of six wild-caught adult females and their prenatal offspring; one litter had been fathered by at least three males. Indeed, for captive breeding programs, Wright (1948) advocated introducing several different males to an estrous female one after another, until her heat subsides, as a means of ensuring success. But since a female does not ovulate until the second or third day after mating, the males that meet her on those days have more chance of fathering her litter than the first male (Amstislavsky & Ternovskaya 2000).

Mating success is all-important to a male. Since he takes no part in rearing his young, fathering many litters is a male's best passport to representation in the next generation (Chapter 14). So, each male attempts to find as many mates as possible each season, and the best way to do that depends on his age and social status (Chapter 8). But for a female, mating is only a dangerous preliminary to the real business, the rearing of the young, and as soon as her short period of heat is over she rejects all males with squeals and savage bites. There is no pair bond of any kind—indeed, the adults appear to have as little to do with each other as possible except when mating.

Even in captivity, where mating encounters can be set up in decent isolation from the possibility of interference from other males, some individual males are consistently uninterested or ineffective in achieving matings and others equally consistently successful (DonCarlos et al. 1986; Sundell 2003). Likewise, Müller (1970) found that some captive female stoats rejected all males, refusing to mate at all, and even the willing females did not necessarily accept all males offered as partners. Of 100 captive matings of least weasels, only 65% led to successful pregnancies. Individual mating success ranged from 0% to 33% in males, and from 0% to 50% in females (Sundell 2003). Breeding weasels is, therefore, an unpredictable business.

Well before her young are due, a mother weasel must find a safe, warm nest for them. She does not make one of her own, since weasels do not burrow, but then, she does not need to. It is a simple matter to find a ready-made nest, and if the owner is still at home it may find itself unwillingly providing board as well as lodging for the weasel family.

The best dens are the ones made by an animal of the right size. Small rodents make good dens for least weasels, but stoats and longtails look for a den made by larger rodents or rabbits. The den needs to be safe from any danger of flooding and thickly insulated with dry grass and leaves. The mother often improves the lining with fur plucked from dead prey; she does not take fur from her own belly, as a doe rabbit will. Nests built under good cover, such as in piles of rocks or logs, are especially attractive since they are safe from the prying paws of larger predators. When she has found a den, the mother weasel collects a store of food and then retires to await the birth (see Figure 14.2).

THE SEXUAL CYCLES IN ADULTS

Common and Least Weasels

The breeding cycles of least and common weasels are broadly similar. The gestation period is about the same (34 to 37 days); both have direct implantation; young females of both subspecies can breed in the summer of their birth when conditions permit (Sundell 2003); and the length of the breeding season is very variable (Figure 9.4A). In Europe, male common weasels can have enlarged testes and are fecund from February until at least early August, occasionally to the beginning of October. The testes regress in autumn, but never relapse into complete quiescence, since the early stages of spermatogenesis can be found in the cells throughout winter. On the other hand, there are no spermatozoa in the epididymis from November to January inclusive, so winter is definitely an infertile period for common weasels. Adult females start coming into heat in February, though they may not necessarily conceive then. They become anestrous in September, or earlier when they are in poor condition.

Implantation in common and least weasels is direct; that is, the fertilized zygote appears to proceed straight through the stages of development without a detectable pause at any stage. The gestation period is, therefore, about the same as the time it takes the embryos to develop, and the ovaries contain corpora lutea only when the female is carrying actively developing young. Pregnancies may be observed at any time from March to early August, and the first litters are born in April. Rearing takes at least 9 weeks, so one full breeding cycle takes 3 to 4 months (see Table 9.3).

When voles are numerous, well-fed adult females may come into estrus again when their first litter of the year has been weaned, at earliest by the end of May. The corpora lutea of the first pregnancy persist well into the second, so females may be found with two sets of corpora lutea in the ovaries from June onward (Deanesly 1944; King 1980c). Second litters are born in July or August. In exceptionally good years some older females may still be suckling in October (Delattre 1983), while the earliest-born of the young females of the season produce their first litter in the year of their own birth (King 1980c; McDonald & Harris 2002).

Common and least weasels can respond to a glut of food by producing two litters in one reproductive season. One group of captive females fed year-round on as many live, breeding voles as they could eat was even persuaded to produce three litters in a season (Frank 1974). In the wild, two litters in a season is about the most any female can manage, and then only when hunting is exceptionally easy. More often, late litters are lost. For example, among 77 female common weasels collected in Britain by McDonald and Harris (2002), the last pregnancy was observed on October 5, and the last postpartum uterus on

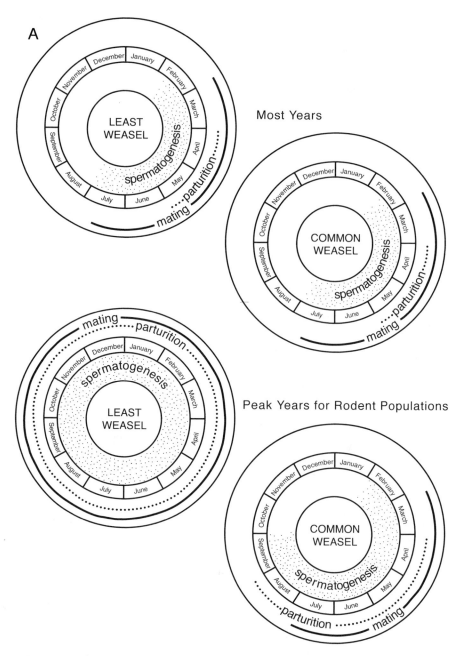

Figure 9.4 The reproductive cycles of weasels in the northern hemisphere. (A) In most years, least weasels and common weasels can rear only a single litter in spring. They remain fertile until midsummer and may mate again but without producing surviving offspring. In years when rodents are abundant, common weasels can rear additional litters in summer and autumn, and least weasels can continue breeding under snow all winter. (B) The reproductive cycles of stoats and long-tailed weasels are fixed by day length, regardless of food supplies.

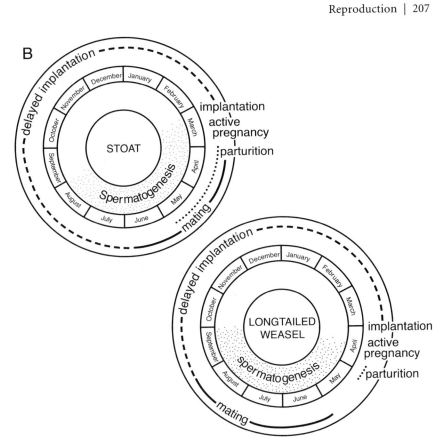

Figure 9.4 (continued)

October 13, but no female captured after August 27 was lactating, suggesting that few late-born young survived.

It is not correct to assume, as in some population models (Chapter 14), that common weasels can successfully produce two litters every year. On the contrary, when food is very scarce, adult females in the wild probably mate as usual but are unable to produce any surviving young even from their first litters. According to Erlinge (1974) and Tapper (1979), common weasels cannot rear young at all unless they have access to a minimum density of some 10 to 14 voles per ha, or about 400 rodents per territory (Jędrzejwski et al. 1995). So pregnancy rates and the lengths of breeding seasons vary enormously in common weasels, from 7 to 8 months in vole peak years to total failure in crash years.

Reproduction by least weasels differs from that of common weasels in two minor, but very interesting, ways. First, when prey are abundant, least weasels in North America can be found pregnant or with small young in most months of the year (Hall 1951; Heidt 1970), whereas common weasels never continue breeding into the winter. In the tundra, least weasels breed under the snow well into the winters of peak lemming years (Fitzgerald 1981). This remarkable ability is actually less surprising than it sounds. The conditions in the far north for breeding of both small rodents and of weasels are, in fact, much more favorable in winter than in spring (Chernov 1985). Once the snow pack is established, the subnivean space provides near constant conditions and a reliable refuge from large predators (Chapter 1). The temperature is near freezing, so female least weasels must keep moving if they leave the safety of their fur-lined dens. Yet, spring conditions are worse. Then, the melting snow often floods the burrows and nests, drowning nestlings, blocking access to food, and exposing small animals to wind chill, late frosts, and hungry predators. The effort these tiny hunters expend on reproduction conveys some idea of the urgency and importance of breeding success to small, short-lived animals.

The second difference between least and common weasels was noted by Frank (1974), who kept both in his lab at Braunschweig, in West Germany. In each of three successive years, a wild-caught least weasel produced three litters, mostly common–least hybrids. The periods of pregnancy (5 weeks) and of rearing (8 to 9 weeks) were the same as he recorded in purebred common weasels kept in the same conditions. The difference was that the female least weasel came into estrus again only 5 weeks after the previous litter had been born, whereas none of the common females did so until 9 to 10 weeks after the births.

This means that the female least weasel was able to start the gestation of a second litter during the rearing of her first, and she completed the production of two litters in 5½ months instead of the 7 to 8 months needed by the female common weasels. Frank emphasized that this happened regularly, and suggested that it might be an adaptation to give least weasels maximum productivity in the short summers of their northern home, especially in periods of rapidly rising numbers of voles. Alternately, if the breeding cycle of least weasels is less closely controlled by season than in other weasels, their ability to breed during lactation would allow female least weasels to respond to a vole peak at any time of year, and to respond more rapidly than other weasels can. The speeding up of the reproduction process in least weasels could be more a matter of competitive advantage (Chapter 14) than of adjustment to short northern summers.

The failure of common weasels to produce surviving young in bad rodent years was already well known when Janne Sundell, a graduate student at the University of Helsinki, Finland, became interested in whether least weasels are also vulnerable to losing their litters in poor years, and, if so, how. The information available did not show whether females attempted to produce young but lost them, or did not bother to try. Not bothering is a strategy common in vole-eating raptors and owls (Southern 1970), but these are longer lived than weasels so have more opportunities to make up for bad years later.

Sundell completed a 5-year study of captive least weasels, following 65 litters of young least weasels averaging five kits per litter (53 litters counted: Sundell 2003). In 1998, some breeding females were kept on a restricted diet until after mating, but were restored to normal rations once pregnant. The four females kept on short rations mated as often as the five fed to excess, but produced fewer young (averaging 5.0 vs. 6.6), which also had a higher mortality rate (31% vs. 2%). Sundell concluded that short-lived species such as least weasels always try to breed at every opportunity, because the distribution of rodents is patchy and the chances of surviving to the next season are poor.

Common and least weasels are easier to breed in captivity than stoats and longtails. They are not strictly confined to one litter a year, and, because they are smaller animals, perhaps they are less likely to be stressed by confinement in small spaces. Experiments on breeding them have illustrated the point, which probably applies to all species of weasels, that the *quality* as well as the quantity of food is a significant determinant of breeding success.

The most likely reason that Fritz Frank was able to breed both subspecies and their hybrids so easily was that he fed them entirely on live *Microtus* from an adjacent breeding colony. The weasels, guaranteed regular supplies of their favorite food, plus the stimulus of killing them, produced 94 young over 10 years (Frank 1974; King 1980e). In the experiments by Sundell (2003), five female least weasels kept on a diet of rodents (85% voles, either live or freshly killed, and the rest defrosted lab mice) produced larger litters (averaging 6.2 kits vs. 4.3) with lower juvenile mortality (17% vs. 58%) compared with four females fed only on defrosted, domestic chicks. These experiments are consistent with the view that, when rodents are scarce, the weasels' chances of successful reproduction are reduced, even if other foods are available.

Stoats

The reproductive cycle in stoats is quite different from those of common and least weasels (Figure 9.4B). First, in both sexes the cycle is strictly controlled by the season, or, more precisely, by the changing ratio of light-to-dark hours (King & Moody 1982; Herbert 1989). This control is fixed, and cannot be overridden even in years when the countryside is teeming with prey. Second, the breeding cycle in stoats includes an obligatory delay in the development of the embryos, which starts about 2 weeks after fertilization. By this stage, each zygote has traveled down the fallopian tube from the ovary, developing as it goes into a hollow ball of 100 to 200 cells, a blastocyst, until it has reached the uterus.

By 38 to 40 days after mating, the tough little blastocysts are visible to the naked eye inside fresh uteri examined via transmitted light (Polkanov 2000). In common and least weasels (and most other mammals) the blastocyst then implants in the wall of the uterus and proceeds to develop to full term. But in stoats and long-tailed weasels, it stops developing and floats freely in the uterus for the next 9 to 10 months. When the days begin to lengthen again in the following spring, it reawakes and continues with its progress as if nothing had happened.

Delayed implantation (also called embryonic diapause) is one of the most fascinating puzzles among the many presented to us by the weasels (Chapter 14).

Immediately after the winter solstice, the first signs of spermatogenesis appear and the males begin to prepare for the spring breeding season. From mid-February to April their testes enlarge rapidly, stimulated by a massive rise in the level of testosterone in their bloodstreams (Gulamhusein & Tam 1974), although they are not yet capable of fertile matings because no spermatozoa have reached the epididymis. In most parts of Europe the fertile season in adults starts in May and lasts until July (in Ireland, March to August: Sleeman 2004). Thereafter, the testes regress, more slowly than they enlarged, and are small from November to early February. The regressed testes of adults in autumn and winter are still distinctly larger than the undeveloped testes of juvenile males at the same season.

The annual cycle in the ovaries (Figure 9.5) is much more complex than that in the testes. Throughout the long period of delay, the corpora lutea are small (0.7 to 0.8 mm across) and produce the small amounts of progesterone needed to maintain the blastocysts. The free blastocysts can be seen in fresh uteri, and there is a 1:1 relationship between the numbers of corpora lutea and of blastocysts. If the ovaries and their corpora lutea are removed during delay, the blastocysts decay (Shelden 1972).

As the spring days gradually lengthen, the increasing ratio of light-to-dark hours passes some more or less definite threshold, and triggers a response in the pituitary. The pituitary then begins to produce larger amounts of LH, the corpora lutea enlarge to 1.2 to 1.3 mm across, and the level of progesterone in the blood suddenly rises. These changes prepare the uterus for implantation, and also stimulate development of other essential organs, such as the mammary glands. About 10 days later the blastocysts implant, spreading themselves evenly between and along the two horns of the uterus (migrating from one to the other if need be). The embryos complete their long-interrupted growth in another 28 days. The number of implanted embryos is often less than the number of corpora lutea, meaning that some potential young have already been lost (Table 9.2).

The control of delayed implantation in stoats, and in the other mammals that have it, is complex (Mead 1989; Renfree & Shaw 2000), and requires a series of mutual signals between the blastocyst and the uterus that must be sent and received correctly. Conventional descriptions cannot account for all these events in terms of the interactions of hormones, which suggests that the most important role in this physiological drama must be played by some additional, unknown actor. Recent research has identified one gene, encoding for leukemia inhibitory factor (*LIF*), that has a critical role in the process of embryo implantation in several species.

LIF is most abundantly expressed in the uterus at the time of implantation, on day 4 after mating in laboratory mice and between days 19 and 25 of the menstrual cycle in humans. Genetically modified mice in which *LIF* is not

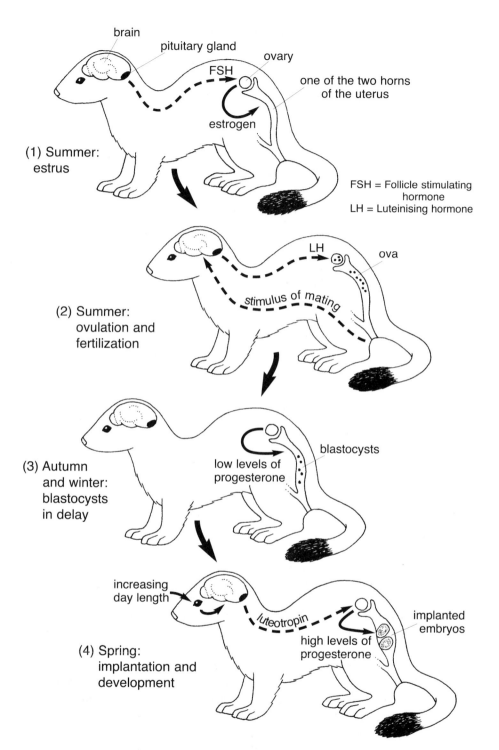

Figure 9.5 Hormonal control of the seasonal reproductive cycle in the female stoat.

211

Table 9.2 Litter Size in Weasels, by Species

	Data from	Country	Mean	Range	Reference
Stoat	CL	Britain	10	6–17	Rowlands (1972)
		New Zealand	10	3–20	King & Moody (1982)
		Sweden	9	5–15	Erlinge (1983)
	Blastocysts	Sweden	7	1–15	Erlinge (1983)
	Embryos	Britain	9	6–13	Deanesly (1935)
		Britain	9	7–10	McDonald & Harris (2002)
		New Zealand	9	6–13	King & Moody (1982)
		Siberia	11	1–17	Ternovsky (1983)
	Births	New York	6	4–9	Hamilton Jr. (1933)
		Germany	6	4–9	Müller (1970)
		Siberia	7	1–14	Ternovsky (1983)
		USSR	9	2–18	Heptner et al. (1967)
Common weasel	CL	Britain	7	4–11	Deanesly (1944)
	Embryos	Britain	6	6–7	Deanesly (1944)
		Poland	5	4–7	Jędrzejwska (1987)
		Britain	6	4–7	King (1980c)
		Britain	6	4–9	McDonald & Harris (2002)
	Births	Britain	5	2–7	East & Lockie (1964, 1965)
		New Zealand	5	3–6	Hartman (1964)
Least weasel	Embryos	Alaska	10	7–16	Fitzgerald (1981)
		Mongolia	12	5–19	Heptner et al. (1967)
		USSR	7	4–10	Danilov & Tumanov (1975)
	Births	Finland	4	4–5	Blomquist et al. (1981)
		Finland	5	1–14	Sundell (2003)
		United States	5	3–10	Hall (1951)
		Michigan	5	1–6	Heidt (1970)
Long-tailed weasel	Births	N. America	7	2–9	Heidt (1970)

CL, corpora lutea

expressed can mate and ovulate normally, but their embryos fail to implant (Stewart et al. 1992). In spotted skunks and in minks, which also delay implantation and are closely related to stoats, maternal expression of *LIF* is required for successful implantation at the end of diapause (Song et al. 1998; Hirzel et al. 1999). This observation may explain the failure of earlier attempts to induce implantation out of season by hormonal manipulation alone (Mead 1989).

Although no research has yet identified the role of *LIF* in stoat reproduction, we predict that *LIF* will be found to be critical. Delayed implantation in stoats is obligate, is prolonged, and shows no apparent natural variation (Chapter 14). Inflexible control of any biological system can only be exerted by a set of genes with almost no variation, like those that govern vital functions such as respiration, because there is never any benefit in behaving differently (Cooper

1999). The virtually zero variance in the normal reproductive cycle of female stoats should make it possible to conduct an experiment and test hypotheses even when sample sizes are small.

The internal coordination of the activities of the reproductive cycle is done by genetically controlled hormones and neurotransmitters, but the external cue that turns the system on and off is day length. Captive stoats can be induced to implant and produce their young at the wrong season by adjusting the ratio of light-to-dark hours in their cages to that typical of spring, but not simply by injecting what appear to be the right hormones (Wright 1963; Mead 1981). Where stoats normally wear a white coat in winter, it is easy to tell when a female is about to produce her young, because the molt and the breeding cycles respond together to changes in day length. Births may be expected some 22 to 25 days after the first brown hairs appear on a female's nose (Ternovsky 1983).

From carcasses, the probable date on which a litter would have been born can be estimated by inserting the average weight of the fetuses into the formula developed by Hugget and Widas (1951):

$$\sqrt[3]{W} = a(t - t_0),$$

where W is the average weight of the fetuses of one litter, a is the specific fetal growth constant, t is the age of the fetuses from conception, and t_0 is the intercept on the time axis. The original formula applies to all mammals, and can be adapted for any species so long as the weight of the newborn young is known and the value of a can be calculated. For example, young stoats in New Zealand are 3 to 4 g at birth (King & Moody 1982). Substituting appropriate values for a and t_0, and rearranging the equation, t can be calculated as a negative value, days before birth, as follows:

$$t = (\sqrt[3]{W}/0.063) + (16 - 40) = (\sqrt[3]{W}/0.063) - 24.$$

Stoats vary tremendously in size across their range, and in some of the local races of smaller adult body weight, the newborn young weigh under 2 g (Table 9.3). For them, this equation can be generalized to

$$t = (24 \cdot \sqrt[3]{W}/\sqrt[3]{W_b}) - 24,$$

where W_b is the average weight at birth (in grams).

The gestation lengths of stoats and longtails are similar, so this equation applies to longtails as well as stoats.

The critical change in day length, which sets off the processes leading to implantation, is reached earlier in places at lower latitudes than at places nearer to the poles; hence, the young are born earlier in warmer climates. For example, in New Zealand, King and Moody (1982) collected stoats from places spanning a range of latitudes from 38°S, where implantation starts at about the end of August, to 45°S, where it starts about 10 to 15 days later. The estimated range of birth dates was from late September to mid-October (the austral spring) in the North Island, and from mid- to late October in the South Island (Figure 9.6).

The roughly equivalent dates in the northern hemisphere would range from late March or early April in California (from south of 35°N to north of 40°N) and France (south of 45°N) to late April or May in Canada (north of 45°N) and Scotland (north of 55°N). In the former USSR the range of birth dates recorded by Aspisov and Popov (1940) was from late March in the Ukraine (45 to 50°N) to early May in Tatary (55°N). Local variation spans about 3 weeks in any one place, but, compared with many other small animals, including both common and least weasels, the season of births in stoats is closely synchronized.

Day length also controls the seasonal development of the testes, so the spring rise in testis weight is conspicuously later in higher latitude males. Since day length also controls the spring molt, weasels change into their summer coats later at higher latitudes (Chapter 3). In fact, the whole suite of related spring activities is closely coordinated by the neuroendocrine system and adjusted to the expected environmental conditions.

While her young are still suckling, a mother stoat comes into heat again. Estrus normally follows littering in adults, and preparations for it are, therefore, also controlled by day length. When the days reach a certain length, follicle-stimulating hormone (FSH) from the pituitary stimulates the ovary to produce estrogen, which prepares the vulva for mating. An estrous female will not ovulate

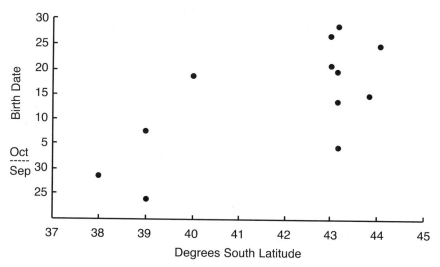

Figure 9.6 The date of implantation in stoats is controlled strictly by day length, so the date of birth of the young is significantly related to latitude. (Note: October is spring in New Zealand.) (Redrawn from King & Moody 1982.)

until she has mated, though, and remains in heat (with swollen vulva) until after mating. The pituitary will not release the LH, necessary for ovulation until it receives the nervous signals from the vulva (Figure 9.5).

It is, therefore, easy to tell the state of any given female because, as soon as she has mated, the swollen vulva subsides and a fresh set of corpora lutea appears in the ovaries. The corpora lutea persist for most of the year, whether or not any young are eventually born, and successive generations of corpora lutea cannot overlap, because each set degenerates before the next estrus.

The male who serves the adult female also fertilizes her precocious female young (p. 225). So, by midsummer practically all wild females of all ages (>99%) are pregnant, carrying a new set of blastocysts and showing no signs of recent estrus. If estrus is brief and synchronized by day length, males must be very good at finding receptive females. The proverbial English description of a particularly persistent human suitor as "a bit of a stoat" (Drabble 1977) is obviously quite a compliment. Meanwhile, in adults the postpartum estrus (extended for as long as is necessary to find a mate) and prolonged delay in implantation mean that almost all female stoats are pregnant almost all the year round. If there were a prize for the sexiest animal, the stoat would surely win it.

According to Ternovsky (1983), Amstislavsky and Ternovskaya (2000), and Polkanov (2000), the estrous season of captive Russian stoats can be months long, and individually variable if a female does not mate early in her season. The gestation period is also variable, depending on the date of mating, and is significantly longer in young females (averaging 317 days) than in adults (averaging 298 days). In New Zealand, three captive females kept in isolation by King and Moody (1982) were still in full heat some 3 months after the normal season, and had no corpora lutea.

How is it, then, that in the wild, estrus is brief, and no prolonged estrus season is ever observed? For example, among 46 adult females collected from all over New Zealand in the southern spring months of September and October, 45 were not yet in estrus, five were in the stage immediately preceding estrus, and only one was in full estrus. Not one of 38 newly independent young females collected in early summer (December) showed any sign of recent estrus (King & Moody 1982:117); all had mated and estrus was over. We conclude that the reported cases of lengthy estrus in caged animals are an artifact of captivity; they might extend the opportunities for breeding stoats for research, but they do not reflect the habits of wild females. Only a lone female with no male for miles and miles around (say, isolated on an island) might prolong her estrus in the wild.

Two persistent errors concerning the breeding of stoats have confused people for many years despite a total lack of evidence that either is true. The first is the idea that females ovulate spontaneously throughout the year but do not conceive until the spring. This story arose because Deanesly (1935), who undertook the first intensive study of the reproductive cycle of stoats, did not know about induced ovulation or delayed implantation. She therefore interpreted the corpora lutea of delay as spontaneous infertile ovulations. Alerted by Wright (1942a), she later re-examined her material and published a correction (Deanesly 1943). Nonetheless, the earlier paper was so good in every other respect, so thorough and so much more widely read than the correction, that the mistake was repeated in general reference books for the following 30 years.

The second error is the idea that the few females that do not conceive during the main breeding season get a second chance the following spring, and that litters conceived then develop directly, with no delay in implantation (Watzka 1940; Kopein 1965). There is no empirical support for this interpretation, either from examination of the reproductive state of large samples of wild-caught females or from extensive observation of the mating behavior of stoats in captivity, but it is an understandable mistake. Females that lose their entire litters for any reason do come into estrus earlier than others, and are ready to mate in spring before any young conceived the previous year could have been born. These females can be fertilized as soon as the males are ready, which could be 6 to 8 weeks before the successful breeders reach postpartum estrus. These early-breeding females must, nonetheless, wait until the *following* year to produce their young, after an unusually long period of delay.

Since shortage of food is a critical factor controlling the survival of young weasels of all species born in the wild, and since every female stoat captured during summer, autumn, and winter is already pregnant, one might expect it to be easy to establish a breeding colony of stoats in captivity, simply by catching females over most of the year and feeding them well. Unfortunately, that is not the case. Although some breeding colonies of stoats have been established (Müller 1970; Ternovsky 1983; DonCarlos et al. 1986), many females fail to produce young when brought into captivity (McDonald & Lariviere 2002; O'Connor et al 2004)). Virtually all wild-caught female stoats of all ages carry fertilized blastocysts, but they seem especially vulnerable to the stress of captivity, especially at first.

What sort of five-star treatment wild-caught females need before they can be persuaded to produce their young in the confinement of a cage is still not entirely clear, although at least one thing seems certain: Female weasels of all species (except, perhaps, the large British stoats) breed best on a diet of small rodents (DonCarlos et al. 1986; Sundell 2003). More important, since well-adjusted and well-fed adult females can live for several years, we need to know how to encourage adults acclimatized to captivity to breed year after year.

Long-Tailed Weasels

The breeding cycle in longtails (Figure 9.4B) is similar to that in stoats (Wright 1947, 1948, 1963), except in one respect. The young are born in April or May, but postpartum estrus in female longtails is inhibited by lactation. Unless she loses her litter, a female longtail does not return into heat until 65 to 104 days after parturition (i.e., in June, July, or August, according to location) (Wright 1948), later in the season than does a female stoat. Females that give birth but lose their litters very early in the season can be ready to mate again earlier, 39 to 71 days after parturition, but still cannot produce the litter until the following spring. Males are fertile from April to August (Wright 1947).

As in stoats, the spring molt is an accurate herald of the reproductive season for longtails. Both males and females are sexually inactive while in their white coats. From the time their brown summer hairs begin to appear, it takes about 21 days to the implantation of the blastocysts and about 47 days to the birth of the young. Any female remaining white unusually late in the season may be confidently assumed to have resorbed her litter. In males the spring rise in the weight of the testes correlates closely with the beginning of the spring molt. For these and other reasons, longtails have proved themselves useful research animals, especially in studies of endocrinology and photoperiodism.

DENS

Den sites may include holes up the trunks and in roots of trees through forest habitat (Murphy & Dowding 1994) or in piles of logs, ditches, and isolated patches of scrub in open habitat. In Ireland, of 19 dens belonging to three radiotracked individuals, 15 were in underground burrows (nine made by rats, four by rabbits, one by a mouse, and one unknown), three in piles of sticks or stones, and one up a tree (Sleeman 1990). One stoat den found in northwest Greenland by Sittler (1995) had been built in the wool of a muskox carcass.

Lisgo (1999) classified seven different types of rest sites or dens used by her stoats, all underground. The dens were in squirrel middens, in tunnels through moss or root systems used by snowshoe hares or chipmunks, under logging slash, under branches or trunks of natural deadfalls, in holes at the bases of trees or snags, in moss hummocks, or in the upturned roots of trees. All stoats used more than one den. Of 42 dens used by male stoats, the order of preference was squirrel middens (55%), holes in trees or snags (24%) and all the rest (<7% each). Of 37 sites used by females, the preferred sites were under logging slash (41%), in holes in trees or snags (30%), in natural deadfalls (16%), and all the rest (<5% each). This difference in rest sites used by the two sexes is just as one would expect from the difference in their preferred hunting areas.

The safest dens are those with the smallest entrances, which keep out unwelcome visitors, including other weasels. Two of the female stoats with young observed by Erlinge (1979a) had chosen dens with access holes too tight to admit a male, and likewise, the entrances to dens of longtails measured by Gehring and Swihart (2000) averaged 30 mm for females and 40 mm for males. Female stoats move about less during the breeding season (Robitaille & Raymond 1995), and females with small young are likely to stay close to their dens except when it is necessary to shift their young between den sites. In New Zealand, Murphy and Dowding (1995) observed an adult female moving her young 500 m to a new den; Dowding and Elliott (2003) identified 14 dens that were used sequentially by more than one collared stoat within 3 months, plus nine dens that were used by both stoats and feral ferrets.

DEVELOPMENT OF THE YOUNG

From Birth to the Opening of the Eyes

Infant weasels look rather alike in all species, both in their appearance at birth and in their early physical development. They are all born completely helpless, and all grow in the same way, but least and common weasels develop more rapidly than do stoats and longtails, and they reach the milestones of development at younger ages (Table 9.3). For example, although young stoats and longtails are born larger than young common weasels, they grow more slowly, and are 6 to 8 weeks old before they are again larger than common weasels of the same age.

Newborn young weigh 1 to 4 g, and at birth, or soon after, their pink skin is covered with a fine, pale natal down, which gives them a silvery appearance. They do not grow the characteristic brown and white fur of adults for several weeks. The eyes can be seen only as indistinct bluish-black dots crossed by the sharp horizontal line of the tightly closed eyelids. The ears are very small, pressed against the side of the head, and often have a waxy deposit in the ear channels. They can easily be told from infant rodents, because they already have the long necks typical of the weasel family, and the front limbs seem to be placed almost halfway down the total length of the head and body.

All four limbs are short, weak, and hardly jointed, but already furnished with broad paws, and the very short toes have fine well-developed little claws. The prominent ribs make the tiny bodies (the width of a pencil) appear to be closely segmented; Ternovsky (1983) commented that "in appearance they are reminiscent of large ants." The tail, little longer than the limbs, is cone-shaped. The urethral opening of the males lies halfway between the navel and the prominent anus; the vulva of the females lies directly in front of the anus.

Newborn young have no teeth, but they do have powerful equipment and instincts for sucking. The jaws are relatively shorter than those of adults, and a massive shovel-shaped tongue takes up most of the space inside the mouth. Left alone, newborn young lie in a heap together, crawling under each other to avoid disturbance; if one is separated it immediately struggles back into touch with the rest. Almost from birth they can produce a fine chirping sound, usually a sign of protest or distress. Most observers report that, at this stage, males and females are roughly the same size, and both are, on average, equally represented.

Table 9.3 Development of Young Weasels

	Least weasel	Common weasel	Stoat	Long-tailed weasel
Total gestation	34–36 days	35–37 days	220–380 days	205–337 days
Birth weight	1–2 g	1–3 g	1–3 g small races 3–4 g Britain, New Zealand	3–4g
Birth coat	Naked	Naked	Fine white hair on back, rest naked	Fine white hair all over
Growth of mane	None	None	14–22 days	none
dorsal fur	18 days	21 days	21 days	28–35 days
black tail tip	None	None	42–49 days[1]	21 days (?)
Teeth, milk	11–18 days	14–21 days	18–28 days	21–28 days
permanent	4–6 weeks	8–10 weeks	10 weeks+	10 weeks+
Opening of eyes	26–30 days	28–32 days	30–42 days	35–37 days
Eat meat	3–4 weeks	3–4 weeks	4–5 weeks	4–5 weeks
Lactation lasts	4–7 weeks	4–12 weeks	4–12 weeks	5–12 weeks
Play outside nest	4 weeks	4 weeks	5–6 weeks	5–6 weeks
Kill prey	6–7 weeks	8 weeks	10–12 weeks	10–12 weeks
Adult size, M	3–6 months	3–6 months	12 months	12 months
Adult size, F	3–4 months	3–4 months	6 months	6 months
Sexually mature, M	3–4 months	3–4 months	11–12 months	11–12 months
Sexually mature, F	3–4 months	3–4 months	4–6 weeks	3–4 months

1. Tail tip begins to darken at about 18–20 days.

(Hamilton Jr. 1933; Sanderson 1949; East & Lockie 1964, 1965; Hartman 1964; Heidt et al. 1968; Heidt 1970; Müller 1970; Blomquist et al. 1981; Ternovsky 1983; Polkanov 2000; Svendsen 2003).

The litters of common and least weasels tend to be somewhat smaller (usually four to eight) than those of stoats and longtails (usually six to 12; Table 9.2), for reasons explored in Chapter 14.

At first, nestling weasels have no proper fur and cannot maintain their own body temperatures. When the mother is not in the den they huddle tightly together for mutual warmth. Even so, if their body temperatures drop below 10°C to 12°C (Segal 1975), the young go into a temporary cold rigor. Their pulses and breathing slow, their metabolism and growth slip into a lower gear, and they become stiff and cold to the touch. The effect is somewhat like hibernation, except that it is rapidly reversible as soon as the mother returns to the nest.

The reason for this curious habit appears to be connected with energy conservation. When the mother is with them, she provides both warmth and food, and they cuddle up close to her and channel as much as possible of the energy she provides into growth (Figure 9.7). When the mother is away, if they tried to maintain normal temperature they would have to draw, from their own meager resources, energy that they need for growth. The advantage to the young of

Figure 9.7 A mother weasel must spend as much time as possible with her young while they are very small. When the young get chilled, they stop growing. Here a female stoat sleeps with her young, who have recently nursed.

not attempting to keep themselves warm when left alone is that they maximize their growth rate when their mother is with them, and when she is away they reduce the chances of running out of energy altogether, even for staying alive, before she comes back. The disadvantage is that, if hunting is bad and the mother has to be away a lot, the young have few chances to grow at all. When that happens, they end up permanently smaller than usual, or dead. In a very large sample of stoats collected from beech forests in New Zealand, where food supply varies substantially between years, we showed that the young of both sexes born in hungry seasons were fewer and grew into smaller adults than those born in good seasons (Powell & King 1997).

Unweaned nestlings spend most of their time asleep. They squeak and chirp in response to disturbances near them, but otherwise they wake up only to suckle and to defecate. The den is kept quite clean because their mother licks up the feces of the young—and, indeed, they apparently perform this service for each other. They make no attempt to leave the den, although the mother may move them if the family is threatened by any interference or by bad hunting in the immediate vicinity. Then she carries each one in turn (Figure 9.8), darting through the undergrowth to the new den. She leaves each of them there, one by one, and returns for the others.

Michael Hitchcock, an English gamekeeper, was once standing by a pile of rotten logs when a common weasel came out, and studied him for a few seconds at a range of hardly more than an arm's length. She then ducked back into the pile of logs and emerged carrying a tiny, blind young one. She carried the young high off the ground as she ran along the hedgerow out of sight. Hitchcock waited, and she returned to repeat the procedure with a total of six young. After the first one, Hitchcock kneeled down by the burrow and watched the operation at close range. He was especially struck by the total fearlessness of the mother weasel in her domestic crisis. The young remained limp and passive in the *tragschlaffe* ("carry-sleep") position while being carried, just as adult females do when being carried about by a male during mating, and made no sound unless the mother grabbed them in a sensitive place.

When the young are 2 to 3 weeks old, their milk teeth begin to erupt, as razor-sharp miniature editions of their adult meat-eater's teeth. As with most mammals, weaning is a gradual process in young weasels. While still blind and deaf (still <4 weeks old; Table 9.3), they chew vigorously on mice that their mother has opened for them, or on small pieces of meat, skin, or bones. Nonetheless, they continue to suckle for several more weeks. They begin to be aware of their surroundings, and they respond to squeaking noises or to human speech by raising their heads, opening their mouths, and hissing faintly. Some make feeble attempts to strike if provoked.

Eventually they try to get to their feet, usually collapsing again immediately. Crawling begins as an unsteady, circular exercise. Before long, the young are

Figure 9.8 The mother weasel carries her kits in the typical carnivore way, grasped gently around the body.

able to crawl short distances, in a more or less straight line, before their mother can retrieve them. From about the time that meat enters the diet at 3 to 5 weeks, depending on the species, youngsters learn to defecate outside the entrance of the den, and from this time onward, the size difference between males and females becomes more and more noticeable.

From the Opening of the Eyes to Independence

After the eyes and ears open, from the fourth week onward, nearly always in the females first, the behavior of the young weasels changes as they begin to perceive the world around them. Within a few days they are actively exploring outside the den, and the mother no longer attempts to keep them all together. They are still a bit wobbly on their feet at first, especially the hind feet, but they can run along at least as quickly as mice.

By about the fifth week of age the young depend much more on the animal prey supplied by their mother than on her milk, and in the wild will soon be weaned. (Captive young ones may continue to suckle for as long as the mother will let them.) The appetites of young weasels are stupendous—each one is soon eating 20% to 40% of its own body weight per day (Sanderson 1949). Their mother may have a struggle to provide such largesse, even though breeding females can increase their hunting effort up to fourfold (Erlinge 1979a). Consequently, there is plenty of incentive for the young, as soon as they attain sufficient coordination, to start providing for themselves.

The milk teeth are usable for chewing on carcasses as soon as the deciduous carnassials (the third upper premolars and the fourth lower premolars; Chapter 2) have erupted. These appear at around the third week, and are replaced by the permanent teeth a couple of months later. Both sets of teeth erupt in a predictable sequence. The permanent carnassials, the fourth upper premolars and the first lower molars, are last in place (Hall 1951).

Killing behavior does not have to be learned, and, although the first attempts are clumsy, the young quickly improve with practice. Some observers maintain that the mother will bring a live but disabled rodent to the den and use it to teach killing technique to the young. We doubt this, since the hunting methods of weasels do not include any complex acquired skills (as in, say, the big cats and wolves), and the family life of weasels is too brief to give the young much chance to improve their proficiency under instruction. Besides, young that have been separated from their mothers at early ages are soon as expert as any others. Of course, the young of nearly all carnivores play with their littermates and with their prey, developing the crafts of their trade, but incidental learning through play is not the same thing as the mother deliberately teaching them.

Within another few weeks the young are more or less fully mobile and their motor coordination is improving daily. Standing up on their hind legs and jumping over obstacles are among the last of the typical weasel skills to develop. By this time the young are fully furred and able to maintain their own body temperatures, at first only inside the den but later also outside it. They follow their mother on short hunting expeditions, dodging from one hiding place to another. If one falls behind it will call loudly with the infantile begging cry. The mother answers by trilling to show the lost one where she is, or, if necessary, she goes back to pick it up. At any potentially dangerous disturbance the mother hisses furiously, which causes the young to "freeze" and keep under cover.

The mother is totally fearless while she has young, and will perform feats of amazing courage, threatening any animal, however large, that gets in her way. Human observers, if unobtrusive, are simply ignored. Michael Hitchcock (p. 221) also reported that he was, on another occasion, standing by a hedge when he heard loud, high-pitched squeaking coming closer. Eventually a family of common weasels appeared, but he could not count the number of young as they were moving about so fast. Some of them actually ran over his boot. Gamekeepers also sometimes see a family of common weasels forming what looks like a "rope", or even "a string of chipolatas"—formed from a parent and a group of young in single file.

Galen Burrell (quoted by Hirschi 1985) watched one mother stoat with her family of young in the alpine meadows of the Colorado Rocky Mountains:

> Grasping the plump vole in her mouth, she carried it about 100 m to the far corner of the rock pile. . . . Here she was greeted by eight young stoats. The nearly full-grown youngsters, mewing and chirping softly, excitedly sniffed their mother, one another, and then the breakfast vole. The stoat family disappeared into their burrow. . . . I . . . waited. . . . The mother reappeared. Standing on her hind legs she first scanned the surroundings, ignoring me, and then chirped. Eight heads popped out of the burrow. In a rushing stream of bodies the young ones followed their mother down the mountainside. . . . They would suddenly pile up behind an obstacle—usually a rock that was just too big for them to leap over. . . . She was forever returning to pick up stragglers. Grasping each one by its neck, she would drag it back to its brothers and sisters. Then the entire family would move on. . . . [They reached] a safe new hiding place. . . . Without pausing to rest she slipped away. [She killed a pika] 50 m away from her young ones. With a tremendous effort she dragged, pushed and sometimes carried the pika (which was twice her size) a short way towards [them, then] she pushed the pika under a rock. Then she headed up the slope towards the place she had hidden [a deer mouse killed earlier the same day]. Despite the time that had passed, she remembered exactly where the mouse was, picked it up and carried it back to her hungry children. Not long after that the mother stoat

moved her brood up to where she had stashed the pika. Darkness fell as the family consumed their meal under the cover of their rock fortress. . . .

This extraordinary account gives a vivid picture of the flexibility of behavior that allows a mother weasel to make choices according to circumstances (e.g., to bring the young to the kill if the kill is too heavy to carry to them). It also underlines the energy invested by the mother in her young, and the enormous commitment of time and effort it takes for her to rear them alone.

Nor is this a unique observation. Sandell (1988) radio tracked a female stoat moving her young from one den to another and rushing around incessantly to provide enough food for them. Female stoats hunt with feverish intensity, and in one case related by Sandell, a female stoat with young ruined a colleague's radio-tracking study of watervoles by killing 15 to 20 voles in less than 2 days. The unfortunate vole researcher "found all of her transmitters in a heap of half-eaten voles. I was excited by this unequivocal proof of a stoat's hunting efficiency; curiously, my colleague wasn't as enthusiastic" (Sandell 1988).

In the wild, family parties of three to six or more stoats and longtails may be seen moving about together in early summer, most often in June and July. As the young get older, they scamper about, chasing each other and making high-pitched squeaking and whistling noises. One eyewitness account described seeing, from a distance of only 4 m, a group of five running across the ground. They included an adult with four young almost as big, running in concert, the adult in front. They "flowed along as if a single animal, accompanied by a wonderfully soft, fluty chirruping . . ." (Ewan Young, personal communication). Usually only one adult is seen with a group of young, presumably the mother, but there are records of family parties accompanied by two adults. We suspect that the second "adult" may be a large young male (maybe a male with only female siblings) or possibly, in longtails, an adult male looking for mates.

By the time the young have hunted together for 2 or 3 weeks, they have their permanent teeth and some experience in using them, and they are ready to set out on their own. Families kept together in captivity have to be separated at this stage, as the young and the mother become increasingly intolerant and irritable with each other.

In New Zealand, as almost everywhere, stoats are distinctive enough to be noticed and rare enough to be remembered by casual observers. In places where many observers aggregate, such as national parks during the summer vacations, this would add up to a lot of information if only it could be organized.

One study done in 1976–1978 attempted to harness this resource and make some sense of it. It involved a simple reporting system whereby visitors to national parks in New Zealand sent in a prepaid record card for every stoat or group of stoats they saw (King 1982). The visitors reported seeing stoats in groups only in late spring and summer (from October to February in New Zealand), most often in December and January. This neatly confirmed what we already knew from trapping studies: Young stoats were seldom caught in any numbers before December, and were most numerous in January when the families break up and the young disperse. Erlinge (1977b) observed the same at the equivalent season (June) in Sweden.

Age and Size at Puberty

Young common weasels born early in the season, especially when food supplies are good, grow very rapidly. Early-born young of both sexes can be physically and sexually mature by the age of about 3 or 4 months. By mid- or late summer (July to August), they may be taking full part in reproductive activities on equal terms with the adults, and in vole peak years their contribution is extremely important (Chapter 10). In most seasons, though, young-of-the-year females either fail to conceive or lose their first litter at a very early stage. The average young common weasel does not breed successfully until its second year.

By contrast, in young stoats and longtails the development of young males and females is radically different (Figure 9.9). Young females reach puberty as nestlings (stoats) or when newly independent (longtails), and then grow to their full adult size by the autumn, when they are about six months of age. Young males stop growing in autumn, and remain immature and distinctly smaller than adults throughout their first winter. In spring, last year's young males reach puberty, and suddenly put on another spurt of growth. They are indistinguishable from older adults by the time they are 12 to 14 months old and the breeding season is well under way.

The precocity of the juvenile females is the most extraordinary thing about the development of young stoats. While they are still in the den, in fact while they are still blind, these tiny, unweaned babies become reproductively mature, and are fertilized by an adult male, probably the same one that mates with their mother. It passes belief that the large and aggressive males can mate with sexually mature but developmentally infant females without damaging them, yet that is exactly what happens.

Many observations of captive matings confirm that adult males do mate with nestling female stoats—not just occasionally but as typical and normal breeding behavior, which leads in the following year to a litter of typical and normal young. Ternovsky (1983) observed 58 young females mating at between 17 and 134 days old, and all of them produced young in the following year, including this remarkable case:

The youngest female was 17 days old on the day of mating (25 May 1980), was 112 mm long, was helpless, deaf, blind and toothless; she was feeding on her mother's milk and could move only slightly by

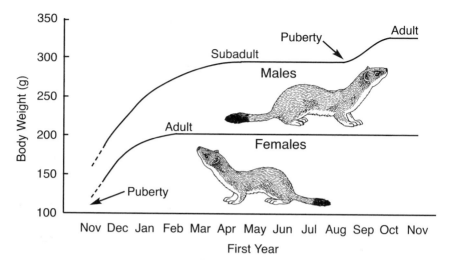

Figure 9.9 The growth patterns of male and female stoats in New Zealand during their first year in life. The two sexes reach both puberty and full size at different ages. (Note: November is the austral late spring, and March is autumn.) The same seasonal pattern applies to stoat populations everywhere, although these body weights are too high for most other stoat populations. (Redrawn from King & Moody 1982.)

crawling. Her body mass (18 g) was 13% of her mother's body weight and only 6% of the body weight of the male. Coitus lasted only for one minute. Spermatozoa were present in a vaginal smear. On 22 May 1981, after 337 days, this female gave birth to 13 young and fed them successfully.

Considering the enormous difference in size and strength between an adult male and a juvenile female, one might regard this behavior as an animal version of rape, but that idea would be quite wrong. Juvenile females are not only willing to cooperate with the male, but they are also positively eager. The arrival of a male (or perhaps more specifically, his scent) stimulates a sexual reaction in blind and deaf female young. They call to him with high-pitched trills and chuckles, grading into the typical adult nuptial cooing signal. They grab onto him as he passes by, or crawl after him, interfering with his mountings with both their mother and sisters (Müller 1970; DonCarlos et al. 1986).

This does not sound like coercion, and, indeed, it is not. It is in a young female's interests to mate with a male that is strong enough to overcome her mother's defenses and enter the den, or careful enough to call when she is absent. The lack of a year-round pair-bond and the short average life span and rapid replacement of locally resident males all minimize the chances that a young female might be mating with her own father.

Adult male stoats can tell the sex of a juvenile as soon as they have grabbed its neck, and they usually drop the males at once. This reaction must presumably be a response to a scent or taste signal. Young stoats develop prominent manes of brown hair on the napes of their necks when about 3 weeks old (Table 9.3; Figure. 9.10). This juvenile mane was first described many years ago (Bishop 1923; Hamilton Jr. 1933), but its function has never been explained. Presumably it is associated with glands that produce a scent or taste distinctly different in young males and females.

Ternovsky (1983) described one case in which the adult male mated with all four young females in a litter of eight, and killed all four males. One might ask why the adult males do not make a habit of this, since, presumably, it would be in their interest to remove potential future rivals. On the other hand, if the old male lived that long, he would be well able to drive any youngsters away and force them to try their luck elsewhere. Meanwhile, restraint might be worthwhile on the off chance that he might be killing his own sons, or that his interference might provoke an aggressive reaction from the mother.

Juvenile female longtails also mature early, though not quite to such a dramatic extent. They are 3 or 4 months old, weaned, and almost full grown before their mother comes into heat between June and August and the adult males are permitted to approach the family. Only then do the young females mate. Young longtails do not develop a mane, as do young stoats, and this difference supports the idea that the function of the mane has something to do with helping adult male stoats to identify their infant mates. One might ask why the young female longtails do not also mate as nestlings; but remember, the two species are not as closely related as they look (Chapter 1).

By the time they leave the family group, virtually all juvenile female stoats and longtails are already fertilized. Even in very large samples of trapped stoats, it is unusual to find even one young female without corpora lutea. The energy required to maintain the blastocysts is slight, so this stage of pregnancy does not add much extra burden to the young females as they complete their own

Figure 9.10 Juvenile stoats of both sexes develop a transitory mane when about 3 weeks old, which seems to be correlated with the extreme sexual precocity of the females. Adult males apparently use a cue located in the mane to distinguish infant females from males, and to mate only with the females.

body growth. The young males of both species are more conventional—they breed first when they reach adult size as yearlings.

DISPERSAL

All weasels are intolerant of crowds, especially crowds of other weasels. Even siblings reared together in captivity, where food supplies are regular and warm nests supplied, have to be separated after they become independent and their squabbles begin to turn nasty (R.A. Powell unpubl.). At that stage or long before, wild-born youngsters will have gone their separate ways.

Most families break up when the young are 3 or 4 months old and fully capable of looking after themselves. Then the young leave the mother's home range to find places of their own to settle. The few observations recorded suggest that young of both sexes are capable of astonishingly rapid travels over long distances before settling down. Most young females remain nearby, usually less than 5 to 6 km away (Erlinge 1977b; Debrot & Mermod 1983). Sixteen of the 18 young females observed by Erlinge (1983) in Sweden stayed in the same area for life.

Young males, however, may travel extraordinary distances in a very short time. In New Zealand in the summer of 1979–1980, six of 65 young males caught in live traps and eartagged were known to have traveled at least 6 km, 8 km, 12 km, 15 km, 20 km, and 23 km. These were only the minimum, straight-line distances between known capture points: The real distances, running in and out and round about as stoats do, must have been much greater. Furthermore, these distances were covered remarkably quickly: 12 km in 27 days plus a further 3 km the next day; 20 km in 5 days plus a further 4 km in the next 2 days; and 23 km in 39 days (King & McMillan 1982). In the same study area a few years later, Murphy and Dowding (1994, 1995) radio tracked an adult male that moved at least 3.7 km in 3 hours and 15 minutes. They also eartagged a young female on December 20, 1990, and were amazed when the same animal (indisputably the same one) turned up in a kill trap outside a rearing facility for rare native birds, 65 km away, on January 13, 1991.

Similar long-distance travels have been recorded even in the smaller weasels. One male stoat tagged in Alaska must certainly have traveled much further than the 35-km straight distance recorded between capture locations in 6 months (Burns 1964). These long treks by young males are no doubt encouraged by intolerance from the established adult males, which tend to move about less once they are established on a home range providing plenty of food. The adult males tolerate the young females, though. The difference in attitude is simply a matter of what serves the best interests of the local resident adult male. He could benefit by driving out the young males but accepting the young females, who will be, next season, his potential rivals and mates.

Most such long treks are made in bumper rodent years when large cohorts of young weasels are born, but are possible any time. The mere ability of such small animals to travel so far explains how least and common weasel populations can spring up where none has existed for years (Chapter 10). So, when populations of any weasel species become extinct, recolonization from long distances away is quite possible. In New Zealand, where reducing stoat populations over large areas is critical to protect native birds, rapid immigration from surrounding uncontrolled populations is a huge problem (Chapter 13).

10 | Populations: Density and Breeding Success

BECAUSE WEASELS ARE predators, they are necessarily rare—much more so than the voles, mice, and rabbits on which they prey. In a meadow or old field, there are from hundreds to thousands of grass plants for every vole, and from scores to hundreds of voles for every weasel. On the other hand, because weasels are the smallest of the warm-blooded predators, they can be much more common than the foxes, feral cats, hawks, and owls that also eat voles. It is therefore often possible to collect large samples from populations of weasels.

One might expect from this that the population dynamics of weasels would be among the best studied of all carnivores, but that is not so. Weasels are too small and secretive to be observed directly (except in occasional lucky glimpses), so systematic study of them has to be done indirectly, by routine trapping or footprint tracking. Suitable field techniques were worked out in the nineteenth century by fur trappers, naturalists, and gamekeepers, but were not applied to scientific studies until the 1960s.

More important, many people do not see weasels as charismatic animals (Chapter 1), and they are unreliable. They are not evenly distributed in all habitats, and even where they do live, they may be abundant one year and scarce the next. These variations make researchers nervous, because it is hard to plan a study on such a slippery, moving target. Many hopeful researchers and students have put in weeks or months of fruitless fieldwork, often in places where weasels were known to have been present at some previous time, before giving up disappointed.

ESTIMATING DENSITY

Of Live Populations

Estimates of absolute density of weasels are rare, because the statistical techniques for calculating the numbers of live animals per unit area impose important conditions on the field data that are hard to meet, including the following:

1. The population must be counted without removing the animals or frightening them so much that they refuse to be counted again. The best available, though not ideal, method for weasels is still the

old one of using live traps to capture, mark, and recapture repeatedly the resident individuals of a given area. But confinement in a live trap is probably unpleasant for weasels, and it tends to discourage at least some of them from risking such an unhappy experience again (King et al. 2003a). Others become "trap happy," especially if wild prey are so scarce that a night in the trap becomes a price worth paying for a free meal.

2. It must be possible to define the area sampled, so as to calculate the catch per unit area, and that is not only very difficult but also has a disproportionately large effect on the results if the boundaries are misjudged.

3. It must be possible to estimate the numbers of animals entering and leaving the population, and the number that are present but not caught. Not every resident will be caught during every trapping session, but if it is caught again in a later session it is usual to assume it was there all along, even though this sometimes seems quite unlikely. For example, one radio-collared common weasel was neither recaptured nor detected by radio for 14 months, until it was again recaptured not far from the site where it was first collared (Delattre et al. 1985)

4. Some individuals will not enter traps at all, so every study of weasels should allow for some unspecified number of shadowy, unidentified figures in the background. The proportion of these probably increases when food is abundant (Teplov 1952; Alterio et al. 1999; King & White 2004).

The problems do not stop once these hurdles are crossed and a set of field data is safely entered into the computer. The simplest way to estimate the population density is the "calendar of captures" or "minimum number alive" method, which compiles regular totals of the number of residents known or assumed to be present each session. Unfortunately, live trapping is very labor intensive, so for very mobile animals such as weasels it is usually impossible to observe more than a few individuals at a time, or to define the boundaries of the sample area. This means that the population estimates are so much affected by statistical errors that they may be quite wildly wrong. More sophisticated methods designed to overcome this problem (e.g., Efford 2004) usually demand far more data than a short-term live-trapping study on weasels can supply.

A few outstanding studies have met this problem by deploying a team of field workers over a large area for several years. The first of these was led by Sam Erlinge (1983), whose group documented the population dynamics and predator–prey relationships of stoats on a 40-km² area of farmland and marshes in southern Sweden over 5 years. Erlinge was the first to test the accuracy of live trap-night indices (live captures per 100 trap nights) on a population of stoats of known density, and he concluded that they are a reasonably reliable guide to changes in relative numbers.

Another team study was done by Jędrzejwski et al. (1995) on a 47.5-km² area of the Białowieża National Park, eastern Poland. The Polish group integrated data from grid-based snow tracking, live trapping, radio tracking, and computer analyses to convert live trap-night density indices into absolute numbers of common weasels with remarkable precision ($r = 0.9$, $P = 0.002$) (Figure 10.1).

One possible way to get density estimates for larger areas is to take reasonably accurate measurements of density in small areas and scale them up. That approach can lead to completely unrealistic results if done uncritically, because it assumes that the whole of the larger area is covered by the same habitat sampled in the smaller one. On the other hand, it is possible to derive a rough estimate of the total population of a large area by multiplying the typical density of a species in different habitats by the area of those habitats available.

Stephen Harris and his team (1995) were the first to attempt a marriage between habitat distribution surveys and habitat-specific density indices, in an attempt to calculate the total national populations of all 63 species of native and introduced mammals in mainland Britain. National biodiversity strategies being developed by countries that are signatories to the agreement made at the United Nations conference at Rio de Janeiro in 1992 need this sort of information, however rough it may be. Under European Union legislation, the monitoring of endangered wild mammal populations is now a statutory responsibility, so the UK Joint Nature Conservation Committee commissioned

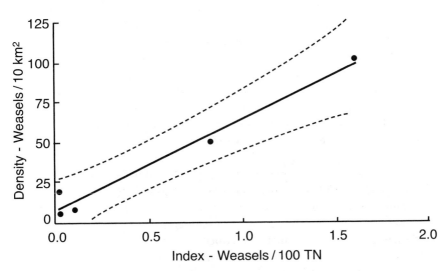

Figure 10.1 The correlation between the absolute density of common weasels and a density index in Białowieża Forest, Poland. (Redrawn from Jędrzejewski et al. 1995.)

Harris and his team to identify which species could be of conservation concern in Britain.

The easy part was to list, from existing survey data, the total national areas of various defined habitats. The hard part was the next step, to get from known figures on the distribution or home ranges of weasels in particular places to a general figure for density per square kilometer of the same habitat. The problem is that home range data can be very misleading if they include, as they usually do (Chapter 8), a lot of spatial overlap and temporal variation. Eventually Harris et al. simply decided to assume the rather generous figures of six stoats per km^2 in all types of woodland, and one to two per km^2 for various types of grassland, which produced a calculated prebreeding national total for Britain of 462,000 (about two per km^2 overall). They calculated the numbers of common weasels from the ratio of common weasels to stoats in gamekeepers' bags, and came up with a national total for them of 450,000.

There is no way to guess how close these estimates are to the mark, but from 1970 to 1995, British gamekeepers working on habitat favorable for wildlife killed about 1.5 to 2.0 stoats per km^2 every year (Tapper 1999), so the prebreeding population must be able to maintain a steady harvest at that level, at least in those habitats. The recent drop in gamekeepers' tallies of stoats and common weasels in Britain is probably due to a decrease in trapping effort, not in the national populations of stoats and common weasels (McDonald & Harris 1999).

In the coastal dunelands of the Netherlands, the prebreeding density of stoats through the 1950s and 1960s was about two to three per km^2 in early spring and six to seven per km^2 in summer; the annual harvest was about four per km^2 (Mulder 1990). In Fennoscandia, numbers of least weasels are reckoned to range from one to 20 per km^2 and stoats from 0.5 to 2.0 per km^2 (Hanski et al. 2001).

Calculating a broad-scale average density for American weasels (all species) would be more difficult, and the only estimate we have found is that given by Craighead and Craighead (1956) for lower Michigan in 1942. They were not specifically studying weasels, which are hard enough to find at any time, so they probably underestimated the numbers of weasels in their study areas. From tracks, observations, and live trapping, rather than home range data, their figures of 27 to 36 per township (36 square miles, 93 km^2) convert to 0.29 to 0.38 per km^2.

In New Zealand, knowledge of the range of absolute densities of stoats before and after control operations is critical to developing effective management programs to save the national icon, the brown kiwi, from extinction. The prospect of losing the kiwi has spurred intense interest in research on the biology and management of stoats there, described in Chapter 13. Independent calculations of stoat densities in New Zealand are few but fall roughly into the same range as for stoats in Britain and the Netherlands.

In a southern beech forest after a productive summer when mice were abundant, Murphy and Dowding (1995) found eight stoats (four male, four female) resident on 150 ha (five per km²) between January and May 1991. In similar conditions but a different place, Alterio et al. (1999) calculated that there could have been a postbreeding population of three to seven (average 4.2) stoats per km², decreasing to two to four per km² (average 2.5) nonbreeding adults in the following winter. After a good summer for mice, densities of stoats can remain higher than normal for another 6 to 12 months in the absence of control measures (Figure 10.4).

Two quite different methods of estimating absolute density from different beech forests have been applied in New Zealand recently. One uses hair tubes (Chapter 8) baited with rabbit meat to collect samples of hairs from stoats, from which individual DNA profiles can be extracted (Gleeson et al. 2003). In the first trial, 60 hair samples were collected each week during a month-long trial, from plastic tunnels placed 250 m apart on a 3×3 km grid (9 km²). DNA profiles were obtained from about 80% of the samples, and from them 30 different stoats were detected (3.3 stoats per km²). A second method uses total removal trapping to calculate absolute density of an undisturbed population. On Te Kakahu (Chalky Island), off the southwest coast of Fiordland, the first forested island where this was tried (Chapter 14 and see Willans 2000), 16 stoats were taken from 514 ha (3.1 per km²) within a few weeks in 1999. Anchor Island (1,100 ha excluding a 200-ha lake) was cleared of 19 stoats in winter 2001 (1.7 per km²) by essentially the same method (M. Willans, unpubl.). On a 750-ha peninsula in a lake, McLennan and his team (McLennan et al. 1996; McLennan, unpubl.) removed 65 stoats in 3 months during a mouse peak year (nine per km²), but many fewer per 3 months (zero to two per km²) in mouse-poor years. Stoats are good swimmers, so the water protecting the peninsula on three sides would only have slowed down immigration, not prevented it.

How many stoats live in New Zealand? Summer irruptions of stoats in beech forests are short-lived, but recur every 3 to 4 years when mice are abundant after a seedfall. In other types of native forest, where absolute stoat densities have never been measured, relative density indices average around the lower end of the range for beech forest (King et al. 1996). The combined area of native forest patches of all types in New Zealand was 62,800 km² in 1993 (Taylor & Smith 1997). If the general prebreeding density of stoats in native forest is two per km², then the total forest population in spring could be about 125,000. Other habitats that might be occupied by stoats, such as exotic forest (14,000 km²), crops (4,800 km²), and tussock grasslands and pastures (135,200 km²) cover another 154,000 km² of New Zealand. The density of stoats in them is unknown but probably low, especially in open country (Keedwell & Brown 2001), but even if it averaged only 0.5 per km², that still makes another 77,000 stoats.

Obviously, it is impossible to estimate the total population of stoats in New Zealand with any confidence. At first glance these estimates look wrong, because they are a lot lower than given above for Britain, a country of roughly the same size. The data for New Zealand are few, but the figures we do have suggest lower average prebreeding densities of stoats in all New Zealand habitats, especially in the much larger total area of forest, than were used for the British estimate. Nevertheless, New Zealand could easily support 200,000 of these easy-to-hide little predators in the spring of a normal year, and for a short period in summer after an especially good breeding season, many more. Half a million stoats could do a lot of damage in a very short time. It is no wonder the New Zealand conservation authorities worry (Chapter 13).

From Kill-Trapping Records

Over the long term, controlled collections made with kill traps can give a lot of information useful to population studies. This method is much less laborious than live trapping, so can be conducted over a wide area, but it can only be used under certain conditions. Kill trapping can affect the density, social relationships, and dynamics of the target population, and, therefore, is inappropriate for many studies, especially those of native species of weasels that are protected. On the other hand, for populations subject to trapping for fur or for control to protect prey populations, calculation of a density index from kill-trapping data corrected for effort can provide important data impossible to obtain any other way.

This method requires that the traps be set out evenly, baited and checked daily, and operated in the same way and for the same number of days per session regularly all the year round. The theory is that, if the trapping operation has been absolutely consistent, or at least if the data are corrected for variation in trapping effort (McDonald & Harris 1999), then changes in the number of animals killed probably reflect real changes in the numbers of animals available to be killed. Likewise, if the living weasels of all ages and both sexes are caught at the same rate (or at least, that differences between them are constant), then changes in the proportions of animals of each age and sex caught should reflect real changes in the structure of the population observed.

It is important to spell out these rather obvious prerequisites, because much depends on them, and they are not always as true as one might expect. For example, if traps are baited, it could be argued that when more weasels are caught it is because they are more hungry, not because there are more of them. If that happens, then changes in the actual numbers of weasels become confused with changes in their willingness to enter traps ("trapability"). Capture-per-unit-effort indices cannot distinguish changes in numbers from changes in detectability (Anderson 2001).

We take seriously the potential errors introduced into any index of captures-per-unit-effort by the expected changes in behavior of weasels in response to bait and natural food supplies, but these errors are not necessarily fatal. On the one hand, although baited traps catch stoats more often than unbaited ones, both do show the seasonal variation in capture rate, which reflects a real variation in density (King & Edgar 1977). On the other hand, Teplov (1952) showed that when voles were scarce in a Russian game reserve, the local population of stoats fell, but the extent of tracks recorded in snow, and the numbers of ermine pelts harvested, both increased. It stands to reason that capture rate should vary with prey density in severe climates: Stoats must avoid chilling whenever possible, and can do so readily when food is abundant by staying in their dens and feeding from a cache. Then stoats will seldom encounter a trap. In mild climates, the risk of thermal stress is less restricting but stoats do still maintain smaller home ranges when food is abundant than when it is scarce (Chapter 8), so variation in capture rate with prey density is still likely (Alterio et al. 1999; King & White 2004).

Kill trapping disrupts a population if the trap sites are too close (Sullivan et al. 2003), so there is much to be said for avoiding removal sampling if alternative methods are available. On the other hand, appropriately designed indexing provides a practical method of monitoring populations that cannot be observed any other way (Caughley & Sinclair 1994). Weasels are among the species that present formidable obstacles to anyone attempting to make conventional population estimates, whereas low-intensity trap-night indices are possible and are consistent with other changes in the population that are also correlated with density. For example, after a productive breeding season, summer density indices and the proportion of juveniles collected do vary together, as expected (King & McMillan 1982). With all their faults, density indices are still useful for handling the readily available data on weasels collected in kill traps, so we will stick with them for the moment.

A simple relative density index for weasels caught on standardized lines is to calculate the number of captures per 100 trap nights, or C per 100 TN (one trap night equals one trap set for 24 hours), allowing for unavailable traps. For example, take a line of 150 traps, checked daily for 3 days. At the end of the session, the results could be tabulated as in Table 10.1. The real density cannot be worked out from the trap-night index without calibrating the index against known densities. Nonetheless, the correlation between them is fairly good, at least until the index exceeds about 20 C per 100 TN (Caughley 1977).

The most extensive data collected this way are for stoats in New Zealand, where the index for 3 months of trapping seldom exceeds 7C per 100 TN (King 1983b), although over shorter periods of a few days in early summer of a mouse peak year it can reach more than 30C per 100 TN (King & McMillan 1982). Jędrzejwski (1995) demonstrated a straight-line correlation between a *live*-trap index and absolute density of common weasels in Białowieża (Figure 10.1), and Erlinge (1983) did the same for live stoats in Sweden, but so far the relative density indices available from kill trapping stoats have not been converted into numbers of stoats on the ground. Although the indices are consistent with how we expect density to vary, both in time and by habitat, they cannot yet

Table 10.1 Calculation of a Density Index from a Hypothetical Set of Controlled Kill-Trapping Results

	a	b	c	d	e	f	g
	Traps untouched	Stoats caught	Rats, etc., caught	Traps sprung, empty	(b+c+d)/2	150 − e	C/100 TN
Day 1	144	2	4	0	3	147	
2	139	5	4	2	5.5	144.5	
3	144	3	2	1	3	147	
Totals		10				438.5	2.28

Although 150 traps were set for three nights, the total number of trap nights is not 3×150 ($= 450$), because every trap that is set off, by a stoat or by any other animal, cannot catch again until it is reset. Assuming that, on average, each sprung trap is out of commission for half a night, half a trap night is subtracted from the total for every trap sprung, for whatever reason (Nelson & Clark 1973). The corrected total number of trap-nights is the sum of column f, that is, 438.5; the total number of stoats caught is 10; the density index is $10/438.5 \times 100 = 2.28$.

be corrected for the ways that the activity, immigration, and trappability of stoats presumably change with food supplies.

A more complex method is available, derived from fisheries management and applied to mustelids by Fryxell et al. (2001). Fryxell and his team were concerned about setting the annual harvest quotas for American martens, an important fur-bearing species in Ontario, as a constant proportion of the available population. They did not have density indices, but martens are similar to their smaller relatives the weasels in that the local population density from year to year is strongly influenced by the success of the previous breeding season, which can be deduced from the age structure of the catch. They knew the age structure of the captured martens and, therefore, could calculate backward to estimate the minimum number of martens that must have been alive at any previous time. Fur harvest managers have used this technique successfully to regulate the marten harvest at close to the maximum sustainable, around 35% pelts harvested each year. The same technique could be applied to estimating the abundance of weasel populations, although no one has tried it, so far as we know.

From Footprint Tracking

Snow provides opportunities to collect data that cannot be collected during warm weather. Weasels travel across the snow with a half bound, the front paws landing nearly simultaneously but with one slightly in front of the other. The hind paws land in the exact prints of the forepaws. Prints of the paired paws may be as close together as 10 to 12 cm for a least weasel, but can reach over a meter apart when a large weasel runs down hill. The tracks of weasels and all their close relatives (ferrets and polecats, minks, martens, wolverines) are the same except for size, and are unmistakable once learned (see Figure 6.1).

The ancient art of reading snow tracks has been developed into a standardized density index, first by Russian game biologists and fur trappers (Aspisov & Popov 1940; Teplov 1952), and more recently by the Scandinavian school of ecologists. For example, on a 28-km² study area at Alajoki, western Finland, E. Korpimaki and K. Norrdahl defined a series of fixed transects across farmland and forest, and checked them systematically for weasel tracks after every snowfall (Korpimäki & Norrdahl 1989a; Korpimäki et al. 1991).

To estimate the density of least weasels, the transects were 60 m apart, equivalent to the mean home range width of both male and female least weasels according to Nyholm (1959b). They measured the footprints to distinguish the sex of each individual (prints averaged 4.6 cm for males and 3.1 cm for females; stride lengths 56.3 cm and 30.2 cm, respectively [Nyholm 1959b]). Only a few females were recorded (the pooled ratio of females to males over the four winters 1984–1987 was 6:30), but that was not surprising since females spend much more time in under-snow burrows (Chapter 2). They assumed a 1:1 sex ratio, and calculated density indices by doubling the number of males, but without correcting for the density of voles. They came up with figures ranging from 2.4 to 13.0 least weasels per km², depending on the density of the vole population (Korpimäki & Norrdahl 1989a:209).

Variations on this method are still widely used in Finland (Aunapuu & Oksanen 2003), but there are three potential problems. The first is that weasels commonly dive down into or pop straight up through the snow blanket. Although it might look reasonable to assume that a track coming up from under the snow was left by the same weasel that went down nearby, that is not always true. Home ranges of males and females overlap, so it is easy to confuse individuals.

The second is that snow conditions affect both the numbers of tracks recorded and apparent sizes of individual footprints. Jędrzejwski et al.(1995) showed that the depth of the snow explained nearly 40% of the variation in numbers of tracks of common weasels counted per km per day, because when snow was more than 40 cm deep the weasels spent most of their time in the subnivean spaces. And, after long experience of tracking mustelids in the snow, R.A. Powell (unpubl.) has records showing that different snow conditions can change the apparent track size observed for the same individual on the same day.

The third problem is that the paw measurements of the two sexes of all mustelids we know overlap to some degree, even for those species with large sexual dimorphism. These three problems combine to make counting individuals and identifying their sex from tracks in the snow very imprecise.

In countries and seasons where snow does not provide a ready-made tracking medium in which to record footprints, it is necessary to provide one. Fortunately, the perpetual curiosity of weasels makes it possible to design a simple and effective method. Artificial tunnels containing tracking papers and ink pads or sooted surfaces (as described in Chapter 8) are set out in grids or lines, and have two main uses. When tracking tunnels were used in remote places as a simple presence/absence survey tool, a durable ink (King & Edgar 1977) was advantageous because it extended the possible period between checks. Alternatively, if routine population monitoring is required, a rough density index can be calculated from the proportion of tunnels recording weasel tracks over a given short period.

Population indices are more accurate if the papers are changed frequently. The longer the period is between checks, the more tracks can appear on the paper, to the point where identification sometimes become difficult. Moreover, if more than one weasel track appears on the paper, there is no way to tell whether they were made by one weasel visiting more than once or more than one weasel. These problems can be reduced by checking the tunnels more often, or reducing the sample period, especially as short-lived, simple dyes such as food coloring can then be used for the "ink."

Tracking tunnels are especially useful because they offer a cheap and easy way to answer one of the most important and difficult questions about any species of weasel: Where are they? Or conversely, are there any here? Students planning to begin a radio-tracking study of weasels, or conservation managers needing to protect especially valued breeding birds from nest predation, need to know where weasels are and how many of them are there.

Tracking tunnels are a simpler method of finding this out than are traps, especially if the weasels are not to be removed when found. Tracking tunnels do not interfere with weasels' normal movements and can be set in arrays of hundreds at a time, and the prints are usually clear and easily measured. The question is, are they reliable enough to be a valuable tool for research and management?

The answer seems to be, as so often, yes and no. There is a correlation between tracking rates and weasel density, but it is not linear. In Britain, Graham (2002) used unbaited tracking tunnels, calibrated against live trapping, to show that the correlation between tracking rates and number of weasels live trapped was close but varied with vole density and with season. The same number of tunnels with common weasel footprints represented a higher density of weasels when voles were abundant than when they were scarce, and more in summer and autumn than in winter and spring. So long as these complications are taken into account, Graham (2002) concluded, tunnel tracking is a reliable means of assessing the abundance of weasels.

In New Zealand, tracking tunnels are now widely used, usually baited (Clapperton et al. 1999), to index population densities of rodents and stoats (Brown et al. 1996; Brown & Miller 1998). Tracking tunnels can document population changes, for example, the reduction in numbers of stoats after a control operation (Murphy et al. 1998, 1999), but less reliably in autumn and spring than at other times of year. When the immigration rate is high, such as in the autumn after a good season for producing young, the residents removed might be replaced too quickly for tracking indices to detect the difference (Dilks & Lawrence 2000).

The ultimate challenge in weasel spotting is to detect the arrival of a single colonizing individual in a large area devoid of them, for example, after a lemming crash in the far north. There, one can exploit a simple index of mustelid activity, such as snow tracking or the spring distribution of lemming nests raided by stoats over winter, to detect as few as two stoats in 1,000 ha (Sittler 1995). Such options, however, are open only to a few. For the rest of us, tracking tunnels are the next best thing.

To see why, consider the model constructed by Choquenot at al. (2001). This model predicted that about 350 tunnels would be needed to detect the presence of a single stoat on an area of 10,000 ha (100 km^2) with 75% confidence (assuming each tunnel samples 1.5 ha, a probability of 0.7 that a stoat encountering a tunnel will enter it, and that home ranges average 50 ha with 20% overlap). If five stoats were present, only about 50 tunnels would be needed to detect at least one of them with the same confidence. Fifty live traps on 100 km^2 (one per 200 ha) would be extremely laborious to operate, assuming they would have to be inspected daily: 350 would be impossible. Conversely, in an area supporting an established population of weasels at a density of say 10 per km^2 (Table 10.2), that is, 1,000 individuals in 100 km^2, the probability of a very small array of tracking tunnels (<50) detecting at least one weasel rises to 99%.

VARIATIONS IN WEASEL DENSITY

The local distribution of weasels is closely related to that of their favorite prey. They haunt places where small rodents or rabbits may be found, such as hedgerows, stone walls, haystacks and brushland, thick forest, or old fields. They avoid places with little cover from hawks or owls for themselves or their quarry, such as ploughed fields and open-floored woodland. Their home ranges are small in places and habitats rich in food, and in these places their local density can be high (Table 10.2). For example, tracks of stoats in the snow can be found much more often in habitats rich in small mammals, such as among the cottonwoods or poplars and willows along river banks and in fields overgrown with scrub, than in open-floored conifer forests or open agricultural fields.

One of the most desirable places for a stoat to live is close to a large colony of breeding seabirds, where food is superabundant, even if only for a short period. Most such colonies are on steep cliffs or offshore islands, or they would not survive the attentions of predators for long, but New Zealand has two remaining very large colonies of burrowing Hutton's shearwaters located high above the treeline on the Kaikoura range of mountains facing the sea (Cuthbert & Sommer 2002). Stoats live in the colonies year round, and are ideal in both shape and behavior to raid the shearwater burrows both day and night, and to escape from the attentions of larger predators whenever necessary.

Surrounded by vastly more food (eggs, chicks, and adults) than they could possibly eat, these stoats can tolerate a local density (17 per km^2) that is high by mustelid standards (Table 10.2)—and even this figure is probably an underestimate. The only higher density figures we could find were for Swedish stoats living in patches of marshland thick with water voles, but Swedish stoats are much smaller than stoats in New Zealand.

Weasel populations are linked to those of their prey in time as well as in space, and their variations in density with time are by far the more dramatic of the two. As in all animals that breed seasonally, including the small mammals on which they depend, the number of weasels in a given place is highest in mid- to

Table 10.2 Some Absolute Density Estimates for Weasel Populations (1 km^2 = 100 ha)

Species	Country	Habitat and density of weasels (n/km^2)	Reference
Stoat	Holland, all year, 1960s	Coastal dunes: Prebreeding 1.6, postbreeding 3.6	van Soest & van Bree (1970)
	Southern Sweden, autumn 1974–1979	Average over rough pasture: 3–10; marshes with abundant water voles: up to 22	Erlinge (1983)
	Ontario, all year 1973–1975	Average, including arable, short pasture, forest: 6; overgrown pasture and shrubby areas: 10	Simms (1979b)
	New Zealand, 1996	Beech forest, summer after seedfall: 4–10, over winter 0–2	Alterio et al. (1999); Basse et al. (1999)
	New Zealand, summer 1997–1999	In breeding petrel colonies: 17	Cuthbert & Sommer (2002)
	New Zealand, winter 1999	On forested island: 3.1	M. Willans (unpubl.)
Longtail	Pennsylvania, January–March 1942	Scrub oak-pitch pine forest: 12–15	Glover (1942b)
	Michigan, January 1937	Farmland: 3	Allen (1938a)
	Kentucky, 1970–1975	Mixed farmland/forest: 2–18	DeVan (1982)
	Indiana, 1997–1999	Mixed farmland: 14	Gehring & Swihart (2003)
Common weasel	Poland, all year 1971–1973	Mixed farmland: 1–7	Goszczynski (1977)
	Poland, all year 1985–1992	Deciduous forest: summer 1.9–10.2; winter 0.5–2.7, depending on forest seeding	Jędrzejwski et al. (1995)
Least weasel	Finland, winter 1983–1987	Farmland and forest: 2.4–13.0 depending on vole cycle	Korpimäki & Norrdahl (1989a)

late summer, at the end of the breeding season. At this time, the young leave their mother's protection and learn to fend for themselves. In a typical year, there will be easy prey available for the young weasels at first—young rabbits and rodents, which, like themselves, are new to independent life and inexperienced in avoiding danger. As the winter approaches and food becomes scarce for everyone, the seasonal crops of young animals—rodents, rabbits, and weasels alike—diminish. By late winter the twin scythes of starvation and predation have cut down the local populations of both predators and prey to their seasonal minimum.

Superimposed upon this regular variation between seasons, there may also be variations among years. Even in a favorable habitat, the local prey species have good seasons and bad ones, so the numbers of prey available in a given season are typically unstable from year to year. The numbers of weasels follow suit. Small prey species affect small weasel species, and larger prey affect larger weasels.

Stoats in Canada

For the small stoats living in open habitats across North America, the most important prey are usually the local species of voles. Changes in the numbers of the small northern stoats are usually linked to widespread changes in abundance of voles, from great scarcity one year through enormous abundance and back again 3 to 5 years later. In high arctic Greenland, the only rodents available are collared lemmings, and there the interactions between lemmings, stoats, arctic foxes, snowy owls, and long-tailed skuas have been worked out in great detail (Gilg et al. 2003). The peaks in abundance of lemmings recur every 4 years, driven by the 1-year delay in numerical response of stoats (see Figure 7.2). In the Kluane area of the Canadian Yukon Territory, Krebs and his team (2001) documented the changes in numbers of predators through the 10 years of a typical snowshoe hare cycle. Systematic counts of snow tracks showed that weasels (mostly the tiny northern stoats, 45 to 106 g) began to increase after a 1991 peak in red-backed voles, and reached high numbers (34 tracks per 100 km per night) in 1994, the year after an irruption of *Microtus* in 1992–1993.

In southern Quebec, a 3-year live-trapping study attempted to document this relationship, on two 90-ha areas of farmland about a kilometer apart. Raymond and Bergeron (1982) estimated the densities of meadow voles and other small rodents, and caught, marked, and released 94 stoats in the years 1978–1980. In each area, changes in the numbers of voles were reflected in the numbers of resident stoats caught in autumn, particularly in the numbers of young. Most of the females had at least attempted to raise a litter every year, because 18 of 21 females showed signs of lactation in early summer. The close link between vole density and the number of young caught in autumn, therefore, suggests that litters did not survive well unless voles were abundant. Other experiments by the same team have shown that the profitability of meadow voles to a hunting female stoat is higher than that of any other prey species (Raymond et al. 1990).

The long-term records of the Hudson's Bay Company fur-trading posts have often been used to ask questions about the population dynamics of northern species, a practice going back to Charles Elton in the 1920s. As mathematical tools have become more sophisticated, analyses can eliminate more of the inevitable "noise" in the data and come to more useful conclusions from historical material. A recent example is the reexamination of records from 45 fur-trading posts in eastern Canada from 1915 to 1940 by Johnson et al. (2000b). The trading posts sampled the populations of ermine (stoats) across a region of comparable habitats occupied by fluctuating species of voles. Such data inevitably are contaminated by unknowable errors, such as whether all the furs counted were the same species (the sample area was north of the range of the long-tailed weasel, but some of the pelts counted could have been least weasels) and how (not whether) pelt price influenced trapping effort. Nevertheless, comparisons of fur returns with data on the local vole populations confirmed that voles and stoat populations fluctuated together in a near-cyclic pattern that was stronger toward the north (Chapter 7).

Stoats in Eurasia

Russia

Ermine was once an important fur resource in Russia, too. Russian scientists concerned with managing the fur harvest invested years of work in the study of population variation in stoats and other mustelids (King 1975b, 1980d; Poddubnaya 1992; Polkanov 2000). They calculated the relationship between the number of ermine skins harvested each year and the fluctuations in the local populations of small rodents and water voles. The pattern was quite clear, even from the somewhat rough and ready density indices they used and the considerable number of complicating factors.

For example, over the whole Kamchatka region, the yield of stoat skins from 1937–1938 to 1963–1964 varied from 4,000 to 12,000 per year. The peak years came about every 3 to 4 years, usually lagging a little behind a peak in the numbers of small rodents. In some local districts within the region, the most productive years exceeded the worst by 15 times, even up to 50 times in others. The best years for stoats and sables (another important fur-bearing mustelid) usually coincided (Vershinin 1972).

On the wide flat floodplains of the Volga and Kama Rivers in central European Russia, the main prey for stoats are water voles, plus a variety of smaller rodents. In this region, the correlation between the numbers of water voles and stoats caught was so close that the yield of ermine pelts each winter could be forecast merely from the number of water voles collected in the previous June (Aspisov & Popov 1940). Only when a low population of water voles coincided with a peak in numbers of smaller rodents was the forecast wrong.

Switzerland

The importance of water voles in determining the population density of stoats in some parts of Europe was confirmed by the work of Claude Mermod and his students Sylvain Debrot and J.M. Weber at the University of Neuchatel in Switzerland. They live trapped stoats in two valleys in the Jura Mountains. The Brévine valley (at >1,000 m altitude) had 1,875 ha of pastures, open fields, and forest. Peat bogs in the valley bottom and 86 km of stone walls offered plenty of shelter for stoats. When Debrot's study began in the summer of 1977, the density of stoats in the surrounding area (estimated from regional hunting statistics) was high (6.8C per 100 TN), but over the following two summers the numbers of stoats in the region fell, reaching 0.7C per 100 TN by 1979.

The number of stoats live trapped on Debrot's study area dropped from more than 50 to only three over the same period (Debrot & Mermod 1983). At first, the resident stoats ranged almost everywhere except into the forest, but as numbers fell, the residents that remained became more or less restricted to the bogs. The reason for the crash was that 1975 had been an extraordinarily good year for water voles, which then disappeared over the following year and remained scarce for several more. When water voles were abundant, the Brévine stoats ate them almost exclusively, and thrived (probably reaching their highest numbers in 1976). When the water voles disappeared, the stoats followed suit. Debrot (1983) showed, from game and fur records covering the surrounding district and extending back to the 1950s, that the correlation between the numbers of stoats and water voles in that area was quite general. Each population peak of water voles (about every 5 years) was followed within a year by a peak in numbers of stoats.

Debrot also had a 616-ha second study area, the Val de Ruz, at 700 m elevation. Here, prey resources were more diverse and less variable, and the density of stoats remained fairly stable around 2.6C per 100 TN over 3 years beginning in April 1978. The stoats' diet in Val de Ruz included more wood mice and birds, and fewer water voles, than at Brévine (Debrot & Mermod 1981).

The study of the stoat population of the Val de Ruz was extended until early 1985 by Weber (1986), using the same "minimum number alive" method of counting resident stoats. The numbers of water voles were again high in 1982, but the response in the Val de Ruz stoats was far less drastic than had been observed at Brévine.

Britain

In Britain, stoats are larger than in Switzerland, and eat rabbits much more often than water voles (see Figure 5.1B). Rabbits were introduced to Britain in

Norman times, but have been common for only the last 200 years or so; stoats are native, and were among the very first postglacial colonizers (Yalden 1999). For almost all the time that stoats have lived there, Britain was largely forested. Until the beginning of deforestation by humans in the Neolithic age, the normal diets of stoats must have been very different from what we see now. Most often, they would have hunted bank voles, wood mice, and red squirrels in the forest, instead of the much more abundant rabbits and field voles of the modern open country. The historical deforestation and increases in agriculture, which have favored the small mammals of fields and hedges at the expense of the mammals and birds of the native woodlands, must also have allowed a considerable historical increase in the average density of stoats.

Many rabbit populations do not show the great ups and downs in numbers that water voles do, so the extent of the British stoats' dependence on them was unknown until 1953–1955. In those years, myxomatosis arrived and spread throughout the islands, and about 99% of all rabbits died (Sumption & Flowerdew 1985). The consequences were sensational—for the vegetation, for small rodents, and for their predators. The upheaval that shook the entire woodland community was very clear in the data of Southern (1970), who documented spectacular swings in the densities of forest rodents and the breeding success of tawny owls in Wytham Wood over the few years following the collapse of the rabbit population. No one was studying stoat populations in Britain at the time, but from another source of information we can see very clearly the impact of the arrival of myxomatosis on stoats.

The "vermin books" of well-organized game estates often contain long sets of figures showing the numbers of predators killed every year. Stoats were always among the greatest enemies of the traditional gamekeeper, for whom every kill was a source of satisfaction. Some keepers used to hang the carcasses along a strategically located fence, accumulating a somewhat macabre "gibbet"—a display intended to demonstrate to their employers the results of their hard work (see Figure 12.2). But in the years immediately after myxomatosis, stoats practically disappeared from the countryside, and from the gamekeepers' gibbets.

Vermin bag records from estates all over the country showed the same remarkable exodus. On one English estate of 9,300 ha with a particularly long record (Figure 10.2), the tallies dropped from 409 to 1,013 a year before the epizootic to 40 to 257 afterward (King 1980c). In the records of another estate in Hampshire of 1,600 ha (Anon. 1960), only 13 to 58 stoats were killed each year from 1954 to 1960, compared with 136 to 302 each year from 1947 to 1953.

Over the whole of Britain, stoats remained scarce until the 1960s, when rabbit populations began to recover, followed by stoat populations. For a decade or so, the bag records for rabbits and the numbers of stoats killed each year by gamekeepers increased together. From the middle 1970s, the national toll of stoats decreased again (Figure 10.2, inset), even though the numbers of rabbits

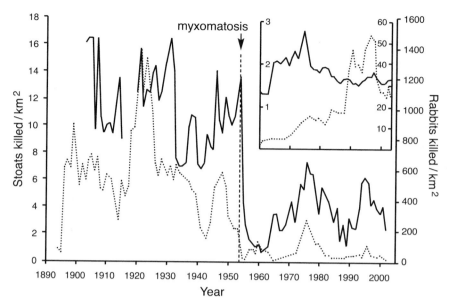

Figure 10.2 The vermin bag records of an English game estate illustrate well the close relationship between the numbers of stoats and rabbits killed on agricultural land under long-term, consistent management, and especially the effect of myxomatosis in 1953. Inset: In Britain generally, the recovery of rabbits accelerated after 1990, but gamekeepers' trapping effort has decreased. The average relationship between stoats and rabbits killed on estates throughout Britain for 1960 to 2004 does not show the close relationship typical of local data. Stoats have solid line; rabbits have dotted line. (Data from the Game Conservancy, courtesy of S. Tapper.)

continued to increase—and after 1990, to soar (Tapper 1999). This time, the decrease in stoat kills has been less to do with stoat population densities than with changes in gamekeeping strategy and reduced trapping effort (McDonald & Harris 1999).

The Netherlands

Drastic reductions in numbers of rabbits have affected stoat populations in other countries, too. For example, stoats decreased and common weasels increased after the arrival of myxomatosis in an area of the coastal dunelands of the Netherlands in 1954, just as in Britain. The difference was that the changes were slower and more gradual on the Dutch dunelands, and lasted only for about 10 years rather than 15 to 20. Mulder (1990) suggested that stoats survived longer in the dunelands because at that time there were no foxes there to compete with stoats for access to the remaining rabbits.

Sweden

In Sweden, stoats are smaller and eat rabbits less often than do stoats in Britain, yet the numbers of stoats still fell after a decrease in rabbit numbers (Erlinge 1983). In this case, the decrease in stoat numbers was attributed to intense competition for voles from the many other generalist predators on the study area (see below).

Stoats in New Zealand

By contrast, New Zealand stoats are large and eat many of the islands' myxomatosis-free rabbits. The effect in New Zealand of successful rabbit control, including the illegal release of a different rabbit disease, rabbit calicivirus disease (RCD) (Jarvis 1999), on the local populations of stoats is similar to that of myxomatosis elsewhere, but on a smaller scale (Marshall 1963; Alterio & Moller 1997b; Norbury et al. 1998).

DEMOGRAPHIC CONTROL OF DENSITY OF STOATS

The general conclusion that the population density of stoats is controlled by variations in the density of their prey is inescapable. Exactly how the ups and downs of vole numbers induced matching effects in stoat numbers was, however, unknown until recently. Popov (1943) reported an early clue, which went unnoticed for decades. In 1937–1938, when stoats were numerous in the Tatar Republic (in the former USSR), the proportion of young in the fur trapper's catch rose to 65%, but when stoats were scarce (1939–1940), it dropped to 19%. The reasons for this shift in age structure remained unexplored at the time.

Since then, long-term attempts to answer that question have been made in two very different environments, using two quite different methods. In Sweden, Erlinge (1983) used live trapping and radio tracking over 6 years to follow the fortunes of the members of an undisturbed population of stoats living on a large area of pastures and marshes. In New Zealand, regular kill trapping was used over 8 years to sample stoats living in three simple forest communities (King 1983b; Powell & King 1997). Radio tracking in these and other study areas added further information later (Murphy & Dowding 1995; Dowding & Elliott 2003; Purdey et al. 2004).

The habitats occupied by the stoats observed, their average body sizes, and the prey resources available to them were all wildly different in the two countries. The most significant contrast, though, is between kill trapping and live trapping as methods of counting the numbers of stoats present. These differences in methodology are also an advantage, because they produce complementary kinds of information. Some data (e.g., on the home ranges and behavior of individuals) can be gained only by watching undisturbed, live animals. Equally important data (e.g., on the age structure, fecundity, and pregnancy rates of populations) can be obtained only from systematic examination of large samples of carcasses. The impressive thing is that the results support each other, and can be integrated to provide a fascinating general view of the way stoats adapt their lives to the resources at hand.

Stoats in Southern Sweden

In the area where Sam Erlinge and his students (Erlinge et al. 1983, 1984) had observed stoats for years, the field voles showed a fairly predictable seasonal variation in numbers, but little variation across years (Chapter 7). Field voles, wood mice, water voles, and rabbits were the main prey available in the 40-km² study area of open fields with interspersed woodlots and marshes. The predators hunting them were stoats and common weasels, foxes, feral cats, badgers, polecats, common buzzards, tawny owls, long-eared owls, and kestrels. Through a gargantuan effort of teamwork, Erlinge et al. (1983, 1984) documented the numbers and annual breeding success of most of these predators. They also estimated the densities of prey available to the predators and, roughly, how many of each prey were taken by which predators (see Table 7.4).

The part of this census work concerned with stoats was done by live trapping in March and April (to estimate the numbers of adults present before each breeding season) and from August to October (to estimate the numbers of young produced). Over the 6 years of the study, 142 individual stoats (75 males, 67 females) were marked and released (Erlinge 1981, 1983). Snow tracking in winter confirmed that the trapping data did reflect the real distributions and numbers of the stoats, even though not all individuals were caught. Erlinge estimated the numbers of stoats present in various ways (not all of which agreed), but the general pattern was very clear. The density of stoats was higher at the beginning of the study than at the end (Figure 10.3). The figures suggested that about 45 to 50 individuals were present in the autumns of 1974 and 1975, and about 35 in 1976. After 1977, the data became too few to calculate actual numbers but, by autumn 1978, there were probably fewer than half the number of individuals that had been present in 1974–1975.

Two important prey were also censused twice a year throughout the study. The numbers of field voles drifted slowly upward, in almost direct opposition to those for stoats. The numbers of rabbits decreased sharply, along with those for stoats. Rabbits were counted because they are a key resource for the larger, generalist predators living in the same area. Rabbits were not often eaten by the stoats, which are rather small in Sweden (though not as small as in Canada) and are rodent specialists. The third most important prey, the water vole, apparently remained fairly stable in population size throughout the study.

Erlinge (1983) calculated the breeding success of the stoat population from his live-trapping records. He could not estimate fecundity, but that did not really

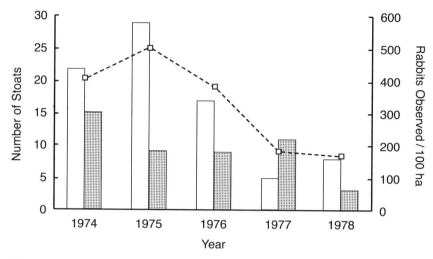

Figure 10.3 The total number of stoats present in Erlinge's study area each year was closely linked with the breeding success of the previous summer, as reflected in the proportion of young in the autumn population, which, in turn, is linked to the abundance of prey. When the rabbit population (*open squares and dashed line*) decreased, the number of young stoats 3 to 5 months old (*white bars*) decreased immediately. The number of adult stoats over 15 months old (*grey bars*) decreased a year later. (Redrawn from Erlinge 1983.)

matter. The corpora lutea counts that were available showed that potential litter size in Swedish stoats is close to the average elsewhere (see Table 9.2), and it is probably safe to assume that mean fecundity is as constant from year to year in southern Sweden as anywhere else. So the start of each year's reproduction saga is almost certainly the same in Sweden as it is everywhere.

The following stages are presumably also the same, because by the time the young stoats appeared in the traps, their numbers had been drastically cut down compared with what was theoretically possible. The vital difference was that there was *no* positive relationship between the numbers of field voles and the production of young stoats each season. In 4 of the 5 years observed, the stoats produced more young in seasons when the field voles started breeding early and fewer when they started late; in the fifth year the voles started early but the stoats had their worst season. If these data are representative, the surprising conclusion must be that field voles in southern Sweden did not influence the breeding success of stoats in those years.

On the other hand, Erlinge did find a relationship between rodent numbers and breeding success, demonstrated in space rather than in time and involving water voles rather than field voles. The isolated water meadows and marshes scattered through the study area, somewhat like islands in a grassy sea, harbored more voles of both kinds than did the open pastures surrounding them. Moreover, the marshes were not all the same. In some, only field voles were common, while others supported water voles too, which were sometimes even more abundant than field voles. Both kinds of voles are among the stoat's favorite prey, but water voles provide larger packets of food for the effort of making a kill. So, given a choice between them, a stoat would probably always prefer a water vole.

Erlinge found that the breeding female stoats living in the marshes with abundant water voles produced more young females (27 young to 11 adults caught, mean 2.45 per adult) than those that had only field voles available (14 young to 16 adults caught, mean 0.88). The difference between these figures was not significant overall, but the trend suggests that breeding success was generally higher in the marshes with plenty of water voles. In the hedgerows on nearby farmland, breeding success was about the same as in marshes without water voles (seven young to seven adults, mean 1:1).

As in all weasels, the age structure of the population in the autumn closely reflected the success of the preceding breeding season. Erlinge's data (Figure 10.3) show that the years when stoat numbers were high were also the years with the highest proportion of young in the autumn catch. The years 1976, 1977, and 1978 were poor years for recruitment, and in fact the population was by then not replacing itself. Decline was inevitable; yet the numbers of field voles, usually regarded as the stoat's most important resource, were if anything increasing. Why didn't the rising numbers of field voles benefit the stoats, as one would expect?

The Swedish stoats lived in a diverse community including many alternative prey, but they had to share their main foods, small rodents, with many other predators. Erlinge's team concluded that the pressure exerted by the whole community of predators on the small rodents was so great throughout the winter, when the rodents were not being replaced by breeding, that by spring there was often little food left for the female stoats. Only in places with augmented supply, such as in the marshes with water voles, were a female's chances significantly improved.

Paradoxically, the general increase in the numbers of field voles through the study was not large enough at the critical times to make any difference. The main reason for this was, they suggested, that as the numbers of rabbits decreased, the other predators had to hunt more intensively for field voles. The decline in rabbits was not itself a crucial problem for the stoats, since the small Swedish female stoats seldom killed rabbits. But other predators sometimes also directly affected the stoat's breeding success (see Figure 2.8). Remains of stoats were found in the pellets of raptors, and two dens of breeding female stoats in meadows were dug up by some predator. Erlinge concluded that the increase in competition for voles, caused by the decrease in rabbits, was the reason for the poor breeding success and, in due course, the decrease in stoat numbers, despite the increase in density of field voles.

Stoats in New Zealand Beech Forests

In New Zealand, the common land mammals in the forests are few and they all belong to introduced species (Chapter 5). Stoats are usually the most common, and often the only, mammalian carnivores. Native raptors (the Australasian harrier hawk, the bush falcon and the morepork, a woodland owl) are scarce. This animal community is not "natural," but at least it is very simple. That makes it easier to see what is happening, and the sequence of events is seldom complicated by competition at any level. One can hardly imagine circumstances for stoats to live in (habitat, prey resources, competitors) more radically different from those in Sweden. Contrasts like these provide exciting opportunities to understand the population dynamics of these adaptable little predators.

King (1983b) was interested in documenting the responses of the forest rodents and mustelids to the masting cycle of the southern beech trees, especially in two mountain valleys in Fiordland National Park, in the far south of the South Island (Figure 10.4). Every 3 to 4 years the beech trees flower synchronously, and when the spent beech flowers fall, the litter-feeding invertebrates get a boost in food supplies (Alley et al. 2001). Several months later, a massive seedfall dumps tons of food onto the forest floor in autumn and early winter (March to April).

The effects are rather like those of a stone dropped in a pond. In both forests, 1976 and 1979 were masting years. Mice benefited from both the additional in-

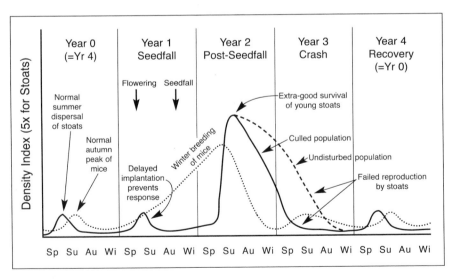

Figure 10.4 The variation from year to year in the numbers of feral house mice and of stoats in southern beech forests in New Zealand. The year in which the beech trees produced a heavy crop of seed is denoted as Year 1. (After King 2002.)

vertebrates and from the seeds. Over the few months following seedfall, the numbers of young mice caught on standard traplines soared, not only because more were born but also because more survived to enter the traps. Further, the adult female mice continued to breed well into the winter, instead of stopping in autumn as usual. By the southern spring (September), an extra large group of young adult mice was already breeding and producing the first of the summer generations. By early summer (November), the mice had reached the relatively high population densities (for the season) of up to 20 mice per 100 TN. The response of rats to the seedfall was modest during the 1970s (King & Moller 1997), but more obvious after a double seedfall in 1999 and 2000 (Dilks et al. 2003).

For the stoats living in the same forests, life was suddenly much easier than usual. Stoats benefited both from the extra mice and from the increased numbers of seed-eating birds (Murphy & Dowding 1995). For the next few months they feasted. Stoats of both sexes and all ages ate mice much more often than usual. But, more important, the increase in the supply of mice in these forests affected the stoats out of all proportion to the increase of mice in the stoats' diet (King 2002; King et al. 2003b).

In nonseed years the density index for stoats ranged from 1 to 2 C per 100 TN in summer down to zero in winter. But in the summer after a big seedfall, the density index for stoats shot up to 5 to 6 C per 100 TN. The increase in numbers of stoats caught in postseedfall summers was directly related to the density index for mice at the same time. The whole chain of events took less than a year, from the March, April, and May of the main seedfall to the following December and January when the season's crop of young stoats dispersed.

The sequence was originally worked out from a large sample of New Zealand stoat carcasses collected in the 1970s (King & Moody 1982; King 1983b; Powell & King 1997; King 2002), and confirmed in the same or similar areas many times since (Murphy & Dowding 1995; O'Donnell & Phillipson 1996; Wilson et al. 1998; King et al. 2003b; Purdey et al. 2004). The data have been used to construct three independent computer models of the relationships between stoats and rodents in this simple, feast-or-famine environment (Blackwell et al. 2001; Barlow & Choquenot 2002; Wittmer et al. unpublished).

The number of young that can be produced by any population of animals in a given season depends on four things: (1) the number of females in breeding condition; (2) their fecundity, or the mean ovulation rate per female; (3) their fertility, or the mean litter size per female; and (4) their productivity, or the number of young reared to independence over the whole local population. For stoats, the number of females in breeding condition can be taken as 100% every year, since almost all females of all ages are fertilized by the end of each breeding season (Chapter 9). We can therefore ignore the first point, but the others are all important.

Fecundity is easy to measure, by counting the corpora lutea in the ovaries or the blastocysts in fresh uteri. By contrast, fertility is very difficult to measure in stoats, because it is impossible to collect large samples of pregnant females or, until very recently, to find breeding dens to observe (see p. 255). Productivity is also very variable, but easily deduced from the age structure of the summer population. So we have the beginning and the end of the story, but there is a gap in the middle. Still, we can see enough to work out the general outline.

Environmental conditions acting on female stoats control the number of young produced each season by a simple, energy-saving, and effective mechanism. The females cannot increase their fecundity in a good year, as common weasels do. The potential number of young stoats born in any given year is already set by the number of ova shed in the previous year, and delayed implantation fixes the cycle regardless of food supplies. The female stoats have no opportunity to increase that number even in a bonanza season. The only thing that can happen is a *decrease* in the mortality of the young at all stages of their lives, from implantation of the blastocysts to independence.

When food is short, some blastocysts fail to implant, or some of the embryos are resorbed before reaching full term, or some of the young born die as nestlings; eventually these losses, or some combination of all of them, reduce litter size to a manageable number. When food is abundant, few potential young are lost at each stage. To follow the process, we need to calculate how many young start off each year as ova, and then how many fall aside at implantation, in the uterus, at birth, in the den, and during the transition to independence.

Fecundity

Of the 451 female stoats collected from all around New Zealand in good condition during the main period of delay in implantation, December to July inclusive, the ovaries of all but two contained corpora lutea of delay. That is, all but these two had mated. The two odd ones had not merely failed to find a mate, because they were not still in heat (Chapter 9). They were, instead, the exceptions that prove how rare any kind of reproductive inefficiency is in stoats. In the rest, the number of corpora lutea, each representing one ovum released during the previous mating season, were counted by serial sectioning. We did not know at the time that it would have been easier to count the free blastocysts, which can be seen in fresh uteri.

The number of corpora lutea per female varied from 0 to 19, but the average number per sample was remarkably constant through the year, in females of all ages (Powell & King 1997), in different years, and in different areas. The local averages varied only from eight to 10, and the general average was 9.7. The number of corpora lutea in the two ovaries of one female were inversely related— when one had more than the average, the other had fewer.

In each area there were some individuals with very high counts in some years, usually in the breeding seasons when mice were abundant, and perhaps this was an effect of good eating. But this higher fecundity had *no* effect on the number of young produced by those individuals in the following season, which is controlled by food supplies at the time of implantation and onward. In some animals, fecundity and productivity are linked, but in stoats they cannot be.

Fertility

Fertility is harder to calculate. There was hardly any information on litter size calculated from embryos, and none at all on nestlings. The whole study turned up only 13 pregnant females, of the 641 collected, and these had from six to 13 embryos, average 8.8. Losses from ovulation (average 9.7) to implantation (8.8), which are common in mammals (Asdell 1964), were modest.

Far greater losses were likely between implantation and birth. In nine of 11 pregnancies where all corpora lutea could be counted, the number of fetuses was fewer than the number of corpora lutea. In addition, some embryos had died and were being resorbed; in 5 out of 12 pregnancies where at least some embryos could be weighed, at least one embryo was resorbing; in one, seven of the eight embryos were reduced to simple swellings, leaving only one normal embryo almost at full term.

If things get very bad, a female can resorb her entire litter, and then appear in the spring with no sign of having produced any young at all. The season of births is quite well synchronized by day length (see Figure 9.6), and more than 99 % of females are fertilized each season, so it is possible to predict the stage in the reproductive cycle that each female should be in at any given time. Females that are not pregnant or lactating at the expected date can be assumed to have lost their litters entirely, either by total resorption or in the den soon after birth. Some are already fertilized for the next season, months before the successful females. These were nearly always found in beech forests in the crash seasons that follow two summers after a good seedfall (Figure 10.4). In one sample taken two summers after a huge peak in numbers of mice, every single one of 28 adult females collected had lost her entire litter: The sample (total $n = 63$) included not a single young one born that season (King et al. 2003b). The mortality of embryos is not always so drastic, but the net result is that fertility is almost always lower than fecundity. Deaths among the small nestlings, which we know nothing about but which surely must happen, must add to these losses.

Productivity

The final step in the long process of producing the annual crop of young stoats is the rearing and training of the young for independence. This phase takes 3 to 4 months, and there is practically no information about it. Biologists studying other kinds of predators, especially raptors, do well at this time, since raptor nests are conspicuous and the young are easily observed. Those who choose to study stoats have, until recently, had to wait until the young are ready to show themselves. The advent of radio tracking and specially trained dogs (Theobald & Coad 2002) have made it easier to find stoat breeding dens, and occasionally the fate of a whole litter can be documented.

In midsummer, young stoats start moving about and exploring the world, including the tempting wooden tunnels that cover traps. The number of young stoats caught per 100 TN is an indirect measure of productivity, not per female but of the population sampled. In poor years, fewer than one young stoat is caught per 100 TN, or even none at all (Murphy & Dowding 1995; King et al. 2003b); in good years, the crop of young can reach more than 6C per 100 TN.

Do these differences reflect real changes in fertility from year to year and not, for example, changes in trappability among the members of a population producing a constant number of young? Yes, because the number of adults caught varies less between years than does the number of young produced (King 1981a). The good years are those when a high proportion of the fertilized ova survive to become live, independent young stoats. Since the energy demands of a lactating female stoat may increase by 200% to 300% (Müller 1970), the chances of the young surviving increase in direct proportion to the mother's chances of finding enough food and still having time to keep her young warm. Those chances are highest when mice are abundant (King 2002).

Conversely, in the bad years, about the same number of potential offspring on average start as ova but few get all the way to independence. Because many females get only one chance to breed, and never "know" whether conditions in the following spring will be good or bad, it makes sense always to start with a large number of fertilized ova and defer adjusting that number for as long as possible. If, by the spring, things look unpromising, a female may be better off saving energy and producing fewer young, rather than taking a chance and losing all of them. Sometimes the chances of raising a few offspring to independence look good to start with but then rapidly deteriorate; then, most of the young will die in the den.

For example, one female radio collared by Dowding and Elliott (2003) on October 9, 2000 was heavily pregnant, and by October 11 had given birth in a rabbit burrow. By October 20 she had moved the kits to a new den under a sheet of roofing iron, and was visited by a male whose home range overlapped with hers, amid much squealing. On October 24, while she was away, Dowding and Elliott lifted the roofing iron and found 10 live kits. On October 26 there were eight live kits and one dead; on October 28, there were six live and one dead; on October 31, there were no kits in the den but possibly one was seen outside with the female on November 1. Rabbits were relatively scarce at the time, and it seems likely that when hunting is bad, the reduction of a litter from 10 to six in a week is not at all unusual. The same process applies to other mustelids with delayed implantation, such as martens; when food is short, young females do not mate, and older ones mate but fail to produce young (Thompson & Colgan 1987).

The Link Between Reproductive Success and Food Supplies

The productivity of stoats is generally held in check by shortage of food, especially protein, for the very young (White 1993). Food resources for stoats in New Zealand beech forests are normally sparse, and determine the level of pre-independence mortality through the breeding cycle. Starting from a constant average of 8–10 blastocysts, the potential young die off in a graded sequence until mortality has matched the number of offspring to food supplies, even reducing them to zero in mouse crash years. Conversely, when mice are abundant, the sudden glut of protein leads to a rapid reduction in mortality of the dependent young stoats. What turns this regulatory mortality on and off?

Juvenile mortality clearly has something to do with the condition of the females, but exactly what the link is, we do not yet know. Perhaps female stoats in forests feeding mainly on rats or mice (or, in nonforest habitats, rabbits) get some kind of stimulus not received by females living on other foods, which in turn reduces intrauterine mortality. Could it be that the body of a whole mouse (skin, bones, guts, and all) contains essential elements not found in the meat taken from larger carcasses? Could it be that a female stoat gets some kind of behavioral cue from the excitement of making frequent kills or of carrying and handling a whole prey animal?

Some results from a captive breeding program for stoats run by DonCarlos et al. (1986) at Minnesota Zoo provide support for the first of these ideas. The female stoats at the zoo were at first fed canned food manufactured specifically to provide captive, wild felids with appropriate nutrition for reproduction, but none of the female stoats produced young. The next year, one female stoat was transferred to a diet of laboratory mice, and she produced young while another one kept on the old diet failed. Later, two more females put on the mouse diet gave no obvious sign of improved condition, but they also produced young. Sundell (2003) observed a similar effect in captive least weasels (Chapter 9).

One possible cue indicating conditions favoring good breeding success could be a reproductive hormone in the diet. If stoats or weasels eating live rodents in breeding condition could absorb rodent gonadotropins in viable form, their own reproductive processes could get an unusual boost that they cannot get from eating nonbreeding rodents or other meat. A similar theory was proposed for red foxes by Lindström (1988), but it has never been tested. According to Rodney Mead (personal communication), one of the most experienced researchers on mustelid reproductive physiology, this mechanism is impossible.

Another candidate for the trigger is lipids from mouse brains, which weasels favor highly. Brain lipids are a highly concentrated source of the nutrition needed for reproduction. Or, perhaps some other mystery ingredient derived from mice is involved. Either way, we remain intrigued, because to us the "mystery ingredient" idea is at least as, or more than, consistent with the data as the other hypotheses explaining higher productivity in good years: improved overall nutritional condition, reduced social stress, or a change in foraging strategy, which allows more time for warming the kits (King et al. 2003b).

If mice are the critical indicator signaling high probability of breeding success for beech forest stoats, then stoat productivity should respond to changes in the abundance of mice—especially if rats are also abundant at the same time. An index of stoat productivity is the ratio of young to adults caught in summer and, indeed, King et al. (2003b) showed that stoat productivity does, indeed, increase with the density index for mice (Figure 10.5). Not only that, the maximum ratios of young-to-adult females predicted by this model are in the low 20s, modestly larger than the maximum recorded ovulation rate for stoats, 20 (see Table 9.2), suggesting that stoats may reach maximum productivity at relatively moderate values of the mouse density index (around 20 C per 100TN). Such high productivity is possible when mice are abundant, because a high population of mice offers an ideal prey resource for nursing stoats: Mice are easy to find, easy and safe to kill, light to carry back to the den, and rapidly replaced. On the other hand, when mice are scarce, home ranges have to be large (Murphy & Dowding 1995), hunting takes much effort and time, and substitute foods such as invertebrates and carrion are insufficient to meet the combined energy demands of hunting over extended areas plus pregnancy and lactation.

Stoats are not native to New Zealand, but they have provided tremendous opportunities to work out these relationships, and the conclusions help us understand the biology of stoats elsewhere.

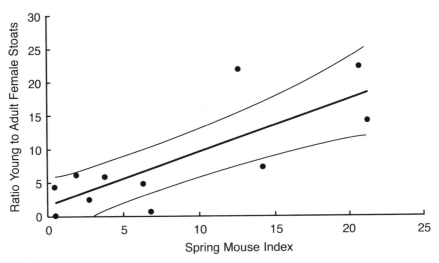

Figure 10.5 The direct relationship between the number of mice and the productivity of stoats (young caught per adult female) in New Zealand beech forests in summer suggests that stoats can reach their maximum reproductive output at mouse density indices of around 20 mice per 100 trap nights. The dark line shows the linear regression for the relationship, and the light lines show the 95% confidence intervals. (Redrawn from King et al. 2003b.)

DENSITY VARIATION IN COMMON WEASELS

Game records show enormous variations in the numbers of common weasels killed each year on English estates. Common weasels are capable of quite startling irruptions, up to a fivefold increase in numbers over the previous year—as illustrated for one estate in Figure 10.6. The best known of these irruptions was clearly related to the huge increase in food supplies for common weasels that followed the arrival of myxomatosis. When the rabbits had gone, grass and herbs previously nibbled constantly short flourished with unprecedented vigor (Sumption & Flowerdew 1985). The densities of small rodents reached record levels, followed by the common weasels. The contrast with the effect on stoats (Figure 10.2, from the same estate as Figure 10.6) is remarkable.

Since myxomatosis, lesser variations in the population densities of common weasels have become clearer. These are also linked to the availability of small rodents, especially voles. Early proof of the connection came from North Farm, the English game estate where Tapper (1979) worked. The numbers of common weasels caught followed the numbers of voles, delayed by a few months because voles and weasels were not counted at the same time of year (Figure 10.7). The same pattern was probably repeated all over England. For example, the records plotted in Figure 10.6 show an extra-large crop every third or fourth

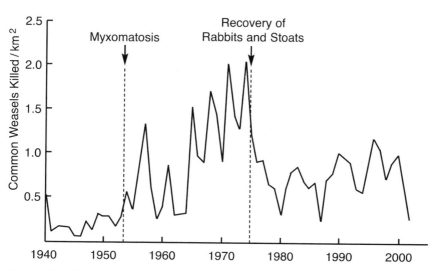

Figure 10.6 The numbers of common weasels killed by gamekeepers increased suddenly after the arrival of myxomatosis on one game estate in England in 1953, and on average remained much higher than previously until the mid-1970s, when rabbit and stoat populations began to recover. (Data from the Game Conservancy, courtesy of S. Tapper.)

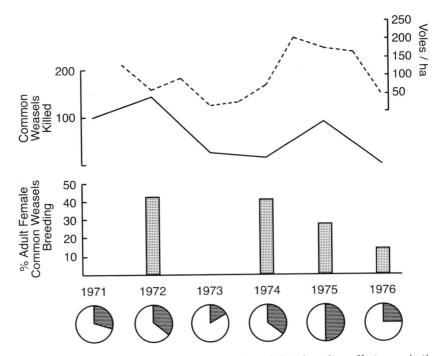

Figure 10.7 The numbers of common weasels and their breeding effort were both closely related to the supply of field voles in Tapper's study area in England. The numbers of common weasels killed by game keepers has a solid line; vole density dashed line; percent voles in diet of weasels black in pie charts; percent adult female common weasels breeding bar graph. (Redrawn from Tapper 1979 and Anon. 1981.)

year (up to twice the number caught the previous year) throughout most of the 1960s and 1970s, suggesting that those were the years when field voles were abundant. More recent data from Białowieża Forest (described below) document similar relationships between common weasels and woodland rodents.

These variations in the densities of common weasels come about because shortage of food has the same controlling effect on their productivity as it has on stoats, although their reproductive physiologies are very different.

Breeding is a hugely expensive undertaking for the small weasels. The average British female common weasel, weighing about 60 g, has to find nearly 1 g of suitable food per hour, or 22 g per day every day, all the year round (Hayward 1983). Her needs increase only modestly (6% to 7%) during gestation, but soar to 80% to 100% extra during lactation, to about 70% of her own weight daily. In cold climates, and when the young are nearing weaning, the additional demand is even higher, reaching up to five to six times her own needs, or up to twice her own weight daily.

Because reproduction is so very energy intensive, there is a minimum density of available prey below which raising a litter is simply not possible. For common weasels, this minimum appears to be about 10 to 15 voles per ha, preferably *Microtus* (Henttonen et al. 1987). More precisely, common weasels need access to at least five reproducing female voles per hectare (Erlinge 1974; Tapper 1979; Delattre 1984).

If the main prey are wood mice, weasels seem to need even more prey to support breeding. For example, in an area of farmland in France where Delattre (1984) was trapping, the combined density of all small rodents was very low, and many of those that were there were yellow-necked and wood mice (see Table 7.5). In the first year, while voles were still in the majority, the common weasels in the area bred well. But in the second and third years, mice made up 80% of the rodents present and the weasels decreased. Delattre suggested that common weasels are so closely dependent on voles that they cannot breed or maintain their populations on mice alone, even if mice are relatively abundant. Later observations showing how vulnerable mice are to being caught within their burrows (Jędrzejwski et al. 1992) casts some doubt on that interpretation, but the general point is valid.

In vole crash years, the breeding success of adult female common weasels is poor, and the few young females that are born do not breed themselves until the following year. For example, in Tapper's study area in Sussex (Figure 10.7), voles comprised between a quarter and a half of the weasels' diet in 1972 and 1974, and at least half the female weasels caught were breeding. In 1973 and 1976, the weasels had to turn to other foods, and few or none managed to produce young.

The total failure of an entire year's reproduction in a small, short-lived animal such as the common weasel has a serious effect on population density. When it happens on more than a very small scale, the result can be local extinction of that population. For example, in Białowieża forest in Poland, in April and May of 1991, the density of female voles was 0 to 1.8 per ha. This density is well below the minimum needed to support breeding by the common weasels (Jędrzejwski et al. 1995). Local extinction of weasels by spring 1992 was inevitable, since almost all the adult weasels present at that time were born in the previous year, and few weasels survive into their third year of life.

On the other hand, common weasels can respond to increases in voles much more rapidly than can stoats. Some idea of the speed of this reaction, and of the following decrease, can be gained from another study done by Delattre (1983) in the foothills on the French side of the Jura Mountains. At 600 to 1,000 m elevation near Levier, field voles staged a population irruption in the summers of 1979 and 1980. Delattre reckoned that in August 1980 the density of field voles was 100 to 200 per ha. The inevitable crash, over the winter of 1980–1981, was so complete that by May 1981 the density of field voles was down to less than a single vole per hectare.

The period of plenty was short, but while it was on, the response of the common weasels was spectacular. Delattre caught none on his 100–ha study area in August 1979, and only two outside it. At the end of 1979, weasels moved in and increased in numbers throughout 1980. In that season, two cohorts of young were produced; the first appeared from May to August, and the second from September to November. Some of the females were still suckling in October 1980. By May of 1981, 19 individuals were living in the area. Two months later, only one was left.

In a bumper season for small rodents, therefore, common weasels increase their reproductive output as do stoats, but by a different and far more effective means (Table 10.3). Potential litter size in common weasels in Britain is generally rather less than in stoats (see Table 9.2). But in good seasons, the best that the stoat can do is to decrease juvenile mortality, whereas the common weasel not only does that but *also* has the option to increase fertility. Increased fertility can produce far more rapid population increases in common weasels, even though their litters are smaller on average than those of stoats (McDonald & Harris 2002). The difference is due to the key part played by young females.

Young common weasels of both sexes can breed in the season of their birth, and delayed implantation does not restrict them to only one litter a year. In a vole peak year, the adults present in spring produce their first litters in about May. Abundant food eases the effort of providing for the litter, and fewer young are lost than usual. In August or September, each mother still alive can produce a second litter, which means that she has doubled her fertility for the season—something a stoat cannot do. This doubled fertility is not in itself, however,

Table 10.3 Effects of Delayed Implantation on the Response by Populations of Stoats and Longtails Versus Common/Least Weasels to Variation in Food Supplies

	Stoat/longtail	Common/least weasel
Delayed implantation?	Yes	No
Fecundity	High (6–20)	Low (4–8)
Earliest possible age of female at first littering	12 months	3–4 months
Life span	1–8 years	1–3 years
Response in good years	Much of the high potential fecundity realized in a single large litter; increase in fecundity impossible	Additional summer litters produced (second litter in adults, first in early-born young females)
Response in bad years	Increase in prenatal and nestling mortality	Decrease in fertility of adults, no summer litters
References	(King 1981a; King & Moody 1982; King 1983b; McDonald & Harris 2002)	(Tapper 1979; King 1980c; Jędrzejwski et al. 1995; McDonald & Harris 2002)

enough to permit the far greater numerical response of common weasels compared with stoats.

The really important difference is that by midsummer the early-born young female common weasels have several times the numbers and only half the mortality rate of the adult females. Yet these young females are fully mature and capable of producing young if well-enough fed. So, the breeding stock in late summer contains many more females 3 to 4 months old than 15 or more months old.

By the time of the autumn peak in weasel numbers in a year when voles are abundant, the majority of the populations of both stoats and common weasels are young animals. In common weasels the great majority of these young animals were produced in midsummer by the early-born young females of the same breeding season. By contrast, all young stoats are born in spring.

Under ideal conditions, a single adult female common weasel in spring can have 30 descendants by autumn, if she bears two litters of six herself (with equal sex ratio) and the three early-born females produce six each: $(2\times6) + (3\times6) =$ 30. This astonishing reproductive capacity is more than sufficient to account for the sudden irruptions of common weasels reported during vole peak years (King 1980c; McDonald & Harris 2002). In fact, during one of the few irruptions actually measured, after a heavy seedfall in Białowieża forest in Poland, the density of common weasels increased fivefold in less than 3 months (from 19 to 102 weasels per 10 km^2) (Jędrzejwski et al. 1995:189).

The Białowieża study documented in great detail for common weasels the same three-stage seedfall–rodent–mustelid interaction as the one that produces massive variations in the densities of stoats in New Zealand southern beech forests. The stone-in-a-pond effect is strikingly similar: Both species of weasels adjust their reproductive effort to the spring density of rodents. The opposite reproductive physiologies of common weasels and stoats, however, make the mechanisms of their responses quite different (Table 10.3).

The Polish study also settled an old argument about whether the weasels' response to an increase in voles is delayed, as suggested by Tapper (1979) from gamekeepers' records (mainly from the spring trapping season), or immediate. The Polish team's intensive year-round monitoring of weasel numbers demonstrated that the effect is immediate. After a superabundant seed crop of oak, hornbeam, and maple in Białowieża forest during the autumn of 1989, voles (*Clethrionomys glareolus*) and mice (*Apodemus flavicollis*) bred throughout the winter of 1989–1990 and rapidly increased in numbers (Jędrzejwski et al. 1995). At their peak in late summer 1990, the forest rodents numbered almost 270 per ha, and supported a postbreeding population of 10 weasels per km^2.

Over the winter of 1990–1991, the rodent and weasel populations crashed together (Figure 10.8). The critical ratio of rodents to weasels required for breeding, about 400 rodents per weasel, was maintained so far as possible by expansion of individual home ranges. Beyond the physiological limit of that adjustment, however, local failure in breeding became inevitable. In spring

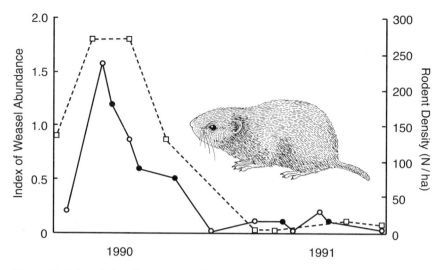

Figure 10.8 Population fluctuations of common weasels (*solid line*) and rodents (*dashed line*) in Białowieża Forest, Poland. Weasel abundance was indexed using live trapping (*open circles*) and direct observations and radiotelemetry (*solid circles*). (Redrawn from Jędrzejewski et al. 1995.)

1991 insufficient numbers of rodents were left to support weasel reproduction on even the largest possible home range. Since very few weasels survive to their third year of life, the local population of common weasels was extinct by spring 1992.

DENSITY VARIATION IN LEAST WEASELS

The least weasel is often considered rather a rare species, since it is usually less well represented in museum collections than the two larger American weasels. Nevertheless, in some years least weasels suddenly become very abundant, only to disappear again just as quickly. The earliest evidence for this erratic variation in numbers came from fur buyers' returns.

Mr. Fryklund, a fur buyer from Minnesota interested in natural history, recorded the species of weasels he handled every year. He offered a better than average price to trappers for whole weasel carcasses, and was well known in Roseau County, so local trappers brought to him all the weasels they caught. From 1895 to 1927, he handled only seven least weasels. In winter 1927–1928, he handled three. But from November 1928 to April 1929, he got 59; from August 1929 to May 1930, 84; then from 1930 to 1935, only three. In his 40 years as a fur buyer, Mr. Fryklund handled 166 least weasels, 143 of them in the 2 years 1928 to 1930 (Swanson & Fryklund 1935).

In North Dakota in 1969–1970, the same thing happened. Wildlife biologists working on the food habits of red foxes noticed that least weasels suddenly began to turn up in the fox dens they excavated. Foxes kill least weasels but seldom eat them, so most of those collected by a hunting fox are left at its den. Also, fur buyers reported more least weasels than usual in those 2 years, and several least weasels were found dead on the roads. Altogether, biologists accounted for six least weasels from November 1968 to June 1969, 54 in the same period in 1969–1970, and eight in 1970–1971 (Lokemoen & Higgins 1972). This irruption of least weasels apparently followed a population increase in meadow voles.

Likewise, no least weasels had ever been recorded in Missouri and southwestern Iowa until 1963. Small rodents were abundant in 1966–1968, and then, in 2 months of intensive trapping in early 1968, Easterla (1970) caught eight least weasels in one 6-ha field.

During these periods of high density, least weasels disperse into new areas. For example, they were unknown in Kansas until 1965, but since then have been dispersing south, presumably stimulated by periodic surges in numbers (Choate et al. 1979). In the late 1980s they reached eastern Oklahoma, some 300 miles further south (Clark & Clark 1988). On the other hand, when the local population of small rodents crashes, least weasels simply disappear. Years of scarcity outnumber years of abundance for least weasels, which explains why sudden changes in the distribution or numbers of these normally seldom-seen little animals are so remarkable.

At Barrow, on the north coast of Alaska, the population fluctuations of the brown lemming have been observed since the early 1950s. Of 66 least weasels collected over the 20 years after 1953, 48 came from lemming peak years (MacLean et al. 1974). The response of least weasels to a lemming peak is presumably the same as that of common weasels to a vole peak, but the results can be even more startling. Not only can least weasels have a postpartum estrus, and produce several litters in very quick succession (Chapter 9), but they are also capable of producing extra large litters (of up to 19 young!) in lemming years (see Table 9.2).

Population fluctuations of least weasels, in relation to the density and distribution of rodents, have also been observed in northern and eastern Eurasia (Nasimovich 1949; Rubina 1960). The estimates of two to 13 least weasels per km^2 over 4 years, calculated by Korpimaki and Nordahl (1989a) from snow tracking, may be a bit rough, but they certainly reflect the way that numbers of weasels follow changes in the numbers of small rodents in the far north.

DENSITY VARIATION IN LONG-TAILED WEASELS

Little information is available on the density and population dynamics of longtails. Simms' (1979b) density index for longtails in his study area in Ontario was much lower (averaging 0.5C per 100 TN) than that of the much smaller stoats (averaging 2.6C per 100 TN). A completely independent study in Indiana also found an overall density index for longtails of 0.5C per 100 TN (Gehring & Swihart 2004). Because longtails have a more generalist diet and a less extremely opportunistic lifestyle, their numbers may be relatively more stable than those of the smaller weasels.

On the other hand, longtails probably still respond to a glut of small rodents just as the other weasels do. Edson (1933) reported that after a heavy crop of vetch seeds in the autumn of 1931, mice were abundant during the winter. In May of 1932 he caught a longtail in a trap set for mountain beavers, followed by nine more before October. Likewise, in summer 1979 both *Clethrionomys* and longtails were common on the Apostle Islands in Lake Superior, but by the following summer, both were gone (R.A. Powell personal observation).

11 | Populations: Survival and Mortality

SIZE AND LIFE span are generally related in wild animals. Mice and voles seldom live more than a few months, whereas bears can live over 25 years and elephants 40 to 50 years or more. Weasels, as a group, are small relative to all other carnivores and, indeed, also to many other mammals. We expect, then, to find that all weasels are relatively short-lived, but that within the group, stoats and longtails will live longer than common and least weasels. The best data we have are from stoats and common weasels, and they confirm the expected pattern.

METHODS OF AGE DETERMINATION

The only method of determining the age of a weasel that can be truly accurate is continuous observation of a marked individual from the time it was recognizably juvenile. Such long-term observation is done best in captivity or in conjunction with a successful live-trapping and radiotelemetry study. For unmarked dead animals collected from the wild, the only methods that are reasonably dependable are those that have been calibrated against a set of specimens of known age.

To our knowledge, two such sets have been collected. The first, the property of Fritz Frank of West Germany, is a set of skulls of 26 male and 18 female common and least weasels and their hybrids bred under seminatural conditions in his laboratory. The other set comprises 22 skulls of stoats from New Zealand, some marked and recaptured from the wild and some held as wild-caught captives. From these two sets of material, two rather different methods of age determination have been developed. One is based on observation of changes in the shape of the skull and the baculum, and the other on the sectioning of canine teeth. Each has its usefulness and complements the other to some extent.

The date-skull-baculum (DSB) method, developed by King (1980e) from Frank's material, is simple, requires no special equipment, and is most useful for distinguishing young animals, not yet full grown, from among a set of skulls and bacula that must not be damaged (see Figures 4.1 and 9.1). The development of the postorbital constriction, allied with features of the baculum and other anatomical features known to be related to age, allow classification of wild-caught common weasels into two, or at most three, year classes. It is very reliable for the first 6 to 8 months after the young begin to appear in the population, though less so in late winter and spring. In Pascal et al.'s (1990) modified version of this method for common weasels in France, the first year class is divided into three stages, and all weasels older than 12 months are grouped into a single class. The skulls of stoats (King 1991a; Sleeman 2004) and longtails (Hamilton Jr. 1933) show the same developmental changes, although they span a longer period than in common weasels, because of the slower development of larger animals (see Table 9.3).

The disadvantage of the DSB method is that it cannot distinguish between the year classes of full-grown animals. This does not matter much for common or least weasels, but for stoats and longtails, if histological machinery is available and destructive processing is permitted, the more complicated but more reliable method of counting cementum layers in the teeth is better.

The teeth of mammals are anchored in their sockets by cementum, a hard substance that is strengthened with new layers year after year. Thin sections of the teeth can be stained to show the cementum lying around their roots. Within the cementum, the edges of the annual layers show up as dark lines in a paler field (Figure 11.1). The technique of counting the cementum lines to estimate an animal's age in years has become routine, provided that it can be proved first that the lines really are annual. For example, the equally obvious lines in the superficial bone of the jaw are also annual in some animals, but not in common weasels (King 1980e).

Young stoats are relatively easy to catch and to distinguish in summer, so King and McMillan (1982) marked and released a cohort of young known to have been born in the southern summer of 1979–1980, and retrieved them over

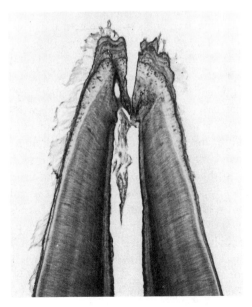

Figure 11.1 Cross section of the canine tooth of a female stoat from Denmark, killed in June at age 5 years and 1 month. Note five annual layers in the cementum. (Photograph courtesy of H. Grue.)

the next few years. By August of 1981, 22 stoats of known or part-known age had been recovered: In every case the number of lines in the cementum corresponded exactly to the number expected (Grue & King 1984). The lines are always formed in autumn and winter, so the date of death must be known before the age of a specimen can be estimated. When that information is available and large samples are classified into year classes, the reliability of many other methods of age estimation can be calibrated (King 1991a). The same method can be applied to common weasels, though for them it still lacks calibration.

The ages of living stoats and longtails can be classified in the summer months when the young are still visibly immature (males have only very small testes, and females have no visible teats), but the ages of living common and least weasels, other than kits, are impossible to estimate with any confidence.

The problems of deciding how to classify the ages of common weasels are a little different from those of classifying stoats and longtails, because of the different reproductive cycles of species with and without delayed implantation. In common weasels during a good year for rodents, the annual cohort of young can be added to the population over a long period. The earliest-born young of the year can be up to 6 months old and reproductively mature while their later-born part-siblings are still helpless nestlings. The physical distinction between young and adult blurs quickly in common weasels, and never can be defined in reproductive terms.

By the end of a productive breeding season the generations of common weasels overlap, although in a big enough sample, a bimodal frequency distribution of age classes may appear (Pascal et al. 1990). The compensation is, however, that these small weasels are short-lived. Young common weasels are usually vastly in the majority, and for most of the year can be distinguished from the smaller numbers of second-year and older adults. In many studies, division into two age classes, while not ideal, may be enough.

By contrast, the annual production of young stoats is closely synchronized by day length, so each cohort is distinct. The young of both sexes can be separated from the adults with confidence until well after the end of the breeding season, and the young males are clearly recognizable until shortly before the next season (see Figure 9.9).

On the other hand, stoats are relatively longer-lived, and for them, separation of the young animals from adults into two simple age classes by cranial features is often not enough. Tooth sectioning of the adults into year classes is usually necessary. This time-consuming and technically demanding operation is best done by someone with experience in cutting and reading the sections.[1]

The results are worth the effort, but it saves money if skull and baculum characters are used to exclude the young of the year from the list of specimens to be sectioned.

---

[1]For example, Matson's Laboratory, P.O. Box 308, Milltown MT 59851 (http://www.matsonslab.com/).

We assume that ages of longtails can also be estimated accurately via cementum annuli, but we know of no one who has done it.

FREQUENCY DISTRIBUTIONS OF AGE CLASSES

Figure 11.2 shows the results of nine studies, done in different countries and circumstances, that have calculated the frequency distribution of annual age classes in a population of stoats. In Russia, the Netherlands, Denmark, Britain, and New Zealand, the ages of stoats killed in the course of routine trapping (for fur, to safeguard protected birds, or for rabies research) were determined; in Sweden and Switzerland, the lives of marked individuals were observed by live trapping. Despite the variation in methods, study areas, habitats, and climates, their results are remarkably consistent.

All the samples agree that by far the largest age class, on average over all years in all habitats, is the first one, that containing the young of the most recent breeding season. The range of variation in the proportion of each age class reflects to some extent the stability or otherwise of the populations.

At one extreme is Debrot's sample from the Val de Ruz, Switzerland, where the density and age structure varied rather little from year to year and it is fair to pool the data from several years. The young of the year always comprised just over half the total, but ranged only from 55% to 67%; the numbers in the later age classes drop away in a regular, smooth curve to the oldest class distinguishable, those over four years old (Figure 11.2b).

At the other extreme is the sample from the southern beech forests in New Zealand, where density varied enormously from year to year (see Figure 10.4). The proportion of young of the year in two areas averaged over 6 years during the 1970s (Figure 11.2g, h) was not so different from that in the Val de Ruz, but it varied from 15% in poor years to 92% during a mouse peak year. In a third area, where the sample was taken during the crash year after a mouse/stoat irruption, the total failure of breeding reduced the proportion of young of the year to zero (Figure 11.2i).

In Erlinge's population in Sweden, the proportion of first-year animals ranged from 31% to 76% (Figure 11.2d), but the proportions of second- and third-year ones were much as in the Val de Ruz. The age structures of samples whose variation in density was unknown (those from Russia, the Netherlands, Denmark, and Britain) are hard to interpret, but give the same general picture.

Huge variations in the number of young stoats produced from one year to the next can often be traced through several following samples. For example, in the New Zealand samples shown in Figure 11.3, the largest cohort of young (almost 90% of the sample of 183 stoats) was produced in the summer of 1976–1977, during a mouse peak, and the smallest in the following crash season of 1977–1978. Normally the 1– to 2–year-old age class is much smaller than the

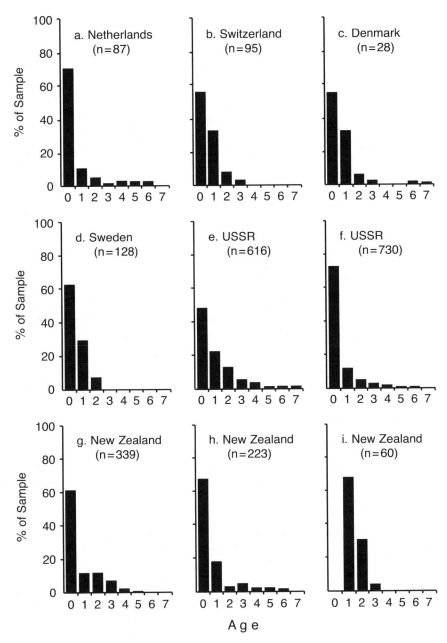

Figure 11.2 Average age structures of some stoat populations. (a) The Netherlands, males only (van Soest & van Bree 1970); (b) Switzerland, live trapped, annual mean (Debrot 1984); (c) Denmark (Jensen 1978); (d) Sweden, live trapped, autumn (Erlinge 1983); (e) USSR, winter only (Stroganov 1937); (f) USSR, winter (Kukarcev 1978); (g) New Zealand, Eglinton Valley (Powell & King 1997); (h) New Zealand, Hollyford Valley (Powell & King 1997); (i) New Zealand beech forest, mouse crash year (note absence of 0 to 1 age class) (Purdey *et al.* 2004.)

0- to 1-year-old class, but in the 1977–1978 sample the 1-year-olds born in 1976 were still almost as numerous as the young of the 1977 season. The earlier cohorts are represented by older animals only, so have already lost most of their members. If we had started trapping in 1971, a year of huge abundance of stoats throughout the country, the 1971 cohort would certainly have been the largest; the bulge of 2- to 3-year-olds is still evident in the 1973–1974 catch.

In severe crash years, when the output of young is zero, a whole annual cohort may be deleted from the population age structure. That extreme result has been documented twice, both times in southern beech forests during the season following a mouse peak. In the summer of 1991–1992, when mice had become very scarce, Murphy and Dowding (1995) caught 37 adults and no young stoats. In similar circumstances in 2000–2001, Purdey et al. (2004) collected 65 adults and no young (Figure 11.2i). In that area, the difference in age structure of samples of stoats collected during and after the mouse peak was very striking (King et al. 2003b).

If the animals were removed for sampling, and especially if the population has been regularly cropped every year, we would expect that the older stoats would disappear sooner than from an undisturbed population. For example, in three samples taken over several years in New Zealand beech forests, the mean age of stoats caught in the early years was significantly higher than in later years, declining from 19 months to 10 months in both Eglinton and Hollyford valleys, and 16 months to 10 months at Craigieburn (Powell and King, unpublished) In another, nonbeech forest, the proportion of stoats older than 12 months old declined over 5 years of sampling, from 52% of 21 collected in summer and autumn of 1983 to 27% of 22 collected in the same two seasons of 1984 and 1987 (King et al. 1996).

So why did samples from regularly culled populations in Russia, the Netherlands, and New Zealand include more older animals than in the undisturbed areas in Switzerland and Sweden (Figure 11.2)? The reason for this apparent contradiction is that kill-trapped samples can be large enough to have a high probability of including some of the rare older individuals, while both of the live-trapping studies quoted above included fewer animals and stopped before the last of the marked young had lived into its undisturbed old age. For the same reason, the effect of trapping history in removing stoats more than 2 or 3 years old from a sample is inconsequential to population density.

There is, in fact, no way to distinguish natural from trap-induced mortality from these figures, and the few older animals make rather little difference to the general age structure of the population and none to its dynamics. By contrast, the variation in the proportion of young of the year is much more significant, and is certainly controlled by food supplies rather than by trapping.

Can we use a list of ages of individuals to estimate how long weasels live? This is not an easy question to answer, because longevity is a tricky concept. It has several meanings depending on the context, and most meanings are not

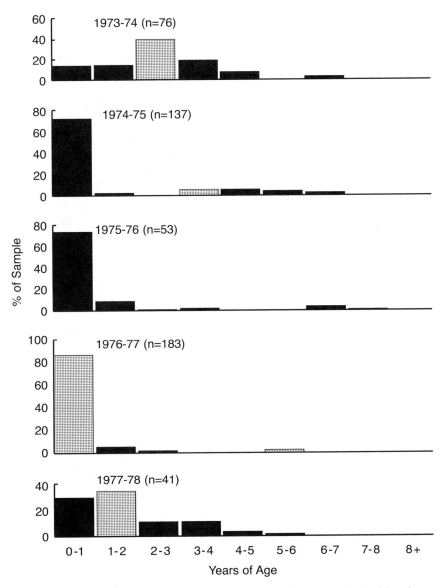

Figure 11.3 When the age distributions for stoats caught in New Zealand beech forests each year are plotted, the extra-large cohorts of young produced after a good seedfall (*gray columns*) remain distinguishable in the following year and beyond. The effect is more pronounced and lasts longer if there was no removal trapping in the seedfall year (e.g., 1971) than in a culled population (1976). (Data from Eglinton and Hollyford Valleys pooled, data from Powell & King unpublished.)

helpful when applied to wild animals such as weasels. One definition, the maximum age a member of a species can reach, is positively misleading, because on average weasels can live much longer in captivity than in the wild (DonCarlos et al. 1986).

The average or median age of a specified wild population is easily calculated, but that tells us nothing about the distribution of the key age groups (juvenile, subadult, adult) across the total range. Even a list of the mean ages of several samples with standard deviations tells us little about the consequences of that age distribution for the dynamics of the populations sampled. What we really want to know is the *probability* of a defined group of animals surviving to particular ages, not just to the oldest age. For a given group, either the total population or a subset of it, we can estimate longevity in terms of survival probabilities, and their converse, mortality rates. We can also see how the probabilities of survival and mortality change under different conditions. To calculate those probabilities, we need to construct a life table.

LIFE TABLES

Life tables were invented by the life insurance industry as a means of estimating risk—that is, how long a certain person can be expected to pay the premiums on a life insurance policy before making the claim. Animal ecologists took over the idea when they saw how useful it was as a means of making standardized comparisons within and among populations. A life table is a useful way to estimate the variation in survival rates (and, conversely, the mortality rates) among the age classes in a population. Many excellent books explain how to construct and analyze the tables, if we can get that far. The trouble with weasels is that few people have been able to get over the difficulties of collecting the valid samples needed for constructing the tables in the first place.

The easiest way to construct a life table requires that the rate of increase of the population is zero (density is not changing) and that the age structure is stable (constant proportions of ages in every generation). When these two conditions are met, a population is said to be stationary, and the standard formulae can be applied. Neither condition is ever met by weasel populations. This does not mean that life tables cannot be constructed for weasels, only that they are harder to calculate and must be interpreted cautiously.

A frequency distribution of age classes for a stationary population is proportional to the probability of living to each age. Two methods of sampling wild populations can be used to collect the age distribution data. Both methods can be used to estimate the probability that a newborn will live to a given age, but different assumptions lie behind each of the two methods.

The first is to catch a group of newborn young representing a particular cohort, mark and release them, and then watch them from a distance to see at what age each dies from natural causes. At some times of year it is not difficult to catch young weasels in live traps and to mark them with eartags. Keeping track of them afterward, however, is very difficult, especially the young males, which may disperse over great distances in their first year (Chapter 9).

The other way is to collect a large sample of dead weasels, taking care to make the traps equally available to both sexes and all ages, and then work out the age of each on the day it was killed. Catching weasels in steel kill traps is usually not difficult, but making the traps equally available to both sexes and all ages is nearly impossible, and estimating weasels' ages has been, until recently, easier said than done, especially in the adults.

Of course, both methods have problems. For example, if kill trapping is used, the complete age structure can be obtained from one or a few samples, but if removal sampling affects the longevity of the older adults in the target population, data from a previously untrapped population may be slightly different from data from regularly trapped populations. The same caution does not apply to the proportion of the young of the year in a sample, because it is controlled almost entirely by the success of the previous breeding season, not by trapping history (Figure 11.3).

Conversely, if live trapping is used, the sampling method has minimum effect on natural age structure, but the ages of the oldest adults can be known only if the study lasts long enough to follow them all to the ends of their natural lives. And one cannot tell if an animal that is not recovered has died or merely moved elsewhere.

Another serious problem for both methods arises from the great instability of all weasel populations. In habitats in which productivity and density vary greatly from year to year, the young entering the population vastly outnumber the breeding adults in good seasons but scarcely match them, or even fail to appear altogether, in poor seasons. Analyses of survival and longevity may then need to separate data from high and low productivity years.

Yet another important source of confusion can be introduced by variation in the season of the year chosen for sampling. Consider two populations, one sampled soon after the end of the breeding season and the other in winter. The first will probably show a far higher proportion of young to adults than the second, because in late summer many more newly independent young are about than if sampling is delayed until overwinter mortality has begun to take its toll. Yet the dynamics and productivity of the two populations could easily be the same.

Finally, all data on age distributions derived from trapping can sample only weasels that survive at least to trapable age; they do not account for young that die at or before birth or in the den. Yet pre-independence mortality is probably the most important factor controlling the age structure of stoats and longtails (Chapter 10). If the total number of weasels born could be known, estimates of the mean ages at death would be much lower, and of the mortality rate of the first year class much higher, than is reported in the current literature. Omission of this information is not serious provided one remembers that all such statistics apply only from the age of independence.

SURVIVAL OF STOATS

In Natural Populations: Sweden and Switzerland

The first method of estimating survival, by following cohorts of living animals, was applied by Erlinge (1983) to 6 years of data on stoats in southern Sweden (Table 11.1). In this type of table, the data are set out as a declining total: so many live animals of age x, of which so many are still alive at age y, so many at age z, and so on. Erlinge had no way to estimate whether any stoats lived for more than 6 years, or to distinguish between death and emigration as reasons for losses of marked stoats. The study area was very large (40 km²), however, with traps distributed throughout, and the stoat population was undisturbed. Consequently, the turnover of resident stoats documented by Erlinge was at least natural, even though he did not know that every stoat that disappeared had died.

The great advantage of this method is that the calculations are unaffected by changes from year to year in the age structure and rate of increase of the population. The disadvantages are that the result is a life table for only a few cohorts in given years, not the population as a whole in all years; that collecting long-term data from marked live animals of known age is a prolonged and time-consuming business; and finally, that it is extremely hard to meet the minimum sample size for a useful table, about 150 animals (Caughley 1977: 95).

Erlinge (1983) started live trapping in the autumn of 1973, and captured, marked, and recaptured 75 males a total of 232 times, and 67 females 171 times, over the years 1974–1979. Those first marked as young born in the years 1973–1976 inclusive, whose ages were known with certainty, were followed for as long as they lived on the study area. Of 47 males present in the study area in their first autumn, 28 were still present a year later, and of these, nine another year later, and so on (Table 11.1). The oldest animals were a male that stayed for 4.5 years and a female that stayed for 3.5 years. Both were born in 1975, the year of highest numbers and most successful reproduction in that population; the longevity of those born in the last 3, lower density years was unknown. The average expectation of further life in this study area for a newly independent stoat aged 3 to 4 months was 1.4 years for males and 1.1 years for females.

The mortality rate among first-year males was 40% (4 out of 10 disappeared between the ages of 0.25–0.5 years and 1.25–1.5 years), and significantly higher (68%) a year later. Male stoats live in the fast lane, so it is not surprising that one male that had been radio tracked through two breeding seasons looked "shabby and worn out" well before his second birthday (Sandell 1988).

Table 11.1 Life Table for Stoats in Sweden from Live Trapping of Marked Individuals

Age class[1]	Number alive	Proportion surviving at start of age class (lx)	Proportion dying at that age	Mortality rate qx%
Males				
3–6 months	47	1.00	0.40	40
1.25–1.5 years	28	0.60	0.41	68
2.25–2.5 years	9	0.19	0.15	78
3.25–3.5 years	2	0.04	0.02	50
4.25–4.5 years	1	0.02	0.02	100
Females				
3–6 months	48	1.00	0.54	54
1.25–1.5 years	22	0.46	0.33	73
2.25–2.5 years	6	0.13	0.11	83
3.25–3.5 years	1	0.02	0.02	100

1. All animals born in the cohorts of 1973, 1974, 1975, and 1976 are considered together. They are grouped as if they had all been born at the same time, and the survivors counted at the end of every year. (From Erlinge 1983.)

For females, the mortality rates were even higher in both their first (54%) and second (73%) years. These figures mean that about half the stoats present in autumn will disappear before spring every year, regardless of the autumn density. There will also be additional losses over the summer.

Another live population was observed in some detail at the Val de Ruz, in Switzerland. The study area was rather small, but the population was quite stable. Debrot (1984) did not calculate a conventional life table, but used a different means of arriving at a similar conclusion. He counted all young of the previous year as adults on January 1, and in the three Januaries of 1979–1981 the total number of newly recruited plus older adults averaged 6.2 on the 616-ha study area. They disappeared at an average rate of 68% a year, which means that two out of every three adults present on any given day will have gone a year later. Conversely, the annual replacement rate of adults was 93%.

Such a high turnover, Debrot admitted, reflects the large dispersal range of the adults compared with the size of the study area. The rapid replacement rate also explains why most of the marked stoats were recaptured infrequently (average 2.3 times each), and why only 7% of adults lived on the study area for more than a year. Some individuals lived (not necessarily only on the study area) to over 4 years old, but the average age of the resident animals was 14.4 months.

This is not the same statistic as expectation of life at independence, but it emphasizes the same point, that most stoats do not live long. In particular, the *average* life spans of stoats in both populations observed by live trapping were not noticeably longer than in the populations observed by removal sampling, described below.

In Culled Populations: New Zealand and Britain

Life tables have also been calculated by the second method, using dead stoats collected in the course of pest control operations in National Parks in New Zealand (Powell & King 1997) and on game estates in Britain (McDonald & Harris 1999). This method also has its problems, but if they can be overcome, a life table can be developed in a shorter time. The validity of the technique depends entirely on three prerequisites:

1. An unbiased method of sampling the population, so that weasels of all ages and both sexes are fairly represented (or if not, that the differences between these groups are at least constant). In addition, sampling effort must either be constant year round or have seasonal variations that are understood and incorporated into the calculations (McDonald & Harris 1999).
2. A reliable method of determining the ages of the dead animals.
3. Confirmation of the important assumption that the distribution of ages among the animals trapped is the same as in the living population (the fact that the animals withdrawn from the population were killed to find out their ages is irrelevant [Caughley 1977:93]).

In a life table calculated this way, the data can be shown as frequency distributions: so many dead animals of age x, so many different ones of age y, so many of age z, and so on. The difference between this type of table and the other is that this one represents the ages of the standing crop (all the members of that population alive at one time), whereas the other represents the life spans of individuals observed over several years.

Because the rate of increase in a real population of weasels is never zero in any one year, and the population age structure varies greatly from year to year, the figures calculated by this method are averages for several cohorts of several years and are only approximate for any given cohort or year. On the other hand, such life tables for weasels collected in different places are remarkably consistent, suggesting that they do represent the general pattern quite well. If the fact that they are only averages is not forgotten, they have their uses.

A preliminary attempt to construct a general life table for stoats, published in the first edition of this book, was calculated by the late Graham Caughley in 1987 by combining samples from three beech forests in New Zealand. In later analyses we (Powell & King 1997) recalculated these data by study site, allowing for the massive annual variations in fertility and age structure (see Figure 10.4; Chapter 10) by compiling two separate life tables representing cohorts of young produced in years of high versus low productivity. A typical result is shown for samples from the beech forest in the Eglinton Valley, Fiordland National Park, New Zealand (Table 11.2 A,B).

Table 11.2 Life Tables for Stoats in New Zealand from Age Determination of Carcasses

A: Eglinton Valley, southern beech forest, seedfall years with abundant mice (159 stoats)

Age class	Number of cohorts sampled	Proportion alive at start of age class (lx)	Mortality rate at that age (qx%)	Survival rate during age class (px)
3 months–1 year	1	1.00	92	0.09
1–2 years	5	0.09	62	0.38
2–3 years		0.03		

B: Eglinton Valley, southern beech forest, nonseedfall years with few mice (147 stoats)

Age class	Number of cohorts sampled	Proportion alive at start of age class (lx)	Mortality rate at that age (qx%)	Survival rate during age class (px)
3 months–1 year	3	1.00	73	0.27
1–2 years	4	0.23	40	0.60
2–3 years	2	0.14	31	0.69
3–4 years	2	0.10	71	0.29
4+ years		0.07		

C: Pureora, podocarp-hardwood forest with few mice, all years (55 stoats)

Age class	Number alive	Proportion alive alive at start of age class (lx)	Mortality rate at that age (qx%)	Survival rate during age class (px)
3 months–1 years	31	1.00	76	0.24
1–2 years	13	0.24	42	0.58
2–3 years	3	0.14	42	0.58
3–4 years	3	0.08	38	0.62
4–5 years	3	0.05	60	0.40
5–6 years	2	0.02		

Animals born in each annual cohort were sampled over several years. The distribution of the ages of the dead animals in the total sample is taken to represent the average distribution of ages of the living animals in the population. The sexes are pooled, because their age distributions did not differ in analyses controlled for seedfall. The Eglinton population had been trapped for several years preceding the dates of sampling; the Pureora population had not. Powell and King (1997) also gave similar tables for two other beech forests, including one (Hollyford) never previously trapped. For cautions, see text. (King et al. 1996; Powell & King 1997.)

The results show clearly the enormous mortality of the young stoats, especially in the high-productivity years during a mouse irruption after a seedfall. In the Eglinton Valley, 92% of young stoats born in a seedfall year died before the age of 12 months, and the whole cohort was gone within 3 years. This was not an exceptional observation; in two other beech forests the mortality rate of first-year stoats after mouse peaks was 91% and 92% (Powell & King 1997).

By contrast, in the intervening years when fewer stoats were born, first-year mortality was lower (but still averaged 73% over four cohorts)—and the remaining few older adults enjoyed relatively good survival over the following 2 years (mortality 30% to 40%). From the fourth year onward, their mortality rate started climbing steeply again. In the only nonbeech forest sampled, at Pureora in the central North Island, the density of mice and the density and population dynamics of stoats were much like Fiordland in nonseedfall years (Table 11.2).

The New Zealand data document a drastic reduction in the prospects of survival of the young stoats born during the stoat "plagues" that coincide with a mouse irruption in these forests (Figure 11.4). If that is true, it leads to the prediction that the oldest members of the population should comprise mostly individuals born in low-density years. When we tested that prediction from our data, we found that, of 70 stoats over 3 years old, only seven males and 13 females had been born in the highly productive mouse peak years (Powell & King 1997).

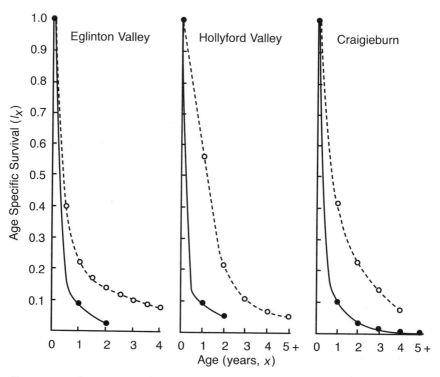

Figure 11.4 The very large numbers of young stoats born after a heavy seedfall in New Zealand beech forests have a consistently lower survival probability than those born in the years between seedfalls. Stoats born in postseedfall years have solid circles and lines; stoats born in nonseedfall years have open circles and dashed lines. (Redrawn from Powell & King 1997.)

McDonald and Harris (2002) made the first large ($n = 822$) collection of stoat carcasses from British gamekeepers. On the farmlands and moorlands they sampled, there are of course annual variations in food supply for stoats, but they seldom reach the feast-or-famine proportions typical of New Zealand beech forests. On the other hand, trapping by gamekeepers is concentrated in spring and early summer, when removal of stoats can best benefit nesting game birds (Chapter 12). Consequently, the samples were affected less by annual variation in resources than by the rather different problem of seasonal variation in trapping effort.

The data show the patterns of age-specific and annual mortality typical of stoats in general, and also, not surprisingly, that the mortality rates of both sexes and all ages of these particular stoats were much higher in spring than at other seasons (Table 11.3). The very youngest stoats just out of the den (0 to 3 months old) comprised the largest single group of stoats caught, but their mortality rate was proportionately much lower (10% to 14%) than that of the smaller group of 12- to 15-month-olds caught in the same season (83%). For both age groups, the normal spring danger period (when rodents are at their seasonal low and before young rabbits become available) was artificially made worse by the management strategy of the gamekeepers.

McDonald and Harris also used contemporary analytical tools to explore parameters not previously considered, such as which of several possible factors

Table 11.3 Variation in Seasonal Mortality Rates (qx%) through the First Year of Life of Stoats and Common Weasels on British Game Estates[1]

	Male stoats	Female stoats	Male weasels	Female weasels
1968–1972 (King 1980c)				
Summer	—	—	14	15
Autumn	—	—	16	11
Winter	—	—	21	22
Spring	—	—	65	58
First year			80	75
1996–1998 (McDonald & Harris 2002)				
Spring	10	14		
Summer	13	27	15	8
Autumn	13	30	24	30
Winter	39	40	16	16
Spring	83	83	73	100
First year	59	74	85	97

1. Ages estimated from carcasses. Median birth dates assumed to be April 1 for stoats and June 1 for weasels. Three monthly seasons for the two species starting from the median birth dates were therefore defined differently: For stoats the first month of each season was July (summer), October, January, and April; for weasels, June, September, December, and March. (McDonald & Harris 2002; King et al. 2003a).

most strongly influence population growth rate. Stoats and weasels in general have very unstable populations, early litters, and short lives (King 1983a; McDonald 2000). For them the population rate of increase varies every year, from a positive value when the population is rising, to a negative one when it is declining.

Species in which the rate of increase (r) fluctuates between high and low values tend to "live fast and die young," as McDonald put it (2000). They are capable of very rapid but temporary variations in density, and are very resistant to control. Stoats and common weasels have long been regarded as so-called r-strategist species (Chapter 14) but until McDonald and Harris' analysis, r had never been calculated for either. McDonald and Harris constructed a mathematical model that suggested that r averaged –0.05 for the British stoat population, at least during the years they sampled.

This negative figure does not mean that British stoats are constantly declining. All such population models demand some conventional but unrealistic assumptions, such as that survival is independent of density and that the population analyzed was closed. Nevertheless, even though we know those assumptions are not accurate in fine detail, the model is useful because it has shown for the first time that the single factor most critically influencing r for stoats was the survival rate of second-year females from April to June, the period when they produce and suckle their first litters.

If a mother stoat dies before her young can kill for themselves (at about 10 to 12 weeks old; see Table 9.3), it is probably safe to assume that the whole litter will be lost as well, so trapping females at that time of year has a disproportionately large effect on the population. If the model's survival rate for that group of females was adjusted by only a small amount, from 48% to 54%, r was lifted from slightly negative to an even zero, and presumably in years or in places where the survival of this group is high, r becomes positive. When those same females are also weaning large litters, because a period of great food abundance has permitted substantial reduction in juvenile mortality, a positive trend could easily turn into an irruption. Take, for example, the irruption that followed the introduction of stoats to Terschelling, off the coast of the Netherlands. The island teemed with water voles when, in 1931, about nine stoats were released on the island. According to van Wijngaarden and Mörzer Bruijns (1961), the colonizing group had multiplied to at least 180 by 1934.

Chapter 10 provides plenty of confirmation that the r statistic for stoat populations frequently does oscillate from positive to negative, producing the unstable numbers typical of stoats everywhere. Clearly there are times when the dynamic equations of local stoat populations can absorb high rates of mortality, both natural and imposed, and times when they cannot. In patchy environments there will always be a mosaic of breeding opportunities for stoats. The high dispersal capabilities of their offspring (Chapter 9) ensure that there is always a supply of young to migrate across the landscape eager to recolonize vacant patches after every local extinction (McDonald & Harris 2002).

Another model explored the dynamics and productivity of stoat populations in New Zealand beech forests (H. Wittmer, R. A. Powell, and C. M. King, unpublished), using the survival data calculated by Powell and King (1997). Wittmer et al. developed age-specific, population projection matrices for stoats living through each phase of a typical beech masting cycle. During a seedfall year, the growth rate of the model population was strongly positive, because the high numbers of mice in late winter enabled the females carrying a normal-sized litter (conceived before the seedfall) to attain such good body condition that intrauterine and nestling mortality was minimized. Therefore, very large cohorts of young were produced at the beginning of the postseedfall year (see Figure 10.4). But these young females emerged (already fertilized) just as the mice were disappearing, and few of them survived to their first littering season as 1-year-olds the next year; those that did lost most or all of their young. Hence, population growth during the postseedfall year was strongly negative. During the transition years between each crash and the next seedfall year, the population was nearly stable. Wittmer et al. linked successive population matrices together to calculate elasticities of the different vital rates at different points in the cycle. This analysis found that the fertility and the survival of 0–1 year old stoats in the post-seedfall year both had great influence on the dynamics of the population. In addition, the analyses demonstrated that the rate of population growth through the cycle differed from that predicted from the stages of the cycle analyzed in isolation. These results illustrate the critical importance of understanding how stages of a population cycle are linked together.

SURVIVAL IN LONGTAILS

Longtails can be as large as are stoats in New Zealand, so probably have similar average longevity. The only data we have found support this reasonable expectation. Linduska (1947) collected 7 years of survival records for various wild mammals on a Michigan farm. Of 73 longtails eartagged, a few were recaptured the year after tagging, but none after 2 years. Svendsen (2003) mentioned that some marked adult longtails lived in his Colorado study area for up to 3 years.

SURVIVAL IN COMMON WEASELS

Populations of common weasels are unpredictable, so no one has attempted to tabulate their age structures by following marked live individuals. The only thing that can be said with certainty is that there is generally a new set of resident common weasels on a study area each season, and that no individual has been known to hold a home range for more than 3 years (Chapter 8). This implies that most common weasels live only for a year or less.

Confirmation of this guess comes from the only three attempts made to construct a frequency distribution of annual age classes from collections of carcasses, one in Denmark and two in Britain (Figure 11.5). The huge preponderance of first-year animals in all three samples confirms our expectations. Various other studies have used age classes defined in different ways, which give results of the same general order but usually are impossible to compare in detail.

The first attempt to construct a life table for common weasels on British game estates, by King (1980c), was based on a sample of 455 dead common weasels, collected all the year round (Table 11.4). All the data were pooled, since there were no significant differences between the age distributions for the five sample areas or for the two sexes. The first-year mortality rate was very high (80% in males, 75% in females), which is not surprising for small predators that live life at such a hectic pace. The expectation of further life for a young weasel at independence was about 10 months in both sexes. The density of a population will therefore depend heavily on the production and survival of young; since this varies drastically according to food supplies, the instability of populations of common weasels needs no further explanation.

As in stoats, and for the same reason (both samples were collected from gamekeepers), there was a well-defined peak in the risk of mortality every year in spring (Table 11.3). On the other hand, spring is a stressful time for weasels anyway, especially if food is short. In Wytham, a protected population never subject to regular trapping, several residents that had been watched for months

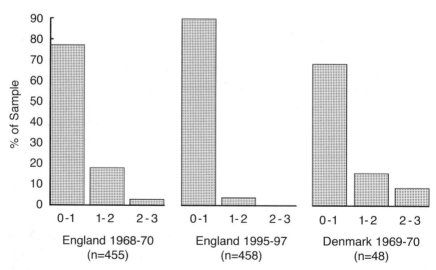

Figure 11.5 The vast majority of common weasels are under a year old, and the oldest of them reach only 3 years of age. England, 1968–1970 (King 1980c); Britain, 1995–1997 (McDonald & Harris 2002); Denmark, 1969–1970 (Jensen 1978).

Table 11.4 Life Table for Common Weasels in Britain in the 1970s and the 1990s from Age Determination of Carcasses

Age class	Number alive	Proportion surviving at start of age class (lx)	Mortality rate at that age (qx%)	Survival rate during age class (px)
1970s (King 1980c)				
Males				
0.25–1 year	339	1.00	0.80	0.20
1–2 years	69	0.20	0.90	0.10
2–3 years	8	0.02		
Females				
0.25–1 year	116	1.00	0.75	0.25
1–2 years	29	0.25	0.80	0.20
2–3 years	6	0.05		
1990s (McDonald & Harris 2002)				
Males				
0.25–1 year	344	1.00	0.85	0.15
1–2 years	50	0.15		
Females				
0.25–1 year	59	1.00	0.97	0.03
1–2 years	0	0.03		

died or disappeared in spring, often after drastic loss of weight (King 1975c), and the age distribution of weasels from Wytham was not different from those on the game estates.

When McDonald and Harris (2002) made a second collection of common weasel carcasses from British gamekeepers, numbering about the same ($n = 458$) but sampling a larger selection of estates, they found similar general patterns (e.g., first-year mortality rate 85% and 97%). But in addition, they contributed two important new details to the story. Their two-species collection strategy enabled them to compare the effects of culling on stoat and common weasel populations sampled in the same areas and by the same means, and their more advanced data analysis techniques allowed a first test of the importance of the late summer litters to the population biology of common weasels.

McDonald and Harris applied to common weasels the same population model that they had developed for stoats (with a slight variation to allow for the different reproductive cycles of the two species). They included the same necessary assumptions (survival independent of density, population closed), except that young common weasels were considered fully independent by the age of 7 weeks.

Again, the model proved extraordinarily useful despite its limitations, because it demonstrated both the similarities and the differences between these two coexisting species. In both, the single factor most critically influencing *r* was the survival of the first-year class, but the difference was that, for the British population of common weasels during the years they sampled, r averaged a *positive* figure, 0.30.

To reduce this positive figure to a negative, that is, to make a breeding population of weasels decline, the survival rate of newborn weasels in the first 3 months of their lives set in the model had to be cut from 0.88 to under 0.56, which might well mimic what happens in less favored habitats when food is very short (Chapter 9). An alternative means of inducing a decline was to reduce the probability of late summer litters (Chapter 10) from 1.0 to 0.4. Conversely, if the survival of newborns and the probability of second litters were both set at high values, as during a rodent irruption, the weasel population growth rate soared.

In reality, both these parameters are strongly affected by food supplies and are probably never the same for 2 years in a row, so the actual numbers are less important than what they tell us about the great capacity of common weasels to respond quickly when rodent numbers increase, and their equally sudden disappearance afterward. Like stoats, common weasel populations are unstable, but they can compensate for the annual harvest imposed by gamekeepers more easily than can stoats.

SEX RATIO

In almost any study of trapped weasels, males will be recorded more often than females. The average proportion of males collected in kill traps is usually around 75% in common weasels, and around 60% to 65% in stoats (King 1975a). These figures are even higher in spring and summer. Live trapping or snow tracking of undisturbed live animals may, over a period (especially in autumn), suggest that males and females are in roughly equal numbers (Chapters 8 and 10). Nonetheless, the individuals most often recaptured are always males (King et al. 2003a). The sex ratio of the young at birth averages 1:1 in all weasels, as best as we can tell, and there is no sign of any difference in the mortality rates of males and females as large as the difference in their capture rates. So what causes the excess of males captured?

One possible reason is that males have more *opportunities* to be caught. If traps are evenly spread out through a suitable habitat, the larger home ranges of the males will include more traps than will the smaller home ranges of females (King 1975a). If the distance between the traps is large, some weasels will not find a trap on their own ground at all, and these are more likely to be females than males (see Table 13.2).

Also, of course, the females of each species are much more likely than males to spend time hunting in burrows or under thick cover where there are no traps. The disparity is increased in spring and summer because many males travel widely then, energetically searching for mates, whereas females tend to be less active, and perhaps also more shy of traps, at that season.

Buskirk and Lindstedt (1989) developed mathematical models to explore these and other possible causes of sex bias in trapping mustelids. They concluded that three factors probably affect the gender balance of captures in all trapping studies: (1) the number of traps on each individual's home range and behavioral differences between the sexes in (2) how they patrol their home ranges and (3) how they approach the traps they find. The "exclusion effect" (some individuals having no traps at all on their home ranges) is a special case applying only if traps are set in a widely spaced grid pattern. Whatever the reasons for the bias, it has important consequences for gamekeepers and conservation managers (Chapters 12 and 13).

Since the enormous fluctuations in population density typical of most weasel populations in good seasons are due to temporarily improved survival of the young, the ratio of young to adults differs distinctly between high and low populations. The effect is similar in both sexes and all ages, however, so density variations have little effect on sex ratio at any age (King & McMillan 1982).

PREDATORS OF WEASELS

Tame weasels of any species, well settled in captivity, can live up to 10 years. Why do most wild weasels achieve only a tenth or less of their potential life span? The most obvious explanation, considering that weasels are obligate predators dependent on wildly variable food supplies, is shortage of food, but there are others.

Stoats and longtails are often killed on the road, especially males in spring (Buchanan 1987; Sleeman 1988a). The active lifestyle of weasels and their insatiable curiosity about dark and inviting-looking holes make them vulnerable to sudden death in kill traps (Chapter 12) and other dangerous places. One unlucky, rain-soaked common weasel crawled into a railway signal box to get out of a storm, short-circuited all the signals, and stopped the trains (Anon. 1998).

In addition, weasels suffer from more natural hazards, including intense persecution from larger predators. Weasels of all species are small enough to be regarded as, or confused with, the normal prey of foxes, coyotes, feral cats, minks and ferrets, plus owls and hawks (Hellstedt & Kallio 2005). Weasels are believed to have somewhat distasteful flesh, so these predators do not necessarily eat a weasel once they have killed it, but that is hardly a comfort. The question is, do these encounters happen often enough to affect the weasel populations?

One of the earliest researchers who believed they do was Latham (1952). He examined the bounty records of the Pennsylvania Game Commission during the 1930s and 1940s, when hunters and trappers were paid to turn in dead foxes, both red and gray, and weasels of all three species. In the early 1930s the total number of weasels killed per year was over 100,000, and of foxes, under 40,000. Then in the late 1930s and early 1940s, the number of weasels decreased, dropping below 20,000 by 1946, while the number of foxes over the same period increased to over 80,000; their relative proportions in the records became nearly reversed. Between 1946 and 1949, the numbers of foxes went down again, which Latham attributed to the heavy toll from the bounty hunters; at the same time, the weasels were apparently recovering.

This inverse relationship was repeated in the relative numbers of foxes and of weasels killed in each county, as shown in Table 11.5. It seems logical to assume, concluded Latham, that foxes reduce and control the numbers of weasels in Pennsylvania.

A single historical event observed and described in unusual detail by Mulder (1990) supports Latham's view. The North Sea coast of the Netherlands from Den Haag to Den Helder is lined with a strip of sand dunes up to 5 km wide, long famous for its rabbits and for the stoats, weasels, and polecats that hunted them. The local game wardens knew their areas and the animals well, and they compiled field notes and annual reports on their activities. The wardens also assisted with several research projects on the stoat populations during the 1960s (Heitkamp & van der Schoot 1966; van Soest & van Bree 1970).

The detailed local knowledge of the wardens documented incidentally the unexpected disappearance of stoats from the dunelands from the mid- to late 1980s onward, following the arrival of foxes between 1968 and 1977. Mulder searched the records, conducted interviews with the wardens, and plotted the acceleration in the recovery of the rabbits from myxomatosis throughout the 1970s and early 1980s. He concluded, on circumstantial but consistent evidence, that the formerly abundant stoats were driven to extinction by foxes throughout the dunelands. The only area not to report this pattern of events, Zwanenwater, apparently proved the rule. There, foxes were shot on sight to protect a nesting colony of spoonbills, and the stoats remained.

Table 11.5 Apparent Reciprocal Distribution of Foxes and Weasels (All Three Species, but Mostly Stoats and Longtails) in Bounty Records From Pennslyvania, 1930–1951

Annual weasel kill per county	Weasel-to-fox ratio
1–99	1:9.33
100–199	1:4.06
200–299	1:2.75
300–399	1:1.86
400–499	1:1.24
500+	1:0.99

(From Latham 1952.)

Another data set, collected before Latham's idea was published and without any special attention to weasel populations, further supports the hypothesis that predation can be an important force, at least sometimes. It stimulated Powell (1973) to explore the situation further. Craighead and Craighead (1956) had studied the ecology and predator–prey relationships of raptors in Michigan in the 1940s. They estimated the densities and food habits of all the local predators, and censused the numbers of their prey available. They reckoned the density of weasels in 1942 was 27 to 36 per township (36 square miles, 93 km²).

Powell took the maximum value of one per square mile, that is, one per 259 ha, or 0.38/km². The Craigheads published extensive tables of data on the contents of nearly 5,000 raptor pellets analyzed, in which weasels appeared consistently as 1% or less of the items listed. Powell calculated that the raptors had killed roughly 70% of the summer (postbreeding) weasel population available in 1942. Powell then constructed a computer simulation model of the weasel–raptor–vole ecosystem, assuming that (1) the area was large enough to operate effectively as a closed system; (2) weasel populations were not limited by food; (3) the reproduction of weasels and of voles was density independent; (4) populations of raptors and of voles were linearly related (this idea was supported by the Craigheads' data); and (5) all mortality of voles and of weasels was due to predation.

If a simulated weasel population could be controlled by predation under these extreme conditions, Powell argued, real predation could probably control real weasel populations. He ran simulations using a variety of starting conditions and ranges of variation in both the weasel and the vole populations. The results always showed that predation kept the weasels below the number that would have been limited by the available voles. Powell concluded that "under certain conditions, a limiting factor for weasel populations is predation by other predators. This point has been given only limited consideration."

Some of the rather obvious problems with this model are explicable. For example, the Craigheads' field data were designed to count raptors, and detected weasels only incidentally, which may be why their estimates of weasel density appears low, compared with those quoted in Table 10.2. But Powell also ran the model starting with weasel densities that were certainly too high, and the results came out the same. He also deliberately included assumptions he well knew were not supported by field evidence, especially that weasel populations are not limited by food and that weasel mortality is due mainly to predation, but only so as to test the question he was asking. The model was designed so that limitation by food and by predation would lead to different outcomes: One set of results would rule out the other, and vice versa. His conclusion was that predation could limit weasel populations, sometimes.

The questions of how often and to what extent raptors might limit weasel populations in real life, as opposed to in a computer, can be addressed only from real data. Fortunately, Erkki Korpimäki and his team in Finland have provided some. They collected detailed information on the diets of 10 species of raptors and 10 of owls, and found that 16 of these species ate, during the breeding season, a range of items including at least 10% small mammals. Their extensive field data also showed that populations of least weasels varied with those of the rodents, and that predatory birds ate weasels along with the rodents, but not to the same extent every year.

Korpimäki and his coworkers became curious when one of us (King 1983c) published a paper suggesting that mustelid populations are controlled mainly by shortage of food, contradicting the raptor predation model proposed by the other (Powell 1973). They used their extensive data (Korpimäki & Norrdahl 1989a, 1989b) to ask which of us was right.

Over the years 1984–1987, the team estimated (1) in late February and March, at the beginning of each breeding season, the densities of least weasels using data collected by snow tracking over 50 km^2; (2) from mid-April to the end of July, the number of least weasels that could have been eaten by the birds of prey; and (3) in May and September, the abundance of small mammals by index snap trapping. Buzzards, kestrels, and three species of owls (eagle, Ural, and short-eared owls) normally ate few small mustelids, on average less than 1% of all the food items identified for these birds. Mustelid skulls did, however, appear in the pellets more often (up to 2%) during the period when voles were decreasing in numbers.

This pattern is quite reasonable, because more weasels are always present after a period of high rodent densities, and predators often have to be content with alternative prey when their favorites are off the menu. Furthermore, both least weasels and stoats might be more exposed to predation in years when rodent populations are decreasing or low, because hunting is harder then and the weasels have to travel further to find the few rodents left (Klemola et al. 1999).

Korpimäki and his coauthors estimated that kestrels took only 20% of 190 available least weasels in the vole peak year of 1985, but kestrels plus short-eared owls took 80% of 650 least weasels available in the vole decline year of 1986. The authors were, of course, quite well aware of the multiple possible sources of error that dog any such exercise. Nonetheless, they listed several good reasons for supposing that their estimates of the predatory impact on weasels were not exaggerated. For example, they sampled the diets only of resident birds of prey, ignoring transient birds of prey and mammalian carnivores.

Other observations in the northern hemisphere confirm that predation is a serious hazard for weasels. Dead weasels or their remains are sometimes found at the dens of foxes and in the nests of raptors; two of nine male longtails radio collared by Gehring and Swihart (2000) were killed by other predators, and Ratz (2000) found a remarkably constant reciprocal distribution of tracking tunnels marked with footprints of stoats versus ferrets. Of eight collared common weasels radio tracked in Białowieża in 1990, one was killed by a fox and a second by another weasel (Jędrzejwski et al. 1995), and the remains of other unmarked weasels were found in the pellets and scats of larger predators. Jędrzejwska and Jędrzejwski (1998) concluded that predation accounted for some 65% of the mortality of common weasels.

Calculations like these suggest that weasel populations can be controlled by larger predators under certain circumstances. Nevertheless, distinguishing between the various possible factors that limit weasel populations, including habitat requirements, predation, and shortage of food, is hard to do, especially when they undoubtedly work together. For example, in Belarus, the invasion of American minks starting in 1988 is staging an as yet unfinished experiment with different results for stoats (a 16-year steady decline in winter snow tracks) compared with common weasels (a corresponding but even steeper increase) (Sidorovich et al. unpublished). Until the arrival of the minks, the Belarus stoats had preferred to hunt water voles in marshlands (as they do in Sweden: p. 251), while weasels preferred forest habitats; since then, stoats have suffered from increased competition with the minks, which weasels can avoid.

King and Moors (1979a) had predicted that body size for weasels is a tradeoff between greater foraging efficiency and aggressive power. Indeed, large weasels can dominate small weasels, but they come off badly against larger carnivores and have fewer escape options than small weasels do. Caught between the two, any increased exposure to predation is a greater risk for the larger of the two weasels of a sympatric pair. The same idea could explain why longtails in northern Quebec were more at risk of predation, compared with stoats, in areas where larger furbearers were protected (St.-Pierre et al. in press-a).

To understand what limits a particular population at a particular time, we need a statistically rigorous model that can eliminate alternative hypotheses under different conditions. Such a model may be ideal in theory, but in practice, extremely difficult to construct.

Raptors certainly kill weasels, perhaps many of them, especially after a vole peak when weasels and raptors are more numerous than usual and voles are becoming scarce. At such times, a weasel population will be decreasing rapidly, whether any are killed by raptors or not. Latham's correlations could well be evidence of active predation on weasels by foxes, but they could equally well be evidence that the habitats in counties supporting large populations of weasels are less favorable for foxes, and vice versa. Stoat populations are prone to local extinction with or without the help of foxes, so we cannot exclude the alternative explanation that Mulder's observations could have been mere coincidence. In part, the question of whether carnivores limit their prey or are limited by them is a matter of scale, and has no simple, general answer (Powell 2001).

The Problem of Black-Tipped Tails

One of the curious quirks of research, and the salvation of many a graduate student, is that it is possible to design a study to ask one question and then be led to a valid answer to a different question. Powell's convictions about the effect of raptor predation on the numbers of weasels in ecological time stimulated him to design a series of simple, elegant experiments with trained red-tailed hawks. They produced a legitimate answer to a long outstanding question in a different field altogether, the effect of raptor predation on the morphology of weasels in evolutionary time. He came up with a simple and convincing explanation for why stoats and longtails have black tips on their tails and least weasels do not.

Whether or not predation by raptors affects populations of weasels, it is a disaster for the individuals caught, and for that reason they avoid open spaces if they can (Figure 11.6). A raptor or an owl can stoop down on a fleeing weasel with terrifying speed, and is unlikely to be deterred by the defensive shriek and "stink-bomb" that might put off a fox or cat.

Only if the talons fail to pierce the thin body of a weasel may the raptor find it has picked up more than it bargained for. Anderson (1966) saw a buzzard swoop, pick up a weasel, and flap away with it, and later he saw it fall to the ground with its captive. When he reached the fallen buzzard, he found the bird lying dead on the ground, with its underparts bloody, and the weasel gripping its breast with meshed teeth. Sometimes, raptors survive such an encounter, although without being able to dislodge the weasel's death grip. Seton (1926) reported an eagle that was found with a bleached weasel skull fixed to its neck.

The white winter fur of the northern weasels is probably itself a defense against attack from the air, even though it does not match the snow exactly. For example, one study area near Robinson Lake, Idaho, supported nine longtails and four stoats during the winter of 1950–1951. The stoats changed to white, but the longtails did not, and two of them were caught by raptors (Musgrove 1951).

Stoats and longtails have another defense against raptors, which is effective all the year round: their black tail tip. Powell pointed out that many small animals, such as fish and butterflies, have spots of contrasting color on their hind ends, which are believed to attract a predator's eye and deflect its strike away from vital body parts such as the head and neck. He asked whether the black tail tip of stoats and longtails might have the same function. The black tip is conspicuous at any time, but especially in winter when it remains black after the rest of the coat has turned white. He decided to test whether the black tail tip is an additional hindrance to raptors hunting white weasels on snow.

Powell had three captive, wild red-tailed hawks, which could be tethered to a running wire laid out between two perches set 30 m apart on the flat, white-painted roof of the old Lion House at Brookfield Zoo outside Chicago. He presented a series of model weasels to each hawk in turn, pulling them across the hawk's flight path attached to a fishing line looped around two pulleys. He estimated each hawk's attack speed, and adjusted the movement of the dummy so that the hawk could only just catch it. The model weasels were made of white artificial fur in two sizes, one 40 cm long, to represent a male longtail, and the

Figure 11.6 A weasel spotted on open ground by a hawk or an owl is nearly defenseless. Its only hope is that it can, by making rapid twists and turns, outmaneuver the predatory bird or, if it is a stoat or longtail, that the bird is confused by the black tip on its tail.

other 17 cm long, representing a male least weasel. Some had a black spot on their backs, some on their tails, and some were plain white. The models were presented in random order to the three hawks until each had seen all six models 12 times (Table 11.6).

The models representing long-tailed weasels with tail spots and least weasels with no tail spots, which were most like the real thing, were missed by the hawks much more often than the other models. Powell's explanation was that the hawks focused their attacks on the black spots. They nearly always caught

Table 11.6 Results of Powell's Experiments on the
Deflective Value of the Black Tail Tip to Dummy White
Weasels Hunted by Hawks Against a White Background

	Number of chances missed			
Model	Hawk #1	Hawk #2	Hawk #3	Total
Longtail, no spot	1	1	0	2
Longtail, tail spot	11	4	9	24
Longtail, body spot	2	0	2	4
Least weasel, no spot	9	7	9	25
Least weasel, tail spot	1	0	1	2
Least weasel, body spot	3	0	0	3

(From Powell 1982.)

either size of model if the spot was placed on the body. If the spot was placed on the long thin tail of the larger models, the hawks failed to grasp it, and they also sometimes checked their attack at the last moment, as if they had not seen the rest of the model until then.

On the other hand, if the spot was placed on the short tail of the smaller model, they usually caught it because the rest of the body was close enough to be within the talons' reach. Larger models with no spots were still visible, even though they were all white against the white-painted concrete roof. But the hawks took fractionally longer to notice and react to the smaller ones with no spots, so often missed them. Powell concluded that the black tail tip on stoats and longtails is a classic predator-deflection mark, and that least weasels do not have it because their tails are too short to hold the mark far enough away from the body.

One might ask, Powell added, why least and common weasels do not have longer tails so that a black tip would be a benefit instead of a liability. He offered the reasonable explanation that they may be too small to keep a longer tail warm during the long northern winters. Perhaps another reason is that they are less exposed to raptors than are larger weasels, because they spend so much more of their time under snow cover.

Snow does not last all year round, however, and, except in the far north, most weasels are brown for more months of the year than they are white. In Switzerland, winter-whitening stoats live alongside non–winter-whitening common weasels in the same areas (Güttinger & Müller 1988). We would like to see Powell's idea tested with models of brown weasels against various natural backgrounds.

Another important factor to consider is the great range in body sizes of weasels with and without black-tipped tails (see Figures 4.5 to 4.7). Some future test should include models representing the races of small stoats that do have short tails with black tips and the races of large European common weasels that have short tails without black tips. In the meantime, Powell's conclusion on the function of the black tail tip as an advantage to individuals seems secure, and it remains valid whether or not predation by raptors has any effect on weasel population dynamics. In nature, predation can have a profound effect on morphology, by determining *which* animals survive, without affecting at all the *number* that survive.

PARASITES AND DISEASES OF WEASELS

Practically all wild mammals, even healthy ones, carry at least a few parasites, internal or external or both. Most are also susceptible to at least some diseases caused by invasive microorganisms. The degree of inconvenience and debilitation accompanying infestation varies a great deal. Some parasites go unnoticed; others can cause intense irritation over a long period; a few are fatal. If parasites or diseases could affect weasel populations we should consider them here.

The best source of detailed information on this subject is a recent review commissioned by the New Zealand Department of Conservation (McDonald & Larivière 2001), as part of its intensive program to develop a means of biological control of stoats (Chapter 13). In general, weasels are susceptible to various diseases, such as tularemia, canine distemper, Aleutian disease of mink, rabies, murine (but not ovine) sarcosporidiosis, and bacterial infections caused by *Bartonella* sp. and *Borrelia burgdorferi*. Another bacterial pathogen, *Helicobacter mustelae*, is widespread in both ferrets and stoats (Forester et al. 2003). Histological signs of some forms of disease in wild stoats in Britain have been described (McDonald et al. 2001), but practically nothing is known about the incidence or effects of any diseases and parasites on wild weasels of any species.

Bovine tuberculosis (TB) is a serious problem for dairy farmers, especially where there are continual "breakdowns" (reinfections of a cleared herd) from contact with infected wild mammals. Removal of the wild species harboring the disease becomes a high priority to farmers in such areas. In Britain the main targets are badgers, especially in southwest England; none of the 33 common weasels nor the 33 stoats examined between 1971 and 1985 was positive for TB (Anon. 1987). In New Zealand the main reservoir of infection is the huge population of introduced Australian brushtail possums. Stoats do pick it up from them, though much less often than do feral ferrets (in one survey only one of 62 stoats collected from a TB area was positive) (Ragg et al. 1995).

McDonald and Larivière (2001) listed the helminth parasites recorded from stoats as *Taenia mustelae*, *T. tenuicollis*, *Alaria mustelae*, *Molineus patens*, *M. mustelae*, *Trichinella spiralis*, *Troglotrema acutum*, *Capillaria putorii*, *Strongyloides martis*, *Aleurostrongylus pridhami*, *Gnathostomus nipponicum*, *Mesocestoides lineatus*, *Filaroides martis*, *Dracunculus* sp., and *Acanthocephalis* sp.

Skin parasites are common in most mammals. Lice, some ticks, and mites are hard to see, especially the larvae, and all we can say is that weasels do have some. The records are sparse and probably grossly inadequate, but for what they are worth we list them here. Common weasels have a specific louse, *Trichodectes mustelae*, and can carry the mites *Demodex* spp., *Haemaphysalis longicornis*, and *Psorergates mustela* (Tenquist & Charleston 1981). Stoats have a specific louse, *Trichodectes ermineae*, recorded in Canada, Ireland, and New Zealand (Jennings et al. 1982; Sleeman 1989a; King 2005b). They also carry the mites *Eulaelaps stabularis* and *Hypoaspis nidicorva* (both normally found on birds), *Demodex erminae*, *Gymnolaelaps annectans*, *Haemophysalis longicornis*, and *Listrophorus mustelae*. Some stoats show symptoms of mange caused by the mites *Sarcoptes*. In Ireland, a comprehensive examination of stoats detected the mite *Neotrombicula autumnalis* (only the larva is parasitic, but one female carried 1,819 of them); the lice *Polyplax spinulosa* (of rats) and *Mysidea picae* (of corvid birds); and the ticks *Ixodes hexagonus* (a nest species; 266 larvae on one female), *I. ricinus* (common on rats), and *I. canisuga* (Sleeman 1989a). A new species of tick *Ixodes gregsoni* has been found on minks, martens, and weasels in Canada (Lindquist et al. 1999).

The only two parasites of weasels that have received considerable attention are the fleas, because they can tell us quite a lot about how weasels hunt and move about their home ranges, and *Skrjabingylus nasicola*, a nematode worm that causes dramatic lesions in the skull.

Fleas

When live-trapped weasels are anesthetized, the fleas in their fur, which are also anesthetized, fall out and can be collected by hand. Several field workers undertaking live-trapping studies for quite other reasons have extended their field routine to collect fleas during the short time that they have the unconscious weasel in their hands. When weasels are collected freshly dead, they often have some fleas still on them, and these are also worth collecting as a routine part of the dissection procedure.

Neither method can be claimed to search the fur thoroughly enough to find all the fleas that might be present, and, of course, lice and mites are harder still to see. Worse, dead weasels collected in kill traps quickly lose many of the fleas they had, since fleas are not confined to the host's body and leave a carcass as it cools (King 1981b). These problems aside, the systematic collection of fleas in various places has produced some intriguing insights into the natural history of weasels.

The larvae of fleas are free-living scavengers of organic material, and they find the best conditions of temperature and food supplies in the nests of small mammals. Fleas have evolved for millennia alongside their hosts, and have developed close relationships with mammals that have substantial nests that are occupied or frequently revisited over a long enough time for the fleas to complete their nonparasitic larval stages. Hence, mammals with small home ranges and permanent dens, such as badgers, moles, shrews, and rabbits, have specific fleas. Those that move around a lot and have only temporary dens or none at all, such as foxes, weasels, and hares, have no flea species of their own. They do carry fleas, but only those normally found on other animals.

Weasels were long assumed to pick up fleas from the bodies of their prey and, perhaps less often, from casual encounters in the grass. But these are probably not the usual ways that weasels acquire their fleas. If they were, the list of the normal hosts of the fleas identified on weasels should closely match the normal diet of weasels, but it does not. The list in Table 11.7 includes fleas specific to hosts that weasels seldom or never eat, such as moles, and some of these in substantial numbers; conversely, there are few fleas specific to hosts frequently eaten, such as rabbits and birds.

The explanation for this apparent contradiction appears to be that weasels normally pick up their fleas from the burrows and nests of their prey (King 1976). These are the places where adult fleas lie in wait, ready to jump onto the first warm furry creature that passes by or stays for a snooze. Weasels feel the cold badly, and since they do not make their own dens, they depend on finding warm nests to sleep in made by other animals. They would be very likely to pick up the fleas specific to hosts that make substantial nests of the right size, even though these hosts are seldom eaten.

Judging by the records of fleas listed in Table 11.7, common weasels must often borrow the nests of moles (see Figure 8.4), and stoats the nests of rats and squirrels, because the fleas of moles, rats, and squirrels are found on common weasels and stoats much more often than their hosts are eaten by them. One flea, *Rhadinopsylla pentacantha*, normally found only in the nests of voles rather than on their bodies, turns up on common weasels remarkably often. Stoats radio tracked in Ireland and in New Zealand liked to sleep in rats' nests and in holes in the ground where rats and rabbits might shelter (Sleeman 1990; Murphy & Dowding 1995).

On the other hand, the nests of many birds offer little protection from the elements. Weasels eat birds often, but seldom rest in birds' nests, and bird fleas rarely turn up on weasels. Irish stoats do eat shrews, yet do not carry shrew fleas. Presumably, a shrew's nest is too tight a fit even for a small Irish stoat.

There are, of course, other considerations. For example, some fleas are specific to one particular host and drop off any other quickly. This may explain why so few rabbit fleas are found on weasels. Less fussy fleas are likely to stay on the "wrong" host for a while, especially if they are hungry and actively searching for a meal. This could be a stronger reason for hopping onto a weasel visiting a nest that has been deserted for days than merely being disturbed by a weasel eating the host they are already on.

Table 11.7 Fleas Found on Common Weasels and Stoats[1]

Flea species	Normal host	On common weasels			On stoats		
		England	Scotland	Switzerland	New Zealand	Britain	Ireland
Megabothris walkeri	Voles	82					
Ctenophthalmus nobilis	Voles, mice	48	67				28
Hystrichopsylla talpae	Rodents, insectivores	25	18	3			
Malareus p. mustelae	Voles	26	11				
C. bisoctodentatus heselhausi	Moles	17	20	3			
Palaeopsylla m. minor	Moles	18	17	2			
Rhadinopsylla pentacantha	Nests of voles	9	16	1			
Dasypsylla g. gallinulae	Birds	5		1		2	3
Megabothris turbidus	Voles	4		1			
Peromyscopsylla spectabilis	Voles	1					
Nosopsyllus fasciatus	Rats				662		17
Palaeopsylla s. soricis	Shrews		1				
Megabothris rectangulatus	Voles		14			1	
Ctenophthalmus agyrtes impavidus				25			
C. b. bisoctodentatus	Moles			47			
C. s. solutus				1			
Monopsyllus s. sciurorum	Squirrels			6			
Peromyscopsylla bidentata				1			
Leptopsylla segnis	House mice			8			
Ceratophyllus gallinulae	Birds			1			
Parapsyllus nestoris	Birds			1			
Orchopeas howardi	Squirrels					2	
Spilopsyllus cuniculus	Rabbits						1
Number of examinations		338	NR	380	1501	NR	122
Reference		(King 1976)	(Mardon & Moors 1977)	(Debrot & Mermod 1982)	(King & Moody 1982)	(King 1976; Mardon & Moors 1977)	(Sleeman 1989a)

1. Columns give total number of fleas found; number of animals inspected, often the same ones on successive days, summed at end.

Skrjabingylosis

Museum curators and researchers dealing with collections of weasel skulls noticed long ago that many specimens were damaged in the postorbital region, immediately behind the eyes. The skulls had what appeared to be dark patches or swellings with thinned walls, and sometimes these had holes in them, even large openings (Figure 11.7).

The cause of the damage became clear when the fresh heads were skinned and the swellings opened. Inside, a mass of bright red round worms could be seen crammed into the tiny sinuses in the nasal bone, coiled over each other and pressing hard against the confining skull. The worms were described in 1842 by Leuckart, and are now known by the tongue-twisting name of *Skrjabingylus nasicola*—after Skrjabin, the great Russian parasitologist, and "nasicola" because of their position. The condition of being infested with these worms is called skrjabingylosis.

These parasites have attracted a lot of attention, for two reasons. First, they are easy to observe in freshly dead mustelids, where they are very dramatic in appearance. Better still, their effects can be studied indirectly from standard museum material. A large collection of skulls allows the incidence and geographical distribution of damage to be calculated with minimum effort. The cheap and simple technique of visual inspection of preexisting material has ensured many more studies of the damage caused by skrjabingylosis than of the parasite itself. Fortunately, the link between the visible damage and the incidence of infestation is secure (Lewis 1978).

Second, there is the economic aspect. In the far north, weasel furs were once a valuable item of trade, and most of the earliest research on skrjabingylosis was concerned with the effect of the disease on the harvest. Looking at an advanced

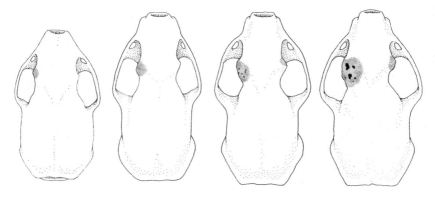

Figure 11.7 The skulls of weasels commonly show more or less severe distortions and perforations in the postorbital area caused by *Skrjabingylus* worms, and their severity generally increases with age (shown left to right). (Redrawn from King 1977.)

case, it is hard to avoid the impression that such severe distortion of the skull bones, and consequent pressure on the brain (Figure 11.8), must have some effect on the general health of the afflicted animal, possibly a fatal one. Hence, in the early 1940s Russian biologists interpreted their data to mean that stoats heavily infested with *S. nasicola* were in poorer condition, were less fertile, and died sooner than uninfested ones, and that the fur harvest was lower in seasons following widespread infestations (Popov 1943; Lavrov 1944).

Conversely, there are other places where any species of weasel is regarded as a pest. People concerned with the protection of birds, usually game or endangered native species (Chapters 12 and 13), would welcome any prospect of biological control of weasel populations using a specific and fatal parasite, and encourage research on means of spreading it (McDonald & Larivière 2001).

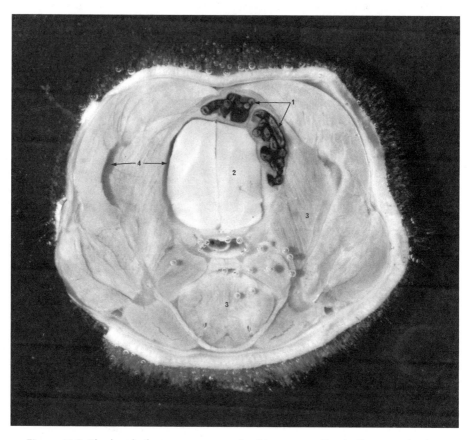

Figure 11.8 The head of a common weasel cut in cross section in the postorbital region. (1) *Skrjabingylus* worms; (2) brain; (3) muscles; (4) skull (to the left of the arrow, the zygomatic arch or cheek bone; to the right, the cranium). (From King 1977.)

The proportion of skulls visibly damaged by skrjabingylosis is not quite the same as the incidence of the disease in a population, since very early infestations are not detected this way. Nevertheless, damage does closely reflect both incidence and severity (the number of worms in each skull). Studies of damaged skulls in museums, therefore, give an idea of local and regional variation in incidence (Table 11.8).

These general figures are not very informative, however. Estimates of incidence are strongly affected by sampling variables, such as the season of collection and the age structure of the target population, and the incidence in any one area may vary from year to year (King 1977; Weber 1986). Moreover, since the worms tend to damage smaller skulls more severely, the disease is probably detected at an earlier stage in females than in males of any species, and more often in common and least weasels than in stoats and longtails. Consequently, estimates of damage are not strictly comparable in skulls of different sizes.

The biology of the parasite was worked out in the laboratory by Dubnitskii (1956). Male and female adult worms are easily distinguished from their size; the females measure 18 to 25 mm long, with a diameter of 0.8 mm, and the males about 8 to 13 mm by 0.5 µm. The first-stage larvae, about 300 µm long, travel from the sinuses down the nasal passages to the back of the throat. From there they pass into the gut and to the outside with the feces. They actively disperse onto nearby grass and leaves, and await their opportunity to invade the soft foot tissues of a slug or a snail.

After 12 to 18 days in a slug or snail, the larvae pass through an obligatory intermediate second stage. When they reach a weasel, the third-stage larvae (now grown to 700 to 750 µm long) pass from the gut into the tissues, through two more

Table 11.8 Regional Variation in Incidence of *Skrjabingylus nasicola* in Weasels[1]

Species	Country	Incidence	References
Stoat	Britain	17–30%	(Lewis 1967; van Soest et al. 1972)
	Eurasia	20–50%	(Lavrov 1944; Vik 1955; Hansson 1970; van Soest et al. 1972; Debrot & Mermod 1981; Sleeman 1988b)
	North America	20–100%	(Dougherty & Hall 1955)
	New Zealand	0–40%	(King & Moody 1982)
Common and least weasels	Britain	70–100%	(King 1977)
	Eurasia	20–60%	(Lavrov 1944; Vik 1955; Hansson 1970; van Soest et al. 1972)
Longtail	North America	0–100%	(Dougherty & Hall 1955)
	Manitoba	100%	(Gamble & Riewe 1982)

1. Ranges of mean incidences given refer to different localities.

molts, and then migrate to the nasal sinuses along the spinal cord as fifth-stage larvae. They grow to adult size and settle in, wriggling around in the confined space and gradually enlarging it.

How they cause the swellings and perforations we see is not quite clear. Simple friction, the rubbing of the worms' hard cuticle against the sensitive bone tissue, could cause the enlargement or, possibly, the worms produce some kind of corrosive agent that interferes with the control of the living bone tissue or erodes it directly. Damage often increases with weasel age, which suggests repeated reinvasions.

The main problem faced by the larvae is how to get from the snail to a weasel. Weasels do not eat snails or, at least, not nearly often enough to account for the high frequency of infestation in some places. Other animals do eat snails regularly, for example, many of the shrews. A larva finding itself inside a shrew will simply retreat into a cyst and wait. The shrew is not needed as part of the parasite's life cycle, but it can act as a paratenic (waiting) host, a bridge to the definitive host where the larvae can complete their development into breeding adult worms.

If an infected paratenic host is eaten by a weasel, the cyst opens in the weasel's gut and the larva continues on its interrupted journey. Hansson (1967) showed that it was possible to transmit skrjabingylosis into a previously uninfested weasel by feeding it shrews. The suggestion that shrews might be a natural paratenic host was strengthened by the fact that incidence rates are particularly high in places where weasels eat shrews more often than usual.

For example, on Terschelling Island, off the coast of the Netherlands, the incidence rate observed by van Soest et al. (1972) was over 90%, much greater than on the nearby mainland coast (23%). As Terschelling has no voles, stoats there had to eat a lot of shrews. North American stoats, too, tend to eat shrews more often than do their European brethren, and they also suffer a generally higher rate of infestation. In western Newfoundland, the incidence of skrjabingylosis in stoats increased dramatically after shrews were introduced to the island (Jennings et al. 1982).

In most places, however, weasels do not normally eat shrews often enough to explain the high incidence of skrjabingylosis. Hansson's experimental animals ate shrews only with the greatest reluctance. Besides, a few weasels in Newfoundland had skrjabingylosis before the shrews arrived, and there are no shrews in New Zealand, where the parasite arrived with its hosts and has persisted for a hundred years (King & Moody 1982).

The solution to the mystery was found by a doctoral student at Neuchatel and published in two joint papers with his supervisor (Weber & Mermod 1983, 1985). Weber discovered encysted third-stage larvae in the salivary and lacrimal glands of wood mice and bank voles, and also in the muscles and connective tissues of their heads. These small rodents, especially wood mice, do eat slugs and snails occasionally (Figure 11.9), not as a main item of diet, but often enough when green food is scarce in late winter that some become carriers of invasive larvae (the strictly vegetarian field vole would rather starve).

Weber fed infected rodents to one common weasel and one stoat, plus six ferrets that were bred in his laboratory and known to be free of the disease. He checked every day for the development of the adult parasites by inspecting the mustelids' scats in water under a binocular microscope. A host carrying both sexes of worms in the same sinus passes a continual stream of first-stage larvae, which can easily be seen swimming about in the water.

Within 30 days, the common weasel, the stoat, and three of the ferrets were infected. Those that had been given whole rodents or only the heads and front ends produced larvae first; one ferret given the hind end produced larvae only much later; and three ferrets given only the viscera (heart, lungs, liver, intestine) remained free. It is a fact that weasels usually begin to eat their prey at the front end (see Figure 2.5), and may leave the viscera. Hence, the larvae not only seem to choose a paratenic host frequently eaten by weasels, but they are also most likely to encyst in the part of the body first eaten by a weasel.

Parasites often show such remarkably close adaptations to the feeding habits of their definitive hosts, for the simple reason that their lives depend on it.

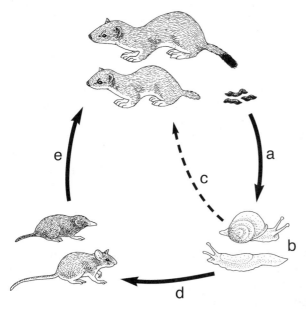

Figure 11.9 The life cycle of *Skrjabingylus nasicola* in weasels. (a) First-stage larvae leave the weasel in the scats. (b) The larvae have to pass through the next two molts in a compulsory intermediate host, a mollusk. (c) Third-stage larvae encysted in the mollusk tissues can reach a weasel directly, but this is rare as weasels seldom eat mollusks. (d) Usually the larvae make use of a paratenic host, a shrew or a mouse that has eaten an infected mollusk. (e) On being eaten by a weasel, the larvae escape from their cysts, pass through another molt, and migrate to the nasal sinuses along the spinal cord. (Redrawn after Weber 1986.)

Nevertheless, the parasitic way of life is hazardous. Weber calculated that, of every 100 first-stage larvae given the chance to invade a slug, only about 24 reached the infectious third stage, and only six became adult worms (Weber & Mermod 1985).

The losses during the life cycle of the parasite are high even in the ideal conditions in the laboratory; in the wild, they are much worse. In Weber's study area at the time of his observations, the infestation rate in stoats was 11%, but only two (0.3%) of 762 molluscs and two (4%) of 48 shrews and small rodents he examined were carrying larvae. The system is wasteful, but it works because the number of larvae produced is so enormous, because small rodents are so frequently eaten by weasels, and because the time the larvae are prepared to wait in a rodent is so long (at least a year).

Geography and climate also play their part in determining incidence. The disease is generally more prevalent in damper habitats, and absent in deserts. The reason is mainly to do with the conditions that favor the survival of the free larvae and their chances of making the hazardous journey from the weasel's scats to a mollusk's foot. The larvae are very susceptible to drying, and to freezing when they are swimming in water (Hansson 1974). The best conditions, both for the larvae and for the mollusks, are found in a mild and humid climate. In different districts of Sweden and Britain (Table 11.8), the frequency of incidence of the disease or the severity of damage it causes increases with the number of rainy days per year.

Strangely, the opposite correlation was observed in New Zealand. In the wettest places sampled, all with mean annual rainfall over 3,000 mm, the incidence of skrjabingylosis in stoats was 0% to 7%, whereas higher local incidences (15% to 60%) were found only where rainfall was under 1,600 mm. This contradiction was completely inexplicable at the time it was published (King & Moody 1982). Since then, however, Weber's demonstration of the role played by wood mice in the transmission of the disease in Europe has suggested a possible solution to the puzzle.

Feral house mice are generally less common in the wetter western mixed podocarp forests of New Zealand, where incidence of skrjabingylosis was low, than in the dryer eastern beech forests and grasslands where incidence was higher. If these mice are the paratenic host transmitting skrjabingylosis to New Zealand stoats, then individual stoats that have passed their whole lives in the period between postseedfall mouse irruptions should be less likely to have picked up the infection than stoats that were born in or lived through a mouse irruption. King (1991c) returned to the original data, and found some support for this hypothesis, especially in females (they eat mice more often than males; Chapter 5). Hence, geographical variation in the relative density of feral house mice could explain the unexpected distribution of skrjabingylosis in New Zealand.

Much work on skrjabingylosis has been done since the early Russian studies that claimed a detrimental effect on infested individuals. When van Soest et al. (1972) measured a sample of stoats from Terschelling Island and found them to be a little smaller in general than those on the mainland, they wondered if the island animals were stunted by the high rate of infestation they suffered. To be valid, such a comparison must be made only between infested and clean individuals of the same age and sex living under the same conditions. Whenever this has been done, there has been no sign that the infested animals were any smaller, were any lighter or leaner, or died sooner, than the others (King 1977; King & Moody 1982).

On the other hand, incidence of skrjabingylosis in Russia and in Switzerland does increase after a population crash. Early writers attributed the decrease in weasel numbers to the increased infestation, but Debrot and Mermod (1981) suggested that the relationship works the other way around. In both countries, stoats reach a high density when water voles are abundant (Chapter 10). Water voles are strictly herbivorous, and are not at all implicated in the transmission of skrjabingylosis. When water voles are abundant, stoats eat them almost exclusively and run little risk of picking up the parasite. When water voles decrease in numbers, however, stoat numbers also decrease and the survivors are forced to turn to alternative prey, including wood mice, which are the most likely of the other rodents to be carrying invasive larvae.

Hence, Debrot documented an increase in infestation during the population decline in the Brévine valley, from 4% when the stoats were feeding almost exclusively on water voles to 50% after the forced change of diet. This attractive idea was confirmed in the Val de Ruz by Weber (1986). Unfortunately, no one seems to have thought of the possible effect on these figures of the change in age structure, which also follows a population decline. The higher proportion of older stoats in the postpeak population would, in itself, cause an increase in incidence, since incidence increases with age and weasels have practically no resistance to infestation and no means of repairing the damage (Lewis 1978). Debrot's hypothesis should be tested; in the meantime, the role of skrjabingylosis in the population dynamics of weasels remains unknown.

Even if *S. nasicola* has no effect on populations, it could still affect the behavior of individuals. The pressure on the brain caused by the distortion of the skull, plus the wriggling of the worms, must be intensely irritating. Weasels observed behaving strangely in the wild, leaping about and somersaulting, are said to be "dancing," either in play or as a clever trick to catch birds (Chapter 6). They are also believed to suffer "fits," or to "play dead" after violent exertion (Chapter 8). Perhaps these gyrations and temporary blackouts are merely an understandable response to the extreme discomfort caused by having to carry such unwelcome and relatively enormous guests in the head. At present, we do not know one way or the other.

12 | Human Attitudes to Weasels in Their Native Environments

HUMAN ATTITUDES TO wild creatures are not, as we would like to think, reasonable. There is hardly any better demonstration of this than the story of how people think about weasels. Contradictions and misconceptions are the stuff of common knowledge, mostly because, until recently, not much was known about weasels. So, people simply projected their own ideas onto the real animals.

There are two main streams of opinion, held by people with different interests. Both start from what seems to be an obvious fact: Weasels appear to be ultra-efficient killing machines specializing on small mammals and birds. Both assume, therefore, that predation by weasels is capable of controlling the populations of these animals, even though in fact it rarely does (Chapter 7). One group concludes that all weasels must be killed on sight to protect birds, especially game birds—even though other predators often take even more birds than weasels do, depending on the habitat. The other group concludes that all weasels must be preserved, and even spread around, to protect farms and plantations from the ravages of pest mammals such as rodents and rabbits.

In our opinion, both views are largely misguided when applied to weasels in their native environments in the northern hemisphere. Weasels introduced as aliens into a totally different environment such as New Zealand are a different matter (Chapter 13), although, even there, some still argue that stoats have a compensating value in controlling rabbits and rats (Fitzgerald & Gibb 2001).

WEASELS AND WILD BIRDS

Bird watching is an immensely popular hobby, and dozens of new books about birds appear year after year. Appeals for conservation funds do well if linked to a particularly attractive bird, at least partly because people tend to feel protective toward birds and respond quickly when birds are under threat. If the threat has reddened teeth and claws, the reaction can be extreme and sometimes irrational.

The natural history literature of North America and Eurasia provides many accounts of weasels raiding birds' nests. Weasels of all three species eat eggs and kill young and adults, especially of small species (see Figures 7.1 and 12.1). One of the earliest and clearest descriptions was given by Hussell (1974), complete with a set of four clear photographs of a stoat raiding a snow bunting nest on Devon Island in July 1969. The stoat was shown leaving the nest entrance with a chick in its jaws at 0755, 0805, 0851, and 1436 hours, and in the early afternoon of the same day a group of at least four weasels was observed in the vicinity, one of which was carrying a young bunting from this nest.

Larger birds such as ducks are not immune either. For example, Teer (1964) observed 59 nests of North American waterfowl, of which three were visited by longtails. The nests themselves were undamaged, but the eggs remaining after the weasel had left were marked with paired punctures, corresponding exactly to the size and position of a weasel's canine teeth. At Union Slough National Wildlife Refuge in Iowa, stoats and longtails were entirely responsible for 27 of 263 upland ducks' nests that failed because of predation in 1984–1985, and contributed to another 11 failures. Weasels also took eggs from at least 5 of 20 more nests that lost up to 7 eggs before the rest hatched (Fleskes 1988).

Over the six summers from 1989 to 1994, Walters and Miller (2001) monitored 239 nests of six species of woodpeckers in montane forest (1,200 m elevation) at Hat Creek in British Columbia. Of the 149 whose fate was known, 22 were destroyed by predators, including 12 (55%) by (probably) longtails. Nine of these 12 nests were concentrated within two small areas (<5 ha) within the 80–ha study area, which might have corresponded with the home ranges of two individuals that learned to spot woodpecker nests.

In two other study areas, longtails were much less often to blame for nest losses than were snakes. In Missouri, longtails contributed only one of 46 predation events on songbird nests in fields (snakes, 33), and none of 15 in forests (snakes, five) (Thompson & Burhans 2003); at Fort Hood, Texas, the predators that destroyed 48 nests of black-capped vireos included 18 snakes but no longtails at all (Stake & Cimprich 2003). In the Californian riparian meadows monitored by Cain et al. (2003), the nesting success of willow flycatchers was correlated with tracking activity indices of stoats.

In northern Norway, stoats are important predators of willow grouse, especially in years when rodents are scarce. Myrberget (1972) calculated the rate of predation, mostly due to stoats, over the 10 years 1960–1969. In 4 years when rodents were abundant, the average loss of eggs was 11% and of chicks 38%. By contrast, in 3 years when rodents were low, the losses of eggs reached 23%, and of chicks 54%. The worst losses were in the vole crash year of 1967, when 36% of grouse eggs were taken, and other birds suffered too. Yet, on a nearby island where there were no stoats, relatively few eggs were lost that year. In a study on a grouse moor in northern England, 45% of curlew nests and 54% of radio-tagged chicks were destroyed by stoats, even though the area was supposedly protected by year-round trapping (Robson 1998).

In the late 1940s in Wytham Wood (near Oxford), researchers from the Edward Grey Institute of Ornithology started mounting hundreds of simple

Figure 12.1 Least weasel raiding the nest of a wood thrush. (Drawn from a remote photograph by Ted Simons.)

wooden nest boxes to tree trunks for a long-term study of the great tit and the blue tit (*Parus* sp). When woodland rodents were scarce, common weasels often raided the boxes (see Figure 7.1). In spring, known residents often turned up in King's (1980b) live traps with yellow yolk stains all down their white chests, and eggshells and feathers were very common in their scats at that time. The damage was eventually eliminated by replacing all the boxes with freely hanging concrete tubes fitted with sloping metal caps that foiled even the agile common weasels (McCleery et al. 1996).

Natural nests are very vulnerable, especially those on the ground. To quote only one example: Clowes (1933) described how, when he was watching gulls on a rugged headland, a female stoat appeared and surveyed the scene. Satisfied, she purred, and three young appeared "as if from the earth." She found a rock pipit's nest, took a nestling and disappeared with it, returning three times and darting off again, each time with a young bird.

Similar stories are legion. In most natural communities and traditional farming areas in the mainland northern hemisphere, these losses were once sustainable over the long term. Since the advent of the intensive agribusiness school of farming, however, the economics of crop growing and livestock rearing have become skewed against wildlife, and changed the equations of life and death for wild birds, to the great alarm of conservationists. The Game Conservancy Trust (Fordingbridge, England) has campaigned against government policies that favor destructive monoculture farming at the expense of wild birds. The Trust also runs extensive education programs to help landowners meet this new emphasis on active protection of wildlife, in addition to their continued interest in game management (Tapper 1999).

Conservation of native species in Britain is now widely seen to be a vital part of game management (Reynolds & Tapper 1996; Tapper 1999). It may seem like a contradiction in terms, but it is not. Most country landowners have always had an interest in conservation, so this suggestion is seen mainly as a shift in emphasis. The Game Conservancy Trust strongly advocates a compromise policy: Run a managed estate including approved forms of predator control, on which shooting generates revenue to fund much-needed but otherwise impossible conservation policies (Reynolds & Tapper 1996).

The habitat protection and modified predator control regime that benefit contemporary game birds also benefit song birds and many other native species. The rare raptors and owls are, rightly, protected everywhere but, on the relatively wide areas of countryside represented by game estates, it is possible to reduce stoats and common weasels (plus foxes and several other common predatory species) to low density at least during the spring nesting season. All these predators are prolific generalists whose national populations are in no danger at all, but localized reduction of them could provide a significant component of a national biodiversity strategy.

To demonstrate this point, a model farm at Loddington in central England is run to show how profitable farming can be compatible with managing land both for conservation and for shooting (mainly wild-bred pheasants and hares). After 7 years of predator control and habitat improvements, several species of birds classed as "nationally declining" have increased in numbers, both over time and in comparison with neighboring farms (Stoate 2002), at a cost of less than 2% of farm profits (Tapper 1999:84).

Unfortunately, many game estates rely on releasing hand-reared birds, a policy that can still produce a shootable harvest but at minimal cost in time and effort spent on predator control. Under that strategy, fewer stoats and common weasels are killed per year than required by the Loddington model, which explains why the national bag for both species has apparently declined over the last 30 years (McDonald & Harris 1999; McDonald & Murphy 2000).

WEASELS AS PESTS OF GAME BIRDS

Game Preservation in Britain, 1870–1914

European farmers and hunters have destroyed predators for centuries (Reynolds & Tapper 1996; McDonald & Murphy 2000), but the most efficient and systematic form of this deadly campaign evolved on privately owned sporting estates, especially in Britain after the 1860s. Game birds were required in large numbers, especially for "driven" shoots using the modern range of sporting firearms. Under British law, all forms of wildlife were then the legal possession of the landowner, who was usually disinclined to share them with natural predators or with people other than invited guests.

The policy of the time was to increase the harvest available to the shooting party by indiscriminate suppression of all predatory mammals and birds and by rigorous enforcement of the laws of trespass, especially against poachers. In Commonwealth countries, such galling social discrimination had ended decades before, with the transfer of legal ownership of wildlife to the state, especially on vast areas of publicly owned land. This option is not available in Britain, most of which is privately owned, even in national parks. The law on British wildlife has merely become much more complicated, while still requiring the goodwill of landowners (Tapper 1999). (In the United States, wildlife species are a publicly owned resource whose management responsibility falls to the state and federal governments.)

Gamekeepers were supplied with guns, poisons, and the newly developed steel kill traps. They proudly displayed their catches on "gibbets" (rows of carcasses hung up by the necks) (Yalden 1999) as evidence of their hard work (Figure 12.2), and some even turned this deadly trade into an art form.

The number of keepers employed increased steadily from 15,000 in 1871 (the earliest available figure) to a peak of over 23,000 in 1911 (Potts 1986). At that time, the heyday of Edwardian country life (Tapper 1992), about half the land area of Britain suitable for shooting was covered by more or less intensive predator control operations, and almost any mammal or bird that got its living even partly by hunting was at risk of its life. The effects on predators were of two main kinds.

Several species of carnivores and raptors, which had previously been widespread, disappeared from most of Britain. The pine marten and the wild cat vanished from England and most of Scotland during the first half of the nineteenth century, and the polecat during the second half (Yalden 1999). Five species of hawks and eagles were completely eliminated by 1916, and three others reduced to remnants.

These extinctions were not due to any contemporary losses of the habitats of these species. Massive deforestation starting in the Neolithic period (5000 years BP) had already reduced the natural forest cover of Britain to its all-time

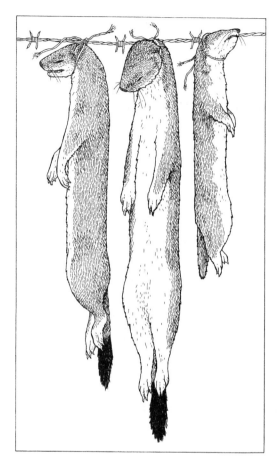

Figure 12.2 Traditional gamekeepers used to display their catches on a "gamekeeper's gibbet" along a fence. The carcasses usually included common weasels and stoats, along with rats, crows, and magpies.

low (probably to under 4% of the total land area) by about 1700. The most serious declines of martens, polecats, and raptors started some 50 to 100 years later, and they coincided rather with the *re*forestation programs, which started about 1750. Even game authorities now generally accept that the disappearance last century of the martens and polecats was due mainly to direct human persecution (Tapper 1992, 1999).

At the same time, huge numbers of foxes, otters, badgers, stoats, and common weasels were also killed, but none of them suffered the same catastrophic population decreases, except perhaps locally and temporarily. This difference arose, at least in part, because foxes and otters were also conserved for hunting, and many badgers could remain secure in their underground setts during the era before power-driven cyanide gassing equipment. Stoats and common weasels were (and still are) simply extremely resistant to control (King & Moors 1979b).

The reason for the resilience of weasel populations in general is one of the consequences of their opportunistic way of life. Weasels have very variable productivity and high natural mortality. Births and deaths are seldom in equilibrium, so local variation in density is expected and frequent local extinctions are normal (Chapter 10). At the same time, weasels are very resistant to total extinction, because a few can always survive hard times in a favorable patch of habitat somewhere, and these holdouts are good at recolonizing abandoned areas when things improve.

By contrast, equilibrium species such as martens, otters, and the large raptors tend to have steadier, lower productivity and, when undisturbed, fairly stable populations with relatively low mortality among the adults. They cannot compensate for any sudden increase in mortality, so they are slow to recover from heavy losses, and are particularly vulnerable to both local and total extinction. They certainly needed the legal protection that was granted just in time to save many of them from oblivion. These ideas explain why it is nearly always the large ones among a given type of animal that first become threatened by persecution (unless they have some refuges or are locally protected), while small ones survive (King & Moors 1979b).

In the conditions of the time, the old policy of enhancing game bags by rigorous predator control alone was highly successful and, although distasteful to modern eyes, it was at least sustainable. The small fields, networks of hedges, regular crop rotations, and absence of chemical pesticides provided the best possible conditions for the nesting and survival of partridges and pheasants. Both flourished greatly, and were shot in numbers that seem unbelievable to today's sportsmen.

Nowadays, game estates are fewer and smaller, and systematic predator control is practiced to only a fraction of its former extent. Pine martens, polecats, wild cats, buzzards, and hen harriers are all extending their ranges as, for reasons summarized above, the twentieth-century concept of conservation has slowly taken over from the nineteenth-century concept of game preservation (McDonald & Birks 2003). Old attitudes toward predators (such as "the only good weasel is a dead one") are changing to a more informed and discriminating assessment; long lines of decomposing vermin on gibbets are now seen as a bad advertisement for the keeper's work, which distracts attention from the real good that keepers do for conservation in the countryside.

In addition, research now plays a key part in determining management policies on game estates. Particularly intensive research has been done on the ecology of the grey partridge in England, including some estimate of the part played in it by stoats and common weasels.

The Partridge Survival Project

The grey partridge is (or was) one of the most important of the English game birds. Bag records from long-established estates, some going back for 150 years, show that populations of grey partridges reached their greatest densities during the heyday of traditional game management by intensive predator control. The early records of the National Game Census, for the peace-time years between 1933 and 1960, showed that about 18 to 20 birds on average were shot on every square kilometer of the estates participating in the recording scheme. Then there was a sudden population crash, so intense that the mean bag for 1971–1979 was only 3.7 birds per km^2. The reasons for this event were sought in a series of studies of partridge ecology and especially by the Game Conservancy's Partridge Survival Project (Potts 1979, 1986).

Among the prime suspects were, of course, the predators. Among nearly 14,000 partridge nests whose fates were known, 1,993 cases of nests lost to predators (36%) were attributed to foxes (Middleton 1967). Stoats and common weasels appeared to be minor players by comparison (7% of the losses were attributed to stoats, and 0.1% to common weasels), even though many keepers can tell stories about stoats cleaning out eggs from partridge nests.

For example, on one estate in Norfolk in 1944, all the losses in one small area (Figure 12.3) were attributed to the work of a single stoat that was seen at one of the nests. On open farmland, stoats usually move along the field boundaries, which effectively channels predator and prey together (Chapter 8). Stoats can also steal whole clutches of eggs and carry them away to cache for later. Cringle (1968), helping two keepers to dig out rabbits in autumn, found a stoat's larder about a foot underground, comprising two or three dozen partridge eggs packed neatly at the end of the burrow. The keepers knew that the nests that had been robbed were all along the hedgebank up to 100 m away, so the eggs had been carried some distance, but not one of them was damaged. A year later, Cringle saw a stoat carrying a tern's egg by walking on three legs and holding the egg against its chin with the fourth.

Studies of all the predators that kill partridges, and of the presumably serious consequences, were regarded as an important part of the Partridge Survival Project right from the beginning. Potts began the work convinced that this presumption was wrong, and set up the study expecting it to show that predation did not control the population density of partridges. Over an area of 13.1 km^2 of the South Downs at North Farm, Sussex, Potts and his team monitored a continuous effort to control all predators of partridges other than the protected raptors. The main predators of partridge eggs, the carrion crows and magpies, were practically eliminated; foxes were removed at an average rate of 3.2 adults per km^2 per year, stoats at 3.7 per km^2 per year, and feral cats at about 1 per km^2 per year (Potts 1980). On other parts of the study area, predators were not systematically controlled.

The team then monitored the reproductive performance of partridges in the areas with and without predators. They also measured various other things, such as the extent of nesting cover and yearly variations in food supplies and in the weather, which could also affect the birds' success.

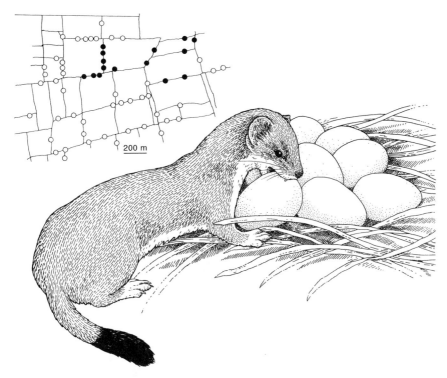

Figure 12.3 Stoats often raid gamebird nests and eat or remove the eggs. Inset: a single stoat was believed to have been responsible for all the partridge nests lost (*filled circles*) in one small area of an estate. (Map from Tapper et al. 1982.)

This effort was systematic and intensive enough to achieve considerable success in removing predators, at least temporarily. For example, although on this estate the traps for stoats were set all year round, the most determined effort was made in spring, when the nests of wild game bird are most vulnerable. The records showed (Figure 12.4) a steep decrease in the number of stoats killed per 100 traps over the first 6 months of the year. The team concluded from this that the resident population was largely removed each spring, although it was replaced later in the year as the season's crop of young stoats dispersed.

Tapper (1976) collected 151 common weasels and 46 stoats from the study area during the critical months of May, June, and July of the years 1971–1974. Of all the items he identified in guts, game birds comprised 2.1% and 6.8% of items eaten by common weasels and stoats, respectively. Tapper calculated, from the densities of chicks in his study area (then about 94 per 100 ha) and from the literature on food consumption and density of weasels (see Tables 2.1 and 10.2), that predation could be an important cause of mortality in game bird chicks in

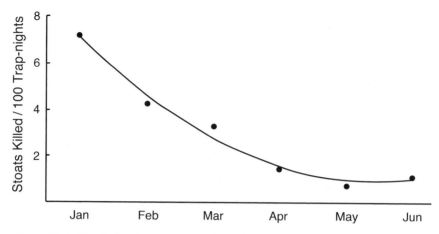

Figure 12.4 The decline in average number of stoats caught per month in the intensive spring campaign at North Farm, Sussex, implies temporary local extinction, but it was usually reversed by immigration within a few months. (Redrawn from Tapper et al. 1982.)

some years, but irrelevant in other years. The difference between years was due to insects, not weasels.

Partridge chicks feed entirely on insects for the first 2 to 3 weeks of their lives. More than 70% of the chicks that fail to reach adult size die during this early period, and this mortality is related both to the abundance of insects and to the weather. In a good year for insects, the young partridges that are not killed by predators are likely to survive, because they have enough food at this critical stage. By contrast, in a bad year for insects, it does not matter if most of the chicks are killed by predators, for they probably would not have survived anyway. Hence, the effects of predation are not the same every year, and calculations of the value of removing predators have to take into account umpteen other factors.

Potts identified three sources of density-dependent losses (those whose action is directly related to the number of animals present): (1) the density of breeding female partridges in spring, (2) the proportion of pairs with a brood in early summer, and (3) the proportion of the total population shot in autumn. The first and, especially, the second of these losses can be reduced by effective removal of predators. Losses of adult birds in early spring are not too serious unless very severe, because those killed can be replaced by immigrants during the partridges' annual competition for territories. By contrast, losses of sitting hens are crucial because by that stage they cannot be replaced. If many sitting hens are killed, productivity for the season will be seriously reduced.

The mortality of the chicks is another important loss, but it is density *inde*pendent, because it is related to the abundance of insects and to the weather, rather than to the density of partridges. Its action does not slacken as the number of partridges declines; it acts blindly, and if it continues at a high rate, nothing can stop it driving the population to extinction. Removal of predators that eat chicks does not improve the survival of chicks in poor years, although density-dependent processes acting later can, in some years, compensate for a poor season for production of chicks, especially if predators are removed.

After 10 years of fieldwork, plus endless patient study of an enormous volume of contemporary and historical data on partridges, Potts set up one of the earliest of the computer models, of the type that we now take for granted, to make sense of these complicated interactions. He deduced that the main reason why grey partridges have become so scarce in Sussex was the high mortality of the chicks, due to the destruction of their food supplies by agricultural pesticides. The only hope of restoring numbers, he suggested, is to reduce the effects of pesticide spraying at the field margins. Nevertheless, this policy would be effective *only* if nesting success was improved at the same time. Why? Because the effects of pesticides were accelerated by predation during the nesting season, which also strongly affects the density of partridges, especially the number of surplus birds available for shooting. Time has shown that Potts' diagnosis was right.

The long-term population decrease in partridges in Britain also coincides with an 80% reduction in gamekeeping since 1911 and a 40% loss of hedges since the 1930s, both of which greatly increase the hazards of nesting for the sitting female partridges. Which of these losses might have disadvantaged the partridges most?

Ever since the work of the influential American ecologist Paul Errington in the 1930s (Errington 1963), game managers have believed that the population density of most animals is set by the extent of their preferred habitat (Chapter 7). Errington was convinced that predation is more often compensatory (substituting for other causes of mortality) than additive (adding to other causes). When predation is compensatory it is relatively unimportant except to "the doomed surplus" individuals unable to find a secure home base. The logical implication of this line of reasoning is that habitat improvement is a much more important means of increasing game than predator control, or even the only means worthwhile. So, Potts' most interesting finding was a completely unexpected interaction between the extent of nesting cover (measured as kilometers of hedgerow per km²) and predator control. Potts argued that, although it is true that predation is not important all year round, it is very important during the nesting season.

Potts concluded that the most urgent and effective way to conserve partridges is to curtail the spraying of toxic chemicals at field margins, since no other measure can have more than a temporary effect if chick mortality remains high. After that, habitat improvements such as the provision of more and better nesting cover will increase the number of partridges only if predators are controlled as well, at least in the nesting season (Figure 12.5).

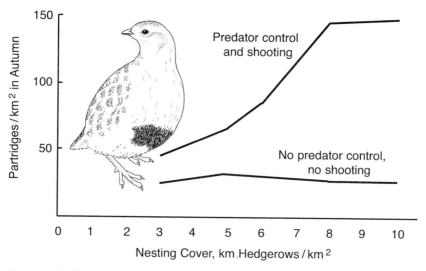

Figure 12.5 G.R. Potts' analysis predicted that the highest numbers of grey partridges in autumn (the shooting season) would be found on estates where nesting cover has been augmented *and* predators have been removed. (Redrawn from Potts 1979.)

The effect of predator control is most decisive where nesting cover is already sufficient to support nests at a density at or above 25 nests per kilometer of hedgerow (Potts 1979). Without gamekeeping, the success of these nests is only about 10%, mostly because the more sitting birds there are, the more easily can predators find and kill them. Gamekeeping in addition to plenty of cover increases nesting success to between 60% and 70%, and results in up to three times more birds in the autumn population. In nature, these extra birds would be the 'doomed surplus' which would be removed by over-wintering mortality or by emigration of birds next spring; but on a game estate nature is managed so as to direct the largest possible proportion of the doomed surplus into the shooters' bags.

The Sussex study area has now been monitored for over 30 years, and provides one of the longest running data sets on the cereal ecosystem in the world (Tapper 1999). The longer it goes on, the more amply it confirms Potts' early conclusions on the interaction between pesticides and predation on farmland birds. It also led to an interesting experiment, done in Wiltshire over 6 years (1984 to 1990).

This research used a classic paired plot design done on two areas of about 5 km² each, and separated by 6 km (Tapper et al. 1996). The common predators were removed every spring for 3 years from one plot, and from the other for the next 3 years. Predator control focused mainly on foxes and corvids, but also removed stoats at an average rate of four to seven per month during each

4-month nesting season. Partridge stocks on the controlled area increased by 75% in the late summer, and were still 36% higher by the start of the following breeding season. Most of this effect was due to the removal of foxes, even though they depended mainly on rabbits and took partridges only incidentally. Stoats played a minor part in this drama, as elsewhere (Chapter 7), but their contribution was still counted.

Common weasels take only chicks, so their effect is short term, and often too minor to be damaging, especially during the years that voles are abundant. On the other hand, artificially reared game birds are especially vulnerable to weasels, because these little predators can run underground through mole tunnels (see Figure 8.4) and come up on the inside of supposedly predator-proof wire enclosures designed to protect the young birds.

Most gamekeepers have a fund of eloquent eyewitness accounts of this problem. For example, one keeper had been unable to prevent continuing losses of pheasant chicks from their wire-mesh rearing coops, or to find what had happened to them, until one morning when he was moving the coops. He went to fire a heap of old straw, but caught a slight movement and, instead, began to turn the heap over while his mate, a first-class shot, stood by with a gun. Under the straw was a mole's nest containing a score of dead chicks, each of which must have been carried or dragged from inside a coop, some 150 m across the grass or through a mole run. While they stood there, a common weasel made a dash for freedom, but the keeper's mate missed it with both barrels.

Stoats also destroy many eggs and chicks from pheasant nests, and they can also kill the sitting hen. Predation by stoats can therefore make a contribution to the problem in the long term, since losses of sitting birds are more directly damaging than losses of eggs or chicks. This is what makes them a pest, less serious than foxes and corvids (Tapper et al. 1996), but more serious than polecats, rats, hedgehogs (Packer & Birks 1999), and common weasels (McDonald & Harris 1999; McDonald & Birks 2003). Studies of upland game birds such as grouse and curlew have reported similar conclusions (McDonald & Murphy 2000).

Modern Game Estates: Controlling Predator Damage, Not Predators

The modern face of game preservation is very different from that of its nineteenth-century forebears. It is still a major participant sport and contributes several million pounds a year to the rural economy in Britain, even though the numbers of keepers is now less than 3,000, and the extent of the area they patrol is reduced to about 27,000 km² — about 12% of the area of Britain (McDonald & Murphy 2000).

Changes in agricultural policy since the 1970s have made the British countryside much less hospitable to wildlife of any sort (Tattersall & Manley 2003), and the large-scale "driven" shoots of Edwardian times, which relied on the millions of wild grey partridges a year bred on each estate under the protection of the gamekeepers, are just a memory. To produce enough game birds for a day's shooting, many estates simply rear large numbers of birds artificially, mostly pheasants (an introduced species in Britain) and release them shortly before the shooting season. The few gamekeepers that are left spend much less time on predator control (McDonald & Murphy 2000), so the total national tallies of stoats and common weasels killed per year are declining (Tapper 1999).

Nevertheless, some estates still take the traditional attitude and attempt to reduce the local density of all unprotected predators of game birds. Fewer than 20% of gamekeepers see stoats and common weasels as major threats now (McDonald & Birks 2003), but both are still included in routine trapping operations, at least during the nesting season. To be effective, more individuals must be removed than can be replaced, at least locally and temporarily, and that means a huge proportion of the target populations have to be removed (for common weasels and stoats at least 80%), even though those that are removed will be replaced very quickly (McDonald & Harris 2002).

Moreover, as Errington saw long ago, even this rate of removal will not have any effect on the density of residents if it only substitutes for, rather than adds to, natural mortality (Reynolds & Tapper 1996). For example, in winter when natural survival rates are low, or in autumn when dispersal is high, trapping can remove large numbers of animals without having any effect on local density. The effects of predator control are seasonal and short, but on intensively keepered estates still specializing on producing wild-bred birds (considered a more sporting target), well worth doing.

On North Farm (Sussex) most resident stoats and common weasels are removed by June. This effort achieves significant protection for game birds and their young during the critical breeding season, even though by September the predators are back, replaced by immigration from surrounding untrapped areas (Tapper 1979; Tapper et al. 1982).

The traps at North Farm are set out at the rate of 300 over 1,300 ha, that is, averaging about 200 m apart in all directions. They are placed along the hedgerows and field margins where common weasels and stoats are bound to find them. The traps are not baited, largely because the sites are so attractive; weasels in general are not only naturally curious, but they also like to keep under cover, and a good site that happens to be on a regular runway under a hedge will go on catching each successive local resident in turn.

When such a system is operated by experts, a real reduction in numbers of stoats is possible, at least while it continues (Figure 12.4). The concentration of the keeper's trapping effort in spring produces a distinct seasonal peak in the mortality rate of stoats and common weasels (see Table 11.3), although it does not necessarily increase the general level of *annual* mortality. Either way, the danger period has passed by June, and in any case affects mostly males. Over the course of a year, many more common weasels and stoats will be taken from an estate than could live there all at once, especially in vole peak years.

In very large areas of less accessible landscapes, where a weasel has a relatively low chance of finding a trap or being shot, attempts at "control" usually turn into harvesting, which takes a yield but does not affect density. For example, on a 50-km^2 stretch of coastal sand dunes in the Netherlands, managed as a game reserve for pheasants, a force of 13 gamekeepers reaped a relatively constant harvest of stoats averaging 210 ± 32 per year, every year from 1952 to 1966. According to Heitkamp and van der Schoot (1966), about 75 stoat territories could have covered the whole reserve, on each of which an average of 4.5 young stoats might be born, giving an annual increment of about 340 stoats a year. Since the keepers removed only 62% of that number, some unknown proportion of the surplus must have been left to die naturally. A century of similar efforts on the game estates of Europe has had no long-term effects on the distribution of stoats and common weasels, for reasons easily understandable from their population biology.

WEASELS AND THE CONTROL OF PEST MAMMALS

Least and common weasels are the smallest and probably the most numerous carnivores in the Holarctic and, weight for weight, they eat the most. The daily food requirement of a biomass of, say, 50 kg of least/common weasels would be much greater than that of an equivalent weight of foxes: 50 kg of least/common weasels eating 35% of their body weight in food per day (see Table 2.1) would need 1,750 g of food (about 88 mice), whereas 50 kg of foxes, at 8% per day, would need only 400 g (20 mice).

Better still, when rodents are abundant, weasels do not kill only what they need for the day: They kill as many as they can and store up the surplus for future use. If the prey were, say, voles causing damage to crops, the weasel would be a much better friend to the farmer than the fox. It is therefore assumed by some that weasels in general have the power to "keep the rodents down" (plus any other small mammals that can cause a nuisance, such as rats and rabbits) and that therefore farmers and foresters should actively encourage weasels to live on their land.

Norman and Stuart Criddle (1925) were among the earliest North American champions of weasels, arguing against the common perception, picked up from the folk lore of Europe and "the game mentality," of the weasel as a savage killer of game birds and poultry. They listed many field observations of all three species of weasels in Canada in the early years of the twentieth century, mostly describing weasels hunting rodents or rabbits. They pointed out that, in 9 years out of 10, a longtail will find sufficient food in the fields and woods, and only on the tenth turn temporarily to domestic fowls. A thousand mice may have been killed in the meantime, but the destruction of half a dozen hens is enough to label the weasel a pest. Surely, they add, this is a remarkably small payment for the great good done by weasels in killing rodents and ample evidence that weasels should be protected.

Field observers can easily find evidence to support this attitude. Lippincott (1940) described how he had tracked a weasel through the snowy woods and farmlands around his home near Philadelphia. All along the two-mile trail, the weasel took no notice of rabbit sign or gray squirrel tracks or roosting pheasants. It stopped only to flush out a rat from a pile of corn shocks and two meadow voles from under a low tree, and to try to get into two other mouse holes, which were too frozen for even its slim body to enter.

On another occasion Lippincott was surprised to see meadow voles running across the lawn beside his house, until he found a small weasel busy flushing them out of nearby weeds. Later the same day he saw three meadow voles scurry across the road, with a brown streak in furious pursuit. He concluded that, even if a weasel does take a few ducklings in the springtime, he still preferred to have a weasel about the place than hordes of rodents.

Other North American authors have also pointed out the absurdity of killing weasels on sight even where there is no bounty on them (Hamilton Jr. 1933; Quick 1944). Weasels probably kill hens only when barnyard rats are scarce, and the rats may well have killed more hens themselves. These authors supported the same idea: Weasels perform a valuable service to farmers.

Whether or not this argument is actually correct has never been tested. In logic and in fact, it is quite untenable, for the same reason that no one attempts to judge the financial position of a company only from the debit columns of its balance sheet. The effect of predation by weasels on rodents is the difference between the rate at which weasels remove rodents and the rate at which rodents are replaced (see Table 7.1), and this is impossible to figure out from casual field observations.

Nevertheless, the idea that weasels "keep the rodents down" remains strong in the public imagination. The logical conclusion is that it should be possible to enlist the help of weasels to reduce high numbers of pest mammals in field crops and forest plantations, or at least to control the damage done by pests. For this purpose, weasels have been released in various places, with mixed results.

In Canada, deer mice and Oregon voles are considered serious pests of forestry, because they eat great quantities of conifer seed and debark young trees, thereby interfering with the regeneration of logged forests. In one short experiment at Maple Ridge, British Columbia, in the autumn of 1978, seven of the small Canadian stoats were captured and released on a 1-ha experimental plot, while the numbers of mice and voles were monitored both there and on a control plot (Sullivan & Sullivan 1980). The hope was that stoats at such a high local density would reduce the number of rodents on the experimental area.

Unfortunately, they did not. Numbers of deer mice remained high in both areas, while numbers of voles decreased in both. On the other hand, there was no evidence that the stoats had stayed where they had been released; seven could not live together on 1 ha at any season, especially not in the fall when the young are normally dispersing. In fact, two of the stoats returned to their original home ranges, up to 4 km away. It is clearly unrealistic to expect predators to stay put and clear rodents from an unfenced area of mainland, just to oblige the owners of the forest.

Weasels seldom persist in hunting prey at low densities unless they have no alternative. The only temperate habitats where the choice is severely limited and emigration impossible are small islands, and weasels arriving on islands can have a substantial impact. One of the earliest known deliberate introductions was in the Shetlands, off the north coast of Scotland, to which stoats were taken in the seventeenth century or earlier (Venables & Venables 1955), but the consequences for the local fauna went unrecorded.

Much more is known about another introduction, on the island of Terschelling off the coast of the Netherlands (680 km²). During an afforestation program in 1910–1930, pines, oaks, and alders were planted on 600 ha of open sand country. The ditches dug to drain the plantations provided ideal conditions for water voles, and after about 1920 the high numbers of voles began to damage trees and gardens. Bounties, mouse typhoid, and poison failed to control them so, in 1931, stoats and common weasels were introduced. By 1937, the stoats had established a fluctuating but substantial population; the water voles (and the common weasels) were extinct (van Wijngaarden & Mörzer Bruijns 1961). Without the water voles, the stoats could not survive either and they, too, are now extinct on Terschelling (Mulder 1990).

This success was repeated in even shorter time on the much smaller Danish island of Strynø Kalv (46 ha), south of Fyn (Kildemoes 1985). High numbers of water voles were destroying the dikes and damaging crops. Beginning in October 1979, the numbers of water voles were assessed by trapping sessions of 5 to 6 days in October, March, May, June, and August for 2 years. In May 1980, six male stoats were released, evenly distributed along a 2-km dike. The numbers of both water voles and of ground-nesting birds on the island were regularly monitored until October 1981. In the first breeding season, 1980, there was no effect on the birds and not much on the voles. In the following year, water voles were very scarce, even though breeding conditions elsewhere in Denmark were favorable in 1981. Some birds were fewer than usual, but could soon be replaced from neighboring islands.

More often, however, weasels fail to do what is expected of them by those who believed that they would "keep the rodents down." The best-documented case is that of New Zealand, where ferrets, stoats, and common weasels were deliberately introduced in the mid-1880s, in the misplaced hope that they would control rabbits (King 1984b). They failed (Chapter 13).

With hindsight, we can see now that the New Zealand experiment could never have worked on the scale required (Trout & Tittensor 1989), but at the time it seemed to be the only way to save the sheep farmers from the ruinous plagues of rabbits. It so happened that one of the most famous examples of successful biological control of pests by introduced predators, the immediate repression of scale insects by the vedalia beetle in the citrus orchards of California, achieved spectacular success almost overnight in 1889. It must have greatly encouraged the continued importation of mustelids into New Zealand, which was going on at the same time. Vedalia beetles saved the citrus farmers, so why did mustelids fail to save the sheep farmers?

The answer is that vedalia beetles have advantages over their prey not shared by mustelids. These advantages include a reproductive rate nearly matching that of the scale insects, efficient aerial searching, no territorial restrictions, and absolutely no risk from killing prey because they are immobile and conveniently clumped. When the scale insects increase in numbers, the vedalia beetles can respond immediately, and when a beetle is stuffed full it simply produces more mouths to continue feeding. These characters make the beetles superefficient shoppers (see Table 7.1).

By contrast, there is a huge disparity in the reproductive rates of stoats and rabbits; searching on and under the ground is energy sapping, inefficient, and confined to the stoat's own home range, and adult rabbits run and hide, and when found are risky prey for stoats (see Figure 6.3). Even when a stoat is stuffed full it cannot produce any more mouths to feed on rabbits until the following year, so it usually goes to sleep. These differences help to explain why stoats never had any hope of controlling the numbers of rabbits over an area as large as the two main islands of New Zealand (114,000 km^2 and 157,000 km^2). Rather, as the history of myxomatosis shows, the converse is true: Rabbits control stoats (see Figure 10.2).

Natural predators are not normally geared to control the numbers of their prey, only to harvest them. A predator can control its prey only if it eats more prey animals than are produced as they increase, and fewer than are still surviving as they decrease (see Table 7.1). This is not what weasels do.

The combined force of all the local predators may have a substantial effect on rodents or birds (Chapter 7) but, in natural ecological communities, weasels usually make only a modest contribution to the total effect. The rate of reproduction of all weasel species is slower than that of any rodent species, and the number of rodents an individual weasel can kill in a day has a practical limit. Consequently, weasels cannot hope to match their consumption to the numbers of prey, so they cannot prevent an increase, nor start a decrease, unless the rodents have already ceased to add recruits to their population. During and after the decrease phase, the "overshoot" of predators caused by the lag of their response ensures that their predation is disproportionately heavy just at the time it should be slackening. This is why predation by weasels has such a destabilizing effect on populations of lemmings and voles such as those described in Chapter 7.

In the real world, the population dynamics of weasels are almost entirely controlled by the density of prey, whereas the density of prey is affected by many other things besides predation—particularly food supplies and social behavior. So, contrary to appearances, weasels do not have the whip hand: Rather, it is the other way about. The very mechanics of weasel predation make it impossible for them to "keep the rodents down," and this is why attempts to use weasels as a control agent on mainland areas have failed.

The only way to get around this is to remove their power of choice. Hunting gets more difficult as the prey become scarce, and the last few are the hardest of all to catch. The weasels' natural inclination to move away when hunting is unrewarding can be frustrated only on small islands. There, it is a matter of hunting the last few or starving; weasels have the same instinct for self-preservation as the rest of us, so, not surprisingly, they oblige.

PROTECTION FOR WEASELS?

After centuries of killing predators to protect human interests, concern for predatory animals as individuals and for their own sake is a very recent idea in Europe (Reynolds & Tapper 1996). European settlers carried their cultural attitudes to North America, including the selective destruction of predators to favor game, although that policy was on its way out in the national parks of the United States by the 1930s (Dunlap 1999). Weasels are small and subject to little rancorous debate compared with large predators, such as wolves, coyotes, and pumas. People wishing the demise of predators and those wishing their protection seldom address weasels, for several reasons. Weasels may kill chickens and other small farm animals, but they do not have the potential to cause large-scale, high-profile damage as do big predators. In addition, weasels are not considered charismatic by most people and, consequently, have few high-profile defenders. Finally, weasels are widespread and generally abundant. Though populations regularly go extinct locally, recolonization is just as regular. Restoration of lost wild populations with captive-bred stock is generally not needed, and unlikely to be practical (Hellstedt & Kallio 2005).

Weasel populations deserve, however, more attention in North America than they have ever received. The actual status is known for few weasel populations, so some could easily go extinct locally and not be reestablished, and no one would know. We know of only one weasel population that might need special protection of its only habitat. The four subspecies of stoats that have been described on the Alexander Archipelago of southeastern Alaska (*Mustela erminea salva, M. e. celenda, M. e. seclusa, M. e. alacensis*) comprise only three distinct lineages, one of which is endemic to the archipelago—the *M. e. celenda-M. e. seclusa* lineage (Cook et al. 2001). Large areas of the forested habitat of that distinct, endemic lineage on Prince of Wales Island are being logged. Island habitats are naturally fragmented anyway, and when that patchwork effect is augmented by extensive fragmentation caused by logging, local extinctions are undoubtedly more common and recolonization slower and more difficult.

In Europe, populations of stoats and common weasels probably do not need protection from human interference. Neither species is protected under any British conservation legislation (they are the only native mustelids that have no conservation status in British law) and neither is being considered for it (McDonald & Murphy 2000), although stoats are protected in Eire. Both species are included in Appendix III of the Bern Convention, which requires only that their status and exploitation be considered by signatory countries.

Culling imposed locally by gamekeepers does not have the capacity to cause a long-term decline in populations of stoats or common weasels, because it is often not heavy enough to reduce density, and even when it is, the effect is always undone within a few months by rapid recolonization. The decrease in numbers killed per year reported by the Game Conservancy is due to declining trapping effort (McDonald & Harris 2002). One could even argue strongly that stoats and common weasels should *not* be protected from trapping, because the loss of access to regular, legal gamekeepers' records would remove the only useful method of monitoring their populations and the vital conservation information such records supply (McDonald & Murphy 2000).

By contrast, small mustelids in general are certainly at risk of secondary poisoning, since around 70% to 90% of arable farms and game estates in Britain use rodenticides both around buildings and in the field (McDonald & Harris 2000). Nine of 40 stoats and three of 10 common weasels examined by McDonald et al. (1998) contained residues of rodenticides. Both species are known to eat moribund and freshly dead rodents affected by poison, with fatal consequences (Chapter 13), although it is not possible to detect from present information whether local populations are affected in Britain.

A more natural hazard is that, since stoats seem to come off worst in encounters with foxes (Mulder 1990), they may also be disadvantaged by the steady increase in British fox populations over the last 35 years (Tapper 1999), depending on how stoats and foxes respond to the equally important parallel increase in rabbits. The same idea has been discussed in North America (see Table 11.5), although it has never been conclusively resolved anywhere. The complex interactions between stoats, foxes, and rabbits remain to be tested.

The one thing small predators do need is protection against the legal use of cruel traps, and in Britain the first attempt to tackle that problem was made more than 50 years ago. The Fenn trap met the definition of a humane trap developed during the 1950s (70% of captured animals must be found dead within a day). The Fenn was efficient and widely used (Anon. 1981) after leghold traps were banned in England under the Pests Act (1954).

In New Zealand, leghold traps were still legal in 1971 when King (1981b) arrived there and persuaded most (not all) national park rangers to use Fenn traps instead (King & Edgar 1977). Analysis of carcasses collected from suppliers that used both confirmed that the Fenn was more humane than the leghold trap, and much better for collecting undamaged carcasses for research (King 1981b).

In New Zealand and many other countries including Canada and Russia, leghold traps are still legal at the time of writing (2006), but perhaps not for much longer. The new definition of a humane trap requires it to guarantee, to a high probability, that captured animals will be rendered irreversibly unconscious within 3 minutes (Powell & Proulx 2003). Fenn traps do not meet these standards (Purdey et al. 2004). New and stricter testing standards for kill traps are now being developed, and when these are adopted, only traps that meet the new standards will be permitted and all other trap types, including the current versions of the Fenn and many others, will be banned.

At the same time, the need to protect native fauna is more urgent than ever, hence the explosion of urgent research on developing new humane traps and alternative, nonlethal control methods (Murphy & Fechney 2003). The hope is that leghold and all other cruel traps can then be outlawed everywhere. Powell and Proulx (2003) reviewed progress made toward this aim in North America.

13 | Stoats as Introduced Pests in New Zealand

ACROSS NORTH AMERICA and Eurasia, both stoats and least/common weasels are native predators that may attack introduced game birds; in New Zealand, they are introduced predators that often attack native birds, many of them endemic. This simple fact makes a world of difference between the attitudes to small mustelids in the northern and southern hemispheres (McDonald & Murphy 2000). Regardless of their sins in chicken coops and on game estates, stoats and common weasels in their native countries have an intrinsic value that they do not have in New Zealand.

Biogeography also makes a difference. The inshore islands of the northern hemisphere are not very different from the mainlands, and those separated by only shallow (<100 m) channels have been connected to the mainland during periods of low sea level in the past. The larger ones are inhabited by weasels, and many of the smaller, closer ones are visited from time to time, since all weasels, especially the large ones, are good and confident swimmers (Chapter 6). The native faunas of those islands have evolved alongside weasels for millennia, so include no species that cannot coexist with them over the long term.

Matters are very different, however, when weasels reach an island further offshore or in the open ocean that has never been connected to a continental mainland. History does not prepare the inhabitants of such a place to deal with predators, and the consequences are likely to be catastrophic. New Zealand is by far the best-documented example of this tragic process.

RAIDERS OF THE LAST ARK

The New Zealand archipelago has a total land area about the same as Britain (270,000 km²), but it has been isolated in the southwest Pacific for some 65 to 80 million years—that is, since before any modern placental mammals evolved. In the absence of terrestrial hunters (mammalian carnivores and snakes), many of the longest established birds of New Zealand found it safe to live, feed, and nest on the ground. In the course of time, many of them became large, flightless, slow breeding, and unique.

New Zealand was the last major land mass on earth to be colonized by people. A series of human invasions, beginning with the first permanent Polynesian settlements about 700 years ago, brought to the islands increasing numbers of colonists plus alien mammals and birds. The intruders arrived with stunning speed, numbers, and superiority, and they overwhelmed the previously undisturbed native fauna, with devastating consequences (Worthy & Holdaway 2002). Twenty-five species of introduced mammals are regarded as pests in New Zealand (Cowan & Tyndale-Biscoe 1997); 50% of New Zealand's breeding bird species have been lost, or 60% if the critically threatened survivors are included (Holdaway 1999).

Stoats and common weasels arrived in large numbers on a series of organized shipments from Britain over about 20 or 30 years after 1884 and liberated on pastures teeming with rabbits. They were part of a huge, unplanned experiment in biological control, intended to save New Zealand pastoral farmers from being ruined by plagues of European rabbits. The assumption was that, in Britain, predators normally kept rabbits under control. That was the only reason people could think of to explain why, in New Zealand, rabbits free of predators reached numbers unknown in Britain, competing with sheep for grass and undermining the pastures with burrows. Therefore, it stood to reason that all that was needed to reduce rabbits in New Zealand to the same nonnuisance level as back in Britain was to bring in predators that could eat rabbits. Foxes were banned, so the task fell to stoats, common weasels, and ferrets.

In the early 1880s, the New Zealand government began advertising for live stoats and common weasels to import. Many of the 17,000 or so gamekeepers then employed on the game estates of Britain (Chapter 12) were glad to take the opportunity to earn some extra pay. The £5 offered for a pair of common weasels was worth the equivalent of about £265 (US $410) today—close to a week's wages. The little animals were easily caught in box traps and shipped in small cages. The mortality rate among them, in the traps and on the long voyage under sail, must have been high. Nevertheless, the scheme ran for at least 20 years, and in the first 2 years, 1885 and 1886, 224 stoats and 592 common weasels reached the other side of the world alive and were released on the rabbit-infested pastures of the new colony (King 1984b).

Stoats certainly could and did kill rabbits, as in Britain, but the farmers' expectation that they would stay on the sheep runs and hunt rabbits indefinitely was asking too much of them. Right next to the pastures were the forests, where birds and wetas (large native insects) nesting in trees were certainly not out of reach of agile stoats. And despite a hundred years' attention from Norway rats and feral cats, semi-wild dogs and human hunters, some of the native birds and lizards that foraged and nested on the ground still survived.

Within a very short period, both species spread throughout the country. Common weasels have always remained scarce, probably because New Zealand has no voles (King & Moors 1979a), so the following account does not include them. By contrast, stoats can now be found almost anywhere, including in all the national parks.

Public attitudes toward stoats in New Zealand are universally hostile, especially where the destruction of the ancient, endemic birds was recent enough to have been witnessed by the parents or grandparents of people still living in the same areas today. Because stoats appear to be such efficient predators, they are sometimes blamed for *all* losses of native fauna, both historical and contemporary. But what stoats might have done in the past and what they can be shown to be doing today are quite different processes, and need to be assessed separately.

THE HISTORIC EXTINCTIONS

Predation has been hugely significant in the historic decline in New Zealand's native fauna (Innes & Hay 1991; Holdaway 1999; McDonald & Murphy 2000; Worthy & Holdaway 2002), but most of the historical extinctions could not have been the work of stoats—not because they were not capable, but because they had no opportunity. Other predators were the first to find the dozens of colonies of tame, defenseless, ground-nesting, and meaty birds that have since disappeared, not only in New Zealand but also all around the Pacific (Steadman 1995).

Melanesian and Polynesian voyagers and their companions (especially Pacific rats) had discovered almost all island groups in Oceania by about 1000 AD. The present Maori population is descended from Polynesian settlers, who arrived in about 1300. European explorers and their companions (Norway and ship rats, plus shipboard cats and assorted livestock, especially goats and pigs) arrived in an accelerating tide from the 1770s onward (Holdaway 1999).

By the time the first stoats arrived in New Zealand in 1884, the native species had been suffering centuries of irreplaceable losses. During Polynesian times, birds and insects that were flightless, or that nested and foraged on the ground, were the first to disappear, along with lizards and frogs. Rats wiped out the small species, spears and clubs the large ones. The medium-sized birds, especially those that nested in trees, fared a little better, but only for a while.

Then, between the first landings by Europeans in 1769 and the first releases of stoats in 1884, more than another century of further damage had been added by cats, by Norway and ship rats and dogs, and by rapid and drastic habitat modification. From data available, *at least* 142 separate island populations of native birds are known to have been extinguished from the main and offshore islands (Holdaway 1999; King 2005c). Of these, 114 (80%) were never in contact with stoats. Conversely, stoats have never reached the most important offshore islands (Stewart, Kapiti, Little Barrier, or any of the outlying islands) on which many endemic birds have survived, sheltered from the disturbances on the mainland.

On the other hand, on parts of the two main islands, especially in the rugged southwest South Island (now the Te Wahipounamu World Heritage Area,

26,000 km²), the ancient fauna was still largely intact when stoats arrived. Some populations of birds that are now lost or severely threatened were still hanging on there, and for them the arrival of stoats was the last straw (King 1984b). Stoats certainly contributed, with the help of ship rats (which arrived at about the same time), to the final disappearance of the South Island subspecies of the bush wren, NZ thrush, laughing owl, saddleback, and kokako (King 1984b). No doubt stoats also immediately began to aid the already advanced and still ongoing losses of the kakapo, both subspecies of takahe, at least three species of kiwi (Figure 13.1), plus the kaka and the yellowhead (Figure 13.2). The periodic increases in predation by stoats after beech mast years (Chapter 10) may explain why the South Island kokako, which lived mainly in beech forest, disappeared while the North Island kokako, which lives mostly in nonbeech forests, has survived into the era of effective predator management (Clout & Hay 1981; Innes et al. 1999).

These grim figures do not mean that the colonizing peoples and their animal companions were unusually rapacious, only that long-isolated island faunas are extremely susceptible to invasion and losses of endemic species cannot be made good from elsewhere. Invasions by commensal rodents were accidental and could not have been prevented, but the deliberate addition of stoats and common weasels to the list of immigrants has arguably made the tragedy far worse. It is all the more sad because the consequences were predicted by conservationists at the time. Their warnings were ignored, and the legislation they proposed to prohibit the import of mustelids was opposed by powerful economic interests (Hill & Hill 1987).

Figure 13.1 Young kiwi (<800 g) are too small to defend themselves from stoats and, consequently, stoat predation on young kiwi is a serious threat to contemporary kiwi populations in New Zealand.

Figure 13.2 Predation by stoats has contributed to the continuing, severe decreases in populations of yellowheads, kakapo, takahe, and saddlebacks from the late nineteenth century onward. Kakapo and saddlebacks no longer survive on the New Zealand mainland at all; yellowheads and takahe do so only in some protected areas.

STOATS AND THE CONTEMPORARY BIRD FAUNA

For many years, the only conservation strategy considered necessary in New Zealand, as in the northern hemisphere, was the establishment of legal protection for national parks and reserves and for native wildlife. Both of these aims were achieved on a large scale and earlier in New Zealand than in many other countries, including legal protection of several large, critically important, uninhabited, and mainly undamaged offshore islands.

Now it is clear that, vital though they were, these measures were not enough (Clout 2001). The fear is that, despite some famous battles won, mostly on offshore islands, conservation action is not halting the ongoing decline in biodiversity on the two main islands (Craig et al. 2000). The survival of many threatened endemic species on the main islands depends on active management, which for some must include control of stoats.

The technology for trapping stoats was easy to learn from British gamekeepers, but reducing local populations of stoats is a much more difficult task in New Zealand than on the average game estate (Chapter 12), for two main reasons. First, effective predator control is always extremely expensive, and therefore must be well justified. Unfortunately, acquiring the necessary data on predator damage is much more difficult for wild birds that nest in tall trees in mountain country than it is for game birds that nest on the ground in accessible farmland under the benevolent eye of a gamekeeper. Second, single-catch traps (the only type available at present) are not cost-effective over the very large areas of rugged, inaccessible mountain forests in New Zealand where the native fauna most needs protection. Widespread control of stoats using existing technologies is currently viewed as "an intractable problem" (McDonald & Murphy 2000).

For both reasons, recent research in New Zealand has concentrated on first, documenting the damage done by stoats; second, developing new and more efficient ways to prevent it; and third, broadening the whole concept of active management from the preservation of high-profile iconic species to the restoration of patches of semioriginal communities in specially designated "mainland islands" (Saunders 2000).

Documenting the Damage

Stoats are probably the most abundant carnivores in the remaining forests of New Zealand, yet the question of the extent to which they threaten the surviving birds was for a long time hard to answer. Early ornithologists often commented on the devastating effects of stoats on native birds (O'Regan 1966), but could not provide the systematic observations needed to balance estimates of predation by stoats against multiple other losses (O'Donnell 1996).

The first and most obvious approach to the question, analysis of what stoats eat, had been done on a large scale by the end of the 1970s (see Figure 5.4). This work confirmed that stoats in New Zealand do indeed eat a great many birds, but that information was no help in trying to assess the effect of the removal of so many birds from a given population.

The only thing that could be said from studying the stoats themselves was that, at certain places and times, stoats eat so many birds that they must be potentially a real threat. New Zealand does not have enough small rodents to "buffer" the birds from the attentions of predators. In the northern hemisphere, good seasons for rodents tend to be easy ones for the birds (Figure 13.3a, b). But in New Zealand beech forests in summer, extra supplies of mice do not significantly reduce the proportion of birds eaten by stoats until the mice reach plague proportions (Figure 13.3c), and that has been observed only twice in 20 years (White & King in press).

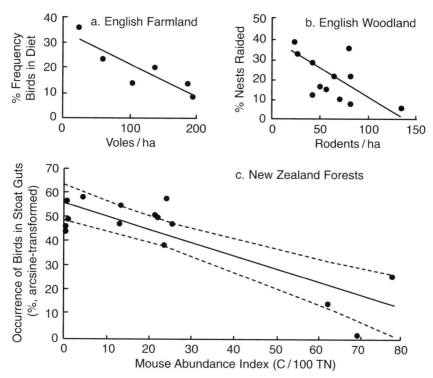

Figure 13.3 In England, as in Europe and North America generally, abundant rodents "buffer" birds against predation by weasels, both in open farmland (a) and in woodland (b). The graphs show that when voles are abundant, birds constitute a small part of weasel diets. The same is usually not true in New Zealand beech forests (c), where the rate at which stoats eat birds is unaffected by modest increases in mice. There, the buffering effect is possible only at extremely high densities of mice, and then only for a short period, because the periods when mice are abundant enough to distract stoats from taking birds are rare and brief (without the three right-hand points, the relationship is not significant). (Redrawn from Dunn 1977, Tapper 1979, and King & White, in press.)

Throughout the 1970s a great effort was made to document the biology of stoats, the necessary prerequisite for developing a workable management policy for them (King & Moody 1982). At that time it was impossible to disentangle the effects on birds of predation versus habitat destruction, and no hard data from New Zealand habitats were available to compare with the international predation literature that was still strongly influenced by the ideas of Paul Errington. King (1983b) suggested that the periodic summer irruptions of stoats she observed in 1976 and 1979 in the southern beech forests of Fiordland (see Figure 10.4) could be periods of greatly increased predation on birds, but confirming evidence derived from measurements of the productivity of the birds did not appear for another decade.

All the surviving native birds of mainland New Zealand live in habitats modified and invaded by a whole range of introduced alien mammals, and reduced and dissected to various degrees by human activities (especially deforestation). From the little that was known in the 1970s about the population dynamics of the remaining mainland bush birds, they appeared either to have accommodated themselves to the new conditions or to be already gone, to extinction or to permanent exile on safe refuges offshore (King 1984b). Doubt remained about the extent to which stoats could still be causing further losses, and it was not clear how we could protect the contemporary species from predation or, indeed, whether protection might now be too late.

Part of the reason for the doubt was that the only way to estimate the impact of predation on birds is to observe the interaction from the birds' point of view, and this is not easy to do. One of the earliest attempts was made in an intensively studied patch of bush near Kaikoura, in the South Island. There, stoats and common weasels together accounted for 77% of known nest losses of bush birds over three seasons, affecting native (101 nests) and introduced (48 nests) species equally (Moors 1983a, 1983b).

It was something of a surprise to learn that the introduced birds, which had evolved with predators and were supposedly better able to deal with them, suffered the same devastating losses as the native birds living in the same forest. Nonetheless, that information on its own does not tell us whether predation affects the numbers of either native or introduced species. Potts' partridge model (Chapter 12) is an example of how much information is needed to understand predator–prey interactions, and nothing like it was available for any New Zealand species until the 1990s.

A pioneering study of the South Island robin, done over several years at Kaikoura and on three mammal-free offshore islands, showed that the mainland robins survived by making a huge effort to compensate for their losses (Flack & Lloyd 1978; Powlesland 1983). By comparison with the island birds, which still live, presumably, in more or less their natural state, the mainland birds studied at Kaikoura for seven summers during the 1970s had much larger breeding territories (1 to 5 ha, instead of 0.2 to 0.6 ha); they started breeding much earlier and finished much later; their productivity was higher (clutch size three instead of two, three broods reared per season instead of one, and three juveniles fledged per pair per season instead of 0.14 to 1.1); nest failure due to predators averaged 55% per season instead of less than 10%; and annual adult mortality ranged from 23% to 37% instead of 17%.

Despite this enormous extra effort, the mainland birds lived at a much lower density (5 to 9 ha per pair, instead of 0.4 ha per pair) and not all available breeding habitat was used every season. The Kaikoura robins were able to compensate at that time, but elsewhere they and many other species could not. Most of those populations, including the one originally studied with such care and intensity at Kaikoura, are now at or beyond the threshold of extinction.

Under New Zealand law it has been illegal to kill or possess any of the threatened native birds since the Wildlife Act 1953, but that did not help them while forests (especially on the lowlands) were still being felled. Therefore, the first and clearest target for the young environmental movement of the 1970s was the fact that many important lowland forests and wetlands still remained unprotected (King 1984b). Some birds are acutely vulnerable to predation, others somewhat less so, but all are vulnerable to the loss of their habitat.

Throughout the 1980s, attention was focused on fierce arguments against large-scale logging of the remaining lowland forests, and the draining of wetlands. The main issue was that the system of reserves and protected lands then established did not adequately represent these lowland habitats, yet they supported the highest diversity of endemic birds. Many native species could not survive the loss or fragmentation of these habitats, with or without stoat control. At that time there *was* still a chance to defend, by radical action if necessary, lowland forests that were then still at risk of clear-felling (Wright 1980), so protesting against habitat destruction was, rightly, top of the conservation agenda.

Much progress was made during the 1980s in establishing legal protection for the lowland forests. Nevertheless, birds continued to disappear even from protected forests. The first edition of this book (1989:227) concluded cautiously that if "habitat protection can be assured, especially in areas where threatened birds still survive, then by analogy with the partridge work we may hope that some carefully planned stoat control work could be worthwhile." On what we know now, that must count as one of the understatements of the century (Wilson 2005).

Over the last few years, intensive field studies allied with computer models have begun to distinguish between the effects of predation and habitat loss. Understanding predation requires rigorous analyses and numerical models based on field-based, quantitative measurements of the mortality definitely due to each species of predator relative to all forms of natural mortality or failures in recruitment. Such measurements have been possible only since the development in the late 1980s of techniques able to measure productivity and mortality of eggs and chicks. The general conclusion is that the prime requirements for protecting native birds in New Zealand are similar to those for protecting partridges in England (p. 317): Habitat protection is absolutely necessary, but not by itself sufficient. Now that the need for protection of lowland forests is widely accepted, attention has turned back to the problem of predation.

Time after time, catastrophic failures in recruitment of young chicks of many of New Zealand's endemic birds have been traced to persistent predation. For some species, stoats are the main threat, and for other species, only an additional one (Table 13.1). For example, young kiwi are especially vulnerable to stoats, but uncontrolled dogs kill many adult kiwi. In beech forests in recent years, heavy seedfalls have been followed by irruptions of ship rats as well as of stoats, doubling the danger for birds that nest in tree holes. The yellowhead (an endemic forest passerine) was once very abundant but has now disappeared from more than 75% of its former range. Yellowheads are at particular risk from agile, tree-climbing stoats and rats (see Figure 6.4), and their productivity and mortality are both significantly affected in years of stoat irruptions (Elliott 1996a). As with the Wytham tits that nested in boxes accessible to common weasels (see Figure 7.1), the sitting females suffer much higher mortality than the males in years when the rates of nest predation are high. In New Zealand beech forests the increased danger can be predicted well in advance (O'Donnell & Phillipson 1996) and can be averted, at least locally, by intensive stoat trapping, except in years when rats are also numerous (O'Donnell et al. 1996; Dilks 1999; Dilks et al. 2003).

Likewise, the survival and nesting success of kaka (an endemic parrot) are seriously affected by predation on eggs, chicks, and nesting females by stoats (Moorhouse et al. 2003). In one forested study area monitored for 11 years, only four young kaka survived to independence out of 20 breeding attempts, and four adult females were killed on their nests (Wilson et al. 1998). The continuing slow decline of the endemic New Zealand long-tailed bat (one of only two surviving native land mammals) is also linked with post-seedfall irruptions of rats, probably accompanied by stoats (Pryde et al. 2005).

Table 13.1 Native Birds in New Zealand Potentially or Actually at Risk of Stoat Predation[1,2]

Species	Study area	Risk	Reference
North Island brown kiwi	Lake Waikaremoana	+++	(McLennan et al. 1996; Basse et al. 1999)
Haast kiwi, Okarito brown kiwi	Westland	+++	(Pyle 2003)
South Island kaka	Nelson Lakes N P	+++	(Wilson et al. 1998)
Yellowhead	Fiordland	+++	(Elliott 1996b)
Black stilt	Canterbury	+	(Dowding & Murphy 2001)
New Zealand dotterel		+	(Dowding & Murphy 2001)
Takahe	Fiordland	+	(Lee & Jamieson 2001)
Yellow-crowned and orange-fronted parakeets	South Island	+++	(Elliott et al. 1996)
North Island kokako	North Island	+	(Basse et al. 2003)
Yellow-eyed penguin	South Island	+	(Alterio et al. 1998)

1. Levels of vulnerability refer to stoats only (+++ very high; ++ high; + other risks higher); all species listed are at serious risk, but for some, factors other than stoats are more important.

2. For a summary of the current status of New Zealand birds, see Wilson 2005.

All species of New Zealand's national icon, the flightless kiwi, are very vulnerable to predation by stoats on chicks up to the age of about 6 months (McLennan et al. 1996, 2004). A model developed from systematic field observations predicted that "the persistence of kiwi on the mainland is now largely dependent on the development of new technology for controlling stoats" (Basse et al. 1999). In the summer of a good mouse year, stoat densities in kiwi areas can reach up to 10 per km^2; the model predicted that the maximum critical threshold density tolerable by kiwi is about two per km^2. More recent work suggests that this figure is, if anything, too high (Barlow & Choquenot 2002), since it takes only a few stoats to do a lot of damage (McDonald & Murphy 2000; Gillies et al. 2003).

Monitoring of kiwi populations has confirmed that effective local reduction in the numbers of stoats can produce increases in counts of kiwi calls (Pierce & Westbrooke 2003) and real improvements in the productivity and survival rates of juveniles. On one study area beside Lake Waikaremoana, in Te Urewera National Park in the eastern North Island (Blackwell et al. 2003), the proportion of young kiwi to reach stoat-resistant size (800 g) increased from 4% to 58% by the third year of stoat trapping. In another area, Trounson Kauri Park north of Auckland, intensive predator control for 6 years, at a cost of more than NZ $1 million, raised the survival of kiwi chicks from (probably) less than 10% to an average of 38% (range 25% to 69%) (Gillies et al. 2003).

Most estimates of stoat densities in New Zealand forests are well above the two per km^2 maximum critical threshold density signaling danger to kiwi (Chapter 10). Where kiwi can be rigorously protected, they can hold their own provided immigration of stoats from surrounding areas can be minimized and dogs can be kept under control. Elsewhere, they are still rapidly disappearing; in some years and places, even intensive trapping cannot always reduce the numbers of stoats enough to save the kiwi chicks. For example, after a good fruiting season for native forest trees (especially rimu) in the winter of 2002, ship rats and stoats became hugely abundant in the Okarito kiwi sanctuary (100 km^2), on the west coast of the South Island. The removal of 353 stoats and 577 rats in the summer after the rimu mast (compared with 124 stoats and 61 rats the year before; J. Crofton unpublished) was not enough to prevent all 14 monitored kiwi chicks from being killed by stoats (Pyle 2003).

Even if they survive, young kiwi have a strong tendency to disperse out of all but the very largest protected areas. This means that the minimum viable area for a kiwi population must be large, at least 100 km^2 (Basse & McLennan 2003)—and controlling stoats over such large areas is difficult. The best prospects are in places protected by natural barriers to immigration, such as in the Moehau kiwi sanctuary at the tip of the Coromandel peninsula. Elsewhere, the ultimate goal of controlling stoats to help the surviving, often aging breeding kiwi replace themselves has been hard to achieve (Gibbs & Clout 2003).

Species that have been long resident in New Zealand are especially vulnerable to predation because they have too few, or inappropriate, antipredator reactions compared with related species that have arrived more recently. New Zealand birds have always had avian predators that hunt during the day by sight, but mammalian hunters such as stoats and rats hunting at night by scent are something entirely new to them. The endemic black stilt and the recently arrived pied stilt have a common Australian ancestor, but whereas the black stilt has lived in New Zealand for millennia, the pied stilt appeared only about 200 years ago. Both nest in the same open gravel riverbed habitats, but the black stilt has lost several antipredator behaviors such as the broken-wing display, and its incubation period has become longer, extending the most vulnerable period of the brood (Pierce 1986).

Likewise, the South Island takahe (Figure 13.2) and the pukeko also have a common Australian ancestor, but the takahe has been in New Zealand since the Pleistocene and is now flightless, while the pukeko arrived within the last 1,000 years. In an experiment reported by Bunin and Jamieson (1996), pukeko responded much more strongly than takahe to a model stoat, and one cross-fostered takahe chick learned this behavior from its pukeko guardians while a parent-reared takahe chick did not.

Young New Zealand robins do not instinctively recognize stoats as a threat, but have to learn to associate the sight of a stoat with the alarm calls made by their parents. Fortunately, they can learn quickly. Wild young robins trained for only 5 minutes with stoat models and robin alarm or distress calls responded with more feather and body displays and stayed away from the nest for longer than other young robins (Maloney & McLean 1995).

Some species of small burrowing seabirds were extinguished from the main islands by predators; others have survived by nesting on predator-free offshore islands. Paradoxically, the Hutton's shearwater still nests in the alpine zone of the Seaward Kaikoura Mountains, in the northwest of the South Island. At this particular site, breeding seabirds are so abundant, and available over such a short period of the year, that the locally resident stoats can make no serious impact on their productivity. Remains of shearwaters were found in 785 of 788 stoat scats collected over three seasons (Cuthbert et al. 2000), but predation by stoats removed on average 0.25% of adults and 12% of chicks per year (Cuthbert & Davis 2002).

The main reason that the future of this particular colony is in no danger from stoats (Cuthbert 2002) is it is so large. At smaller colonies, the toll taken by predators can easily exceed the birds' production, just as colony size is important for Scottish seabirds visited by introduced American minks (Craik 1997). No doubt many small breeding groups of seabirds around the coasts of New Zealand have been wiped out in the past, unrecorded. For example, nesting colonies of New Zealand dotterels and yellow-eyed penguins are often threatened by predators including stoats (Moller & Alterio 1999; Dowding & Murphy 2001).

The effects of stoat predation can become very complex, and can extend across more than one trophic level. At Craigieburn, an isolated patch of mountain forest, low pollination rates and reduced fruit set of the native mistletoe is attributed to low densities of bellbirds, induced by stoat predation. Removal of stoats benefited the bellbirds: Nests in the trapped area in 2000–2001 were four times more likely to succeed than in the untrapped area (64% compared with 16%), although the increase in bellbird density was not enough to make any difference to the mistletoes (Kelly et al. 2005).

CONTROL OF STOATS IN NEW ZEALAND

The recent documentation of the damage to native fauna due to stoats (and, to a much lesser extent, common weasels) fully justifies the conclusion that the introduction of stoats was one of the worst mistakes ever made by European colonists in New Zealand. However greatly one might admire stoats as individual hunters making a living against all the odds, or as examples of evolutionary adaptation at its most impressive (and we both do admire them, as is clear from previous chapters of this book), they do not belong in New Zealand, and conservation authorities are under great pressure to find better ways to control them.

The danger from stoats to important threatened native fauna in New Zealand is taken very seriously. Throughout the 1990s, the Department of Conservation (DoC), the government agency responsible for protecting native fauna and landscapes, increased its investment into research on stoat biology. The groundwork had been laid in the 1970s but, at that time, the information on stoats could not be coordinated with the necessary observations on the birds. Only in the mid- to late 1990s were several key papers published on the impact of stoat predation on threatened species (Elliott 1996b; McLennan et al. 1996; Wilson et al. 1998; Basse et al. 1999).

As a direct result, in 1999 the New Zealand government accepted the argument that only an extensive, nationally funded campaign could protect endangered ground and hole-nesting birds from stoats (Hackwell & Bertram 1999). It therefore granted NZ $6.6 million to instigate a 5-year program to find more cost-effective and sustainable approaches to stoat control. The new funding stimulated an explosion of new research, as summarized in the annual reports of the Stoat Technical Advisory Group and published by the Department of Conservation (Murphy & Fechney 2003). For a short period, far more money was being spent on stoat research there than in any other country in the world, for example, NZ $1,352,000 in 2000–2001 (Parkes & Murphy 2003).

Obviously, effective control of any animal population is not merely efficient killing, although it may involve that. It is the translation of ecology into management policy. A rational stoat control program must (1) ensure that stoats are at least one of the main reasons for the observed damage and remove all cause for suspicion that any improvement in the birds could be due to something else; (2) understand the normal population ecology of stoats and how to disrupt it by the most humane and efficient method possible; and (3) monitor the effectiveness of a program and abandon it if it has no impact on the target population. These requirements are easy to understand but hard to meet in practice, especially as they require different strategies in different habitats.

For example, in North Island mixed forests Brown (1997) filmed 65 nests of New Zealand robins and tomtits in the 1993–1994 breeding season, of which 72% were lost to ship rats. No stoats were seen at the nests during this study, so Brown concluded that stoats were probably not important predators of forest passerines in nonbeech forests when rats were present. Likewise, intensive monitoring of nests of another species living in nonbeech forest, the North Island kokako, recorded many rats and possums, but only one stoat (Innes et al. 1999). Stoat control, therefore, needs to be focused on protecting species where stoats are usually the important predators of eggs or young (Table 13.1), such as yellowheads (Dilks 1999), kaka (Moorhouse et al. 2003), and robins (Etheridge & Powlesland 2001) in beech forests, and kiwi (Gillies et al. 2003) anywhere.

Stoat Control in Theory

Control of a population can be indirect, by manipulating its habitat and food supplies, or direct, by removing more animals each season than can be replaced. Indirect control is usually a side effect of some other process. For example, myxomatosis removed the main prey of British stoats, and thereby achieved a spectacular reduction in their numbers (see Figure 10.2). A similar though more moderate effect was noticed in New Zealand when real control over rabbits was first achieved (by aerial distribution of poison baits) from the 1950s onward (Marshall 1963).

Unfortunately, deliberate habitat manipulation usually does not work for predators that, like stoats, normally live almost anywhere they can find food and cover. For example, at one stage it was suggested that a "grass wall" surrounding the nests of yellow-eyed penguins in southern New Zealand might deter predators from reaching them. In fact, not only did the tall grass fail to keep the stoats out, but it also supported so many mice that it actually attracted stoats to hunt there (Alterio et al. 1998; Ratz 2000).

Actively removing the local rats or rabbits might work better, but may have dangerous side effects if it is done so quickly that hungry predators then switch their attention to birds (Alterio & Moller 1997b; Murphy et al. 1998). Where indirect control is inadvisable or not possible, direct control (what Graham Caughley [1977] called "frontal assault") is the only option.

The great difficulty of achieving effective, long-term control of small, short-lived mammals with high annual productivity like stoats ("r" strategists, see

Chapter 14), is that the control measure(s) must add to, not replace, the natural mortality rate. Weasels of all species are naturally resistant to control, because of their normally high mortality rates—around 70% in stoats (Chapter 11)—and their skill at recolonization. Productivity of stoats is variable but can also be very high, depending on food supplies during the previous breeding season (Chapter 9).

Ideally, the females should be brought to very low numbers before they have given birth, but that is at present very difficult or, in many places, unattainable. One promising alternative is to use trained "stoat dogs" to find breeding dens, and then, depending on the location and construction of the den and the number of exit holes, the whole litter can be humanely dispatched (Theobald & Coad 2002). This is much more efficient than waiting until they are independent and having to catch them all individually, but such a skilled resource is not available everywhere.

These aspects of their population dynamics make the large-scale management of stoats a severe challenge, but a lot of progress has been made in the last decade. Biologists in New Zealand are finding practical ways to reduce stoat numbers, to monitor the results with statistical rigor (Brown & Miller 1998), and to model both current and future control options (Barlow & Choquenot 2002; Barlow & Barron 2005).

New Zealand may be the only country in the world where stoats have become, by a historic human misjudgment, such a widespread and serious invasive pest, but the rapid advances in understanding of stoat biology there are of interest to students of small mustelids everywhere. The earliest days of stoat control to protect wildlife in New Zealand benefited from the ecological knowledge and practical skills developed over centuries both by North American fur trappers (Seton 1926) and by British gamekeepers (Anon. 1981). Now, New Zealand is the world center for research on the ecology and management of small mustelids, and New Zealanders are beginning to reverse the previous flow of information.

Stoat Control in Practice

Trapping

Most official stoat trapping operations since 1972 have used Fenn traps (pp. 326, 327), set in tunnels and baited with hens' eggs or meat. Fenns are practical, easily operated, and well documented (Anon. 1981), and still offer the only proven and safe method of removing stoats over wide areas (Dilks et al. 1996, 2003; Lawrence & O'Donnell 1999). Vertical bars set across tunnel entrances (King & Edgar 1977) at 30 mm spacing discourage entry by ground-feeding birds while allowing stoats to pass through at full speed (Short & Reynolds 2001). Conibear traps and various other types have been tried, but are not widely used.

In recent years more attention has been given to the question of balancing the need of native populations for protection with the need of individual pest animals for humane treatment (Littin et al. 2004). New standards for kill traps have been developed, including the requirement that captured animals should be rendered irreversibly unconscious within 3 minutes. The Fenn trap does not pass this test (Purdey et al. 2004), so it will eventually be replaced by more recent trap designs.

On the other hand, the design of a single-catch trap is only one component of effective control by trapping, for two reasons. First, the average probability of capture is relatively low, often less than 0.2/day, and is zero for at least some individuals (King et al. 2003a). In areas where traps are set for years at a time, some stoats learn to become very wary of traps and resistant to capture even if dozens of traps are concentrated over a limited area (Chapter 8). Second, even if efficiency is maximized by selecting the correct bait, layout, seasonal timing, and length of operation to suit the purpose (King et al. 1994; O'Donnell et al. 1996), trapping is always labor intensive, and therefore costly.

Take, for example, the effect of trap spacing on the sex ratio of the catch. In an early experiment setting out an array of kill traps at different spacings, the wider the distance between them, the fewer stoats were caught per kilometer of trap line and the higher the proportion of males (Table 13.2). The removal of even large numbers of males makes little difference to the number of young born the next spring, since stoats are polygynous and males are very efficient at finding estrous females (Chapter 9). All females are already pregnant by midsummer, including the young females that are fertilized before they leave the den, so any that escape trapping can produce a litter even if every male in the area is killed. Put another way, wide-spaced traps may do more harm than good, because they remove mostly males. Removing males leaves more food for the surviving females, and may even enhance the ability of the local stoat population to recover. Therefore, although a widely spaced array of traps may be less effort to operate (Chapter 8), it will have little effect on the next year's population if it does not remove enough females.

Research on improving baits and lures has been intense in recent years. The most effective lures for stoats are fresh meats or eggs, but many new baits are being tested (Spurr 1999; Montague 2002; Spurr et al. 2002). In response to field reports of weasels hunting by sound (Willey 1970), electronically produced sound lures have been tried, but with equivocal results (Spurr & O'Connor 1999). Synthetic lures incorporating components of stoat musk, a natural scent marker of great significance to stoats (Chapter 8), have been, so far, not very successful (Clapperton et al. 1994). Natural lures made from real anal scent glands help (Clapperton et al. 1999), but obtaining large supplies of the natural product would be logistically difficult.

All other artificial odors and flavors offered to stoats in pen trials (e.g., commercial trappers' lures, food flavors, trimethylamine, and synthetic fermented

Table 13.2 The Effect of the Spacing Between Kill Traps on the Number of Male and Female Stoats Caught Over 17 Months

	Distance between traps, m				
	100	200	400	800	2100
Number of traps	22	22	22	22	19
Length of line (km)	1.8	4.0	8.0	16.0	42
Total number of stoats caught	14	25	36	65	123
Sex ratio (% males)	29	32	44	42	76
Stoats caught per 100 trap nights	0.47	0.59	0.87	1.45	1.13
Stoats caught per km of trapline	7.8	6.3	4.5	4.1	2.9

(Data from experimental studies in southern New Zealand in 1972–1976, reported by King 1980a.)

egg) were unattractive (Spurr 1999). Perhaps young predators learn what prey are worth hunting and safe to eat during a critical period of their development, so they might not recognize artificial baits as potential food. If so, the best baits to offer may depend on the natural diet of stoats in each location. Bait markers help in field trials of new baits, because they can identify which animals ate the bait and which did not. Markers such as rhodamine B and iophenoxic acid are tasteless and can be detected in the whiskers and blood, respectively, of stoats that had eaten marked bait up to several weeks previously (Spurr 2002a, 2002b; Purdey et al. 2003).

One early set of field trials showed that capture rates of stoats were higher early in a trapping session than later, and higher in baited than in unbaited traps (King 1980a). The most efficient way to run a trap line in that situation was, therefore, to set the traps close together, bait them well, inspect them daily for a few days, and then spring them closed and do something else for the rest of the month.

In that trial, the number of stoats caught fell by about 60% by the end of each 14-day session, although immigrants soon replaced those caught by the beginning of the next month's trapping. It seemed likely that less than half the number of stoats present in the forest had been caught, even though the greatest practicable effort had been expended. Ratz (1997) calculated a linear regression comparing capture rate with cumulative number of stoats removed, which gave median estimates (with very wide confidence intervals) of 58% of stoats removed after 8 days and 87% after 12 days. Unfortunately, it takes very few stoats to make a lot of difference to a local population of threatened birds, so the risk remains that removal rates like these may not be sufficient to achieve a useful level of protection.

In an unstable population, a decline in the numbers of stoats after all that effort is not necessarily proof of the effectiveness of control, because trapping often does not add to, rather than replace, natural mortality (McDonald & Harris 2002). Moreover, young stoats are able to travel very long distances (20 km or more) within a few weeks of independence (Chapter 9), and can repopulate cleared areas within 2 to 3 months (Murphy & Dowding 1994; Alterio 1996; Murphy et al. 1999).

On the main islands of New Zealand, therefore, the intensity of trapping required to remove more stoats than die from natural causes necessarily involves a huge commitment of labor for a temporary result. The Department of Conservation does not conclude that stoat trapping is worthless; rather, they see it as "rather like cleaning toilets, you just have to keep on doing it" (E.C. Murphy, personal communication). But the incentive to find more cost-effective methods is very, very strong.

Only on the offshore islands can rapid recolonization by stoats be prevented, and then only if they are isolated by more than about 2 km of water. The story of Maud Island (309 ha), an important rodent-free sanctuary in the Marlborough Sounds (at the northern end of the South Island), illustrates two of the most extraordinary abilities of stoats: swimming and evading traps.

Crouchley (1994) described how Maud has been invaded three times in a decade by stoats swimming across the 900-m salt-water channel from the nearest mainland. In April 1982, the first stoat was discovered because it responded to taped calls of the saddleback, an endemic forest bird that had recently been introduced to the island. Traps (up to 80) were immediately set all over the island, but it was 7 months before the first of seven young stoats was caught, and it took 16 months to get the last of them. The first one seen must have been a female, carrying (as usual) a litter of unimplanted blastocysts (Chapter 9), and the seven stoats captured were presumably the young she produced on the island. She herself was never caught.

A similar story was repeated on the same island in 1989, and this time the litter included at least five young. Like the previous immigrant adult female, this one was also never trapped despite the intense efforts of the Maud Island staff. She lived on the island for 18 months until she was found dead in February 1991. By then, sibling matings between her offspring had begun to produce the next generation, of whom at least three were included among the total of 16 stoats caught up to July 1994. In March 2003 a stoat was discovered on the island for the third time—fortunately, this time it was a young male. Obviously, small islands lying as close to the main islands as Maud will never be safe from these repeated incursions.

By contrast, on Te Kakahu (Chalky Island, 514 ha) off the southwest coast of Fiordland, stoats had lived for many years free of all contact with humans. The island supported no rodents, but several large breeding colonies of burrowing seabirds and of New Zealand fur seals provided plenty of eggs, chicks, and carrion from early spring to late autumn. During the winter their primary diet was bush birds, supplemented by carcasses of adult seabirds and perhaps also cached chicks and eggs. Like all stoats, however, they no doubt still preferred fresh meat if they could get it.

On that assumption, the Department of Conservation planned an intensive trapping campaign for the winter of 1999 (Willans 2000). The plan was to avoid the Maud experience by timing the start of trapping just at the time that resident stoats had few live prey to hunt, and by making sure that every stoat had access to several traps and could not learn to avoid them until too late. After the last of the young seabirds had fledged in April and May of 1999, 140 double trap tunnels were placed at 100–m intervals all over the island and prebaited with fresh eggs and day-old chicks for 2 weeks. Most of these were quickly taken, presumably by hungry stoats.

When the traps were set on June 29, 15 stoats were killed in the first 2 weeks (11 on the first night!), and one more was found when the team next visited the traps in October. Further offerings of fresh baits remained untouched, trained stoat dogs could find nothing, and stoat footprints were no longer seen on the sandy beaches of the island. Te Kakahu is now available for restocking with threatened species from the mainland, and the numbers of small bush birds are recovering. The once-only cost of removing the stoats (NZ $62,000 over 14 months to June 2000; M. Willans personal communication) is trivial beside the present and future value of the island as a predator-free reserve.

It is beyond belief that the two female stoats that invaded Maud could have avoided traps for months, when the whole resident population on Te Kakahu was eradicated within weeks. Te Kakahu was not a fluke, either: Similar operations have been repeated on other Fiordland islands (in Anchor Island, 2001, and Bauza Island in 2002) (M. Willans, unpublished). The contrast between these stories illustrates how well stoats can learn to avoid traps (Chapter 8). Unfortunately, textbook operations like the ones in Fiordland are often logistically impossible to repeat on the main islands, but they certainly add to other pressures to find means other than trapping for protecting native fauna from stoats.

Poisoning

New Zealand legislation controls but does not prohibit primary poisoning programs targeted directly at stoats. No toxins are presently registered for use against stoats, although some (including 1080, diphacinone, and cholecalciferol) may be used under experimental research permits; trials so far show that all have drawbacks (Lawrence & Dilks 2000; Spurr 2000; Spurr et al. 2002). Poisoning is unpleasant to think about, but the looming extinction of irreplaceable endemic birds is worse. Consequently, in New Zealand toxins have remained a permissible tool for predator control long after most other countries have banned them. The toxins used, the research programs testing them, and the people handling the most dangerous of them, are all subject to stringent supervision by regulatory agencies concerned as much to ensure humane treatment of the pest animals as to protect nontarget species and human interests.

One toxin commonly used is 1080 (sodium monofluroacetate), a natural plant toxin widespread in native vegetation in Western Australia, which is very effective and breaks down quickly in soil and natural water. Stoats are very susceptible to minute amounts of 1080 (1 mg is a fatal dose for stoats of all sizes), and it acts quickly and apparently humanely (Potter & King unpublished). When good delivery methods have been developed that can fully protect nontarget species, 1080 could become a viable alternative to trapping for targeted use against stoats.

Rodenticides are widely used to control rats and voles on farms worldwide (Chapter 12), and stoats are vulnerable to secondary poisoning when they eat dead or sublethally poisoned rodents (McDonald et al. 1998; Shore et al. 1999). In New Zealand, where the only native mammals that could be affected are rare, semiterrestrial short-tailed bats (King 2005a), 1080 has been an important and routine weapon against rabbits, rats, and possums for several decades.

Secondary poisoning has probably had local and temporary effects on stoat populations for much of that time (Gillies & Pierce 1999; Murphy et al. 1999; Alterio 2000). Since most introduced mammals are regarded as undesirable in New Zealand, secondary poisoning is less a cause for concern to managers than it is elsewhere. The argument is that the ecological costs of using toxins against introduced mammals are much less than the damage costs to native fauna if they are not used, although we still need to learn more about the ecosystem-level consequences (Innes & Barker 1999).

Nonlethal Control Methods

Arguments about the ethics and principles of lethal predator control and its use in conservation management have accelerated in recent years, and seem likely to continue (Littin et al. 2004). In this context, animal welfare is concerned with *individual* pests, whereas conservation is concerned with protection of *populations* in the wild (Eggleston et al. 2003). Both concepts are important but, unfortunately, sometimes incompatible. If we accept that all animals have the same rights to protection without consideration of the damage some may do to others, then we are ignoring our responsibility to vulnerable species and to biodiversity generally (Fulton & Ford 2001).

It could be argued that lethal control of exotic predators is justifiable to protect threatened species, especially species that are disadvantaged by human activities such as intensive agriculture and chemical sprays. Effective eradication using humane lethal methods not only is cost-effective, but also makes all further killing unnecessary. This process is well advanced, especially on the off-shore islands of New Zealand (Towns et al. 1997; Towns & Broome 2003). Islands will always be the most important refuges, and progress in clearing stoats and rodents from all suitable offshore islands continues apace (Parkes & Murphy 2003). An alternative strategy would be to develop nonlethal means of protecting vulnerable prey.

Nonlethal control is possible if the aim is to prevent damage rather than reduce predator numbers. For example, an obvious solution to protect nesting game birds is a predator-proof fence. In North America, several designs for electrified predator-proof fences kept out predators of the size of foxes, skunks, and raccoons, but not stoats or raptors. The cost per additional successful nest was over $100 in two cases reviewed by Reynolds and Tapper (1996), and the consequences of even a single fox breaching the fence were so catastrophic that predator removal continues to be the preferred policy.

New Zealand has a smaller suite of predators and, most significantly, no foxes. There, continued research has produced a genuinely practical nonelectrified design effective against stoats (Clapperton & Day 2001), and even mice. New Zealand is leading the world in the technology of creating predator-proof enclosures. At the time of writing (2006), some 15 of these enclosures have been built around New Zealand, and another five are imminent, at massive cost but with enthusiastic community support (R. McGibbon, personal communication). Some of the earliest enclosures already support protected species. Other nonlethal methods, such as repellents (Spurr 1997) and fertility control (Cowan & Tyndale-Biscoe 1997; Norbury 2000), are also being developed. Potential new techniques involving genetic engineering of vectors to carry stoat-specific biocontrol agents is not ruled out, but would require extensive public consultation (Fitzgerald et al. 2002), and in general a better strategy is to improve current methods first.

COMMUNITY RESTORATION

The trend toward implementing more humane treatment of all animals, including pests (Chapter 12), has stimulated intensive research not only on more humane traps but also on cost-effective, nonlethal methodology. Interest in this subject is worldwide, but there is probably more concern about it in New Zealand than anywhere else in the world. What happens on those remote islands may well affect how pest mammals are treated in many other countries in the future.

There are good reasons to be cautious about lethal control policies. Invasive pests cannot be treated in isolation from the rest of the ecosystem with which they interact. When pests are removed without reference to the other species in the same community, unexpected consequences tend to follow (Zavaleta et al. 2001).

For example, unless the two most serious pests of nesting birds (rats and stoats) are removed together, there is a risk that removal of stoats alone might permit rats to increase in numbers (King 1984b; Barlow & Choquenot 2002). Rats are also serious predators of birds, and live at much higher densities than do stoats. In the Eglinton Valley after two heavy seedfalls in consecutive years (1999, 2000), stoats and rats increased rapidly in numbers together (Dilks et al.

2003). Even though modeling suggests that stoats cannot prevent rodent irruptions (Blackwell et al. 2003; Ruscoe et al. 2003), the possibility that the unprecedented response of rats in the Eglinton could have been helped by a decade of stoat trapping has not yet been eliminated. In the same years, the Eglinton population of native long-tailed bats suffered greater predation and lower survival than in nonseedfall years, correlated with variations in the density index for rats (Pryde et al. 2005).

Conversely, after successful control operations in North Island forests that affected rats more than stoats, stoats switched to eating more birds (Murphy et al. 1998). Another interaction that complicates the picture is that mice increase in numbers whenever rats are removed (Innes et al. 1995), which is a concern since mice are important food for stoats and, therefore, affect their numbers (Chapter 10).

As confidence and technology improve, management policies in New Zealand are looking toward more ambitious goals—beyond mere eradication of pests, including stoats, to broader ecosystem restoration, at least in patches (Towns et al. 1997; Atkinson 2001). Stoats, common weasels, and ferrets might not have been among the aliens that must now be tackled if the upper chamber of the New Zealand Parliament had not allowed to lapse in 1876 the bill (already passed by the lower chamber) that would have prohibited the introduction of mustelids (Galbreath 1989:126). But, so long as the modern political will supports the public yearning to put right that historic mistake, some hope remains for the threatened native fauna.

14 | Puzzles: Sexual Dimorphism, Delayed Implantation, and Coexistence among Weasel Species

TO SPECIALIZE IN hunting small rodents, a successful predator must have not only obvious characteristics, such as the ability to enter very small spaces, but also less obvious ones, such as some means of surviving and breeding year after year despite huge variations in food supply. Weasels have in fact evolved a whole set of special adaptations to deal with these conditions, which make the weasel way of life possible. Most of these adaptations have been introduced in the preceding chapters.

Of course, no one means to imply that animals are capable of sitting down and thinking out a plan for survival. It is just that certain sets of characteristics are likely to favor the animals that happen to have them in a given set of conditions. Over the long term, only the favorable sets of characteristics survive to be observed.

Some of the adaptations of weasels are long term, involving physical characteristics unchangeable in an individual's lifetime, such as the long thin shape of the weasel body or the details of the reproductive cycle. Some are short term, including the flexibility of behavior and reproductive response appropriate to different and changeable circumstances. And some (such as delayed implantation) have caused biologists considerable consternation, because it is not obvious how certain weasel characters could have evolved or what adaptive value they now have.

To understand the ways in which weasels have evolved to deal with the hazards of their lives, and to understand the puzzles they still present to us, we must draw on practically everything we know about weasels discussed in this book so far.

HOW SIZE AFFECTS POPULATION DYNAMICS

The characteristics that determine the demography of a population collectively make up its life history pattern. They include, for example, age at first breeding, mating system, litter size, average number of litters per season and per lifetime, amount of parental care invested in each litter, expectation of life at birth, and so on.

Life history patterns are determined mainly by two things: body size and the permanence of the habitat relative to the lifetime of an individual. Small animals can survive in small patches of the environment that cannot support large animals, which tend to range across many different patches. However, since small patches tend to be unreliable and often short-lived, small animals must also be good at dispersing from one patch to the next as local food supplies run out.

For example, an overgrown corner of a field containing many voles provides an excellent temporary home for a common or least weasel, although after a while the weasel and its young will probably have to move on to find another field. By contrast, the whole field and a large area around it may be the stable and permanent home of generations of foxes or badgers (King 1983c). On the other hand, because small animals generally have small home ranges, they can, reach greater population densities than larger animals of similar food habits. For example, common weasels are much more abundant than wolves, and vary in numbers from year to year much more. In Białowieża Forest in Poland, during an irruption of forest rodents, the density of common weasels went from two per km^2 in spring to ten per km^2 in summer 1990, falling to less than one per km^2 in spring 1991; but the density of wolves averaged two to three per 100 km^2 throughout (Jędrzejwska & Jędrzejewski 1998).

The suite of characteristics typical of small, opportunistic animals living in variable and unstable habitats make up the "r-strategy," so named decades ago because the net effect of that combination of characteristics is to maximize r, the rate of population increase. The opposite characteristics typical of large animals living near carrying capacity (K) in relatively stable and reliable habitats make up the "K-strategy," which tends to maximize sustainable population density. For K-strategists, the rate of increase is far less important than producing a few high-quality young, well developed and prepared to compete with others for a place in the adult population. Between the two extremes are intermediates that have these characteristics developed to different degrees.

The idea of the r–K dichotomy is obviously a gross oversimplification of how real animals live. Nevertheless, it is a robust generalization that has been immensely helpful in the past, and it does capture some important general truths explaining broad-scale variations in population dynamics. For example, a set of related species can be arranged along a spectrum, from extreme opportunists at one end through all shades of intermediates to extreme equilibrium species at the other. Their positions on the spectrum are determined from the relative importance for each species of the characteristics that most augment the rate of population increase, which are early maturity, more than one litter per year, large litter size, and more than one litter per female's lifetime, in that order.

Mustelids demonstrate this idea well (King & Moors 1979b). Least and common weasels are the most variable in numbers of all the mustelid family, and stoats and longtails are next. In Britain, Europe, and Asia, the badger is the largest mustelid species, well adapted to relatively stable woodland conditions. In North

America, the wolverine is the largest mustelid, and populations of wolverines, where undisturbed, remain much the same from year to year.

Common and least weasels have all the characters of determined opportunists. They mature early and bear small young with a short life expectancy. In seasons of abundant food, spring-born young can breed in the season in which they are born, while adults can produce a second litter. Their populations fluctuate wildly from year to year and are governed mainly by food supplies. The parent animals most likely to pass their genes to the next generation tend to be those that produce and rear to independence the most young most quickly.

Common and least weasels concentrate on quantity rather than on quality. They spend the least possible time on parental care, because, if food is abundant, the effort spent on training one litter could better be spent on producing another one. Besides, training individual young is not a good investment of parental energy for the small weasels—when food resources are unpredictable, chance is probably as important as hunting or fighting prowess in determining whether a young common or least weasel finds a suitable home range to settle and hunt. The more young are produced, the greater the chances are that one of them will survive and breed in its turn. On top of that, the mortality rate of adults is so high that few live through more than one breeding season, so they have every reason to make the most of the present one.

Stoats and longtails have similar traits, developed to a lesser extent. Females mature and are mated at a few months old (Chapter 9), but then are prevented by a long period of compulsory delayed implantation from producing the litter until the following season. Young males cannot mate until 13 or 14 months old for stoats, and about 15 months old for longtails.

Individual stoats and longtails of both sexes are incapable of producing young in their first year, and only one litter a year is possible in older ones, however abundant the food supplies. Hence, the potential rate of population increase for stoats and longtails is considerably less than that for common and least weasels. On the other hand, although the mortality of the first-year young is very high for stoats (and probably also for longtails), those that survive to their second year in an untrapped population have some chance of living for 3 to 6 years, and breeding several times (see Tables 11.1 and 11.2) .

Populations of stoats and longtails are unstable, but not to the same extent as those of common and least weasels, not only because stoats and longtails are longer lived but also because they can catch a wider range of prey, switching from one to another if necessary. They spend rather longer on parental care, partly because they cannot produce another litter until the following year anyway, and perhaps also partly because greater investment at this stage has some chance of being repaid with greater success later.

For example, larger than average size is an advantage to a male stoat or longtail, because large males are more likely to be dominant and to be successful breeders (Erlinge 1977a). Among the things that determine whether a male will grow to his full potential size is whether his mother feeds him well (Ralls & Harvey 1985). In litters of weasels raised in captivity on guaranteed food supplies, the young males grow larger than wild-reared young of the same age (East & Lockie 1965; Hayward 1983). In habitats with feast-or-famine food supplies, young born in years of abundant food grow larger than young born in other years (Powell & King 1997).

As the theory predicts, the differences in population dynamics and parental care among weasel species, and between all weasels and the larger predators, are related to their small size and fast-paced lives.

SEXUAL DIMORPHISM

Male weasels of all species are substantially larger than females (see Table 4.1; Figure 14.1). The reason for this difference was at first considered to have something to do with food (Brown & Lasiewski 1972). Because the two sexes are so different in size, they tend to eat different things; so, the argument ran, the difference must have arisen so that each could avoid trespassing on the other's food supplies. In times of food shortage, this trick might be valuable to both. Indeed, the overlap in the diets of males and females is substantial at all times, especially when food is short (Chapter 5).

Reasonable as this argument sounds, there is no evidence to support it. Fortunately, there is a much better hypothesis, which proposes that males and females are different in size for reasons to do with that old driving force, reproduction (Erlinge 1979a; Moors 1980). Large and small predators tend to choose prey appropriate to their size, for good reasons (Chapter 7); therefore, the observed differences between male and female weasels in their diets are the consequence, not the cause, of the difference in their body size.

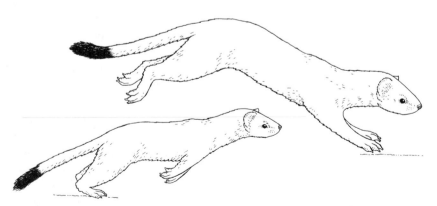

Figure 14.1 Male weasels are significantly larger than females. (See also Figure 9.9.)

The fluctuations of populations of voles and lemmings over a 3 or 4-year period create alternate feasts and famines for weasels. Across these fluctuations, the chances of a female weasel producing young that survive to breed in their turn range from practically nil to very high. The population densities of the weasels themselves vary over the same period, during which the chances of a male weasel gaining several mates, or any at all, also range from good to bad.

The probability of a given individual weasel of either sex meeting a disastrous season, in which successful breeding is impossible, is high; and because weasels have relatively short life spans, the probability of surviving to a better season is low. The common problem for weasels of both sexes, living in such a variable environment, is the risk of failing to leave any young at all. The two sexes have responded to this problem in different ways.

A female weasel is an energetic little hunter, and she always raises her offspring with no help from a male. In good years the food resources available are sufficient in quantity, distribution, and rate of renewal for the females to manage very well alone. A female breeding in a good year can harvest enough prey to feed her entire litter, and those breeding in the best of the intervening years can usually manage to produce at least some young. Since the effort of raising a litter is so great, a female with her first litter has a rather small chance of having another one later in life (except when food is very abundant), so her best chance of leaving surviving offspring is to invest the maximum possible effort into the one she has. Occasional lucky observations of weasel mothers with their young show that they certainly do take this advice seriously (Chapter 9).

Of course, the task would be easier for her if the father stayed with her and helped protect and provide for the litter. But from his point of view, that would not be a good policy. In good years, the male who spreads his genes liberally stands to gain many more descendants than the male who has stayed with one mate. The best bet for a male is, therefore, to invest the minimum possible time in each of as many litters as possible, and to rely on the females to work, for their own reasons, to raise the young he fathered.

Mammals that normally produce relatively large litters of blind and helpless young tend to have stable pair-bonds among the adults, whereas those that produce a few, well-developed young, born fully furred and with eyes open, are often polygamous. Weasels, however, have relatively large litters of helpless young, but no pair-bonds at all. Why are weasels so different?

The explanation has yet again to do with the consequences of the weasels' specialization as rodent hunters. A female weasel cannot afford to accommodate a growing litter internally for too long after they begin to ruin her slim figure (Figure 14.2) (Gliwicz 1988), but neither can she afford to reduce the number of young she could produce when conditions for breeding are favorable. Her solution is to divide her resources between a large number of small kits, and to avoid carrying them about with her by dropping them in a safe place as soon as they are viable. She is then free to concentrate on bringing food to them.

Figure 14.2 A pregnant female weasel builds up caches of extra food for as long as she can still squeeze into small spaces to search for rodents. When hunting becomes difficult, she retreats to her chosen birthing den and waits. By producing extremely altricial young, born as soon as they can live outside the uterus, a mother weasel minimizes the period when the alteration of her svelte outline reduces her hunting success. Once the kits are born, their mother can leave them safe in their natal den while she hunts.

People do sometimes report seeing male stoats, longtails, or common weasels taking food to a den, or families of weasels accompanied by both female and male adults. These males are more likely to be motivated by sex than by paternal responsibility. In fact, two breeding dens observed by Erlinge (1979a) had such narrow entrances that no male could enter. In addition, a female weasel with small young protects her family so fiercely that she becomes temporarily dominant over even the larger males. Adult female weasels of all species may have postpartum estrus, so a male bringing food to the den is most likely to be attempting to placate the female's hostility and to curry her favor to gain a mating.

Male stoats and longtails are especially likely to bring presents, but only for strictly sexual reasons. Because of delayed implantation, the young born this year were fathered last year, very likely by another male, and it is not in any male's interests to invest time and effort into feeding another male's young. On the other hand, he might be willing to invest in the female young as sexual partners for himself, and the potential rewards of gaining access to them are great enough to be worth a considerable investment. Perhaps this explains why male longtails sometimes do, apparently, help provide for a litter (Hamilton Jr. 1933; Gamble 1980). Young female longtails have to be well grown before they are ready for mating, and a male might increase the number of young females becoming independent by bringing food to them in their den. Gehring and Swihart (2004) reported that, in Indiana in May, one radio-collared male was twice relocated within a natal den containing a litter of 6- to 8-week-old young.

Because the roles of male and female weasels in reproduction are totally different, it is not surprising to find that the adaptations that each has evolved are also quite different. Females need to be highly efficient providers of prey for their young, and males need to dominate other males in competition for access to the most females. Both aims are critically affected by body size.

Small size increases hunting efficiency for a female feeding her young on voles and mice, and small size allows her to minimize the food she must eat to maintain herself, leaving more of what she catches for her young (Moors 1977, 1980). On the other hand, large size increases a male's chances of success in confrontations with other males (Erlinge 1977a) and, perhaps, somewhat lessens his chances of being attacked by a swooping raptor (Korpimäki & Norrdahl 1989a). Hence, the balance of size-related advantages favors different average body sizes for male and female weasels.

Efficiency in hunting may mean either minimizing the time taken to collect a certain amount of food or maximizing the amount of food that can be collected in a given time. Female weasels could do either, according to the need and the conditions. For example, when the young are very small, they need the mother with them for as much of the day as possible, to keep them warm and to feed them milk. She would do best to avoid letting them chill while she is out of the den by minimizing the time she spends hunting at that stage. Later, as the young grow and their demands escalate, their growth and future prospects may depend on the extent to which she can fulfill their needs. They can keep each other warm by then, so the mother would do best to maximize the amount of food she can collect during the time she has the energy to go out to hunt. It would be interesting to know whether female weasels do switch from the one hunting method to the other as their young grow.

While the explanation of sexual dimorphism that we have presented appears sound and has broad support, the evolution of sexual dimorphism in mustelids in general, and in weasels in particular, is controversial. Tamar

Dayan and Dan Simberloff (1994, 1996) argued from patterns in canine sizes that sexual dimorphism in mustelids evolved to reduce competition between the sexes for food.

Because males of each mustelid species in a community are smaller than the females of the next larger species, the animals form a series of "morpho-species" that get bigger in size as one goes from common or least weasel female to male, to stoat or longtail female to male, to polecat or marten (or longtail where stoats and longtails both live) female to male, and so forth. The body sizes of these animals do, indeed, form such a series, and Dayan and Simberloff have shown that the largest diameters of their canines also form a stepwise pattern of in-

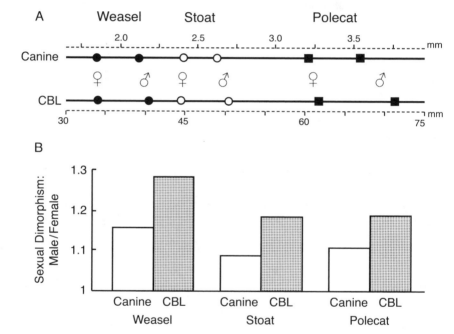

Figure 14.3 (A) The mean largest diameter of canine teeth and the mean condylo-basal lengths (CBLs) of skulls for female and male common weasels, stoats, and polecats in Britain. These six morpho-species form a controversial example of character displacement. Although the diameters of canine teeth of weasels, stoats, and polecats form a stepwise pattern from female to male common weasel to female to male stoat to female to male polecat in Britain, we note (B) that the diameters of the canines for males and females within each species are actually more similar (the sexual dimorphism is smaller) than are skull lengths. Holmes (1987) and Holmes & Powell (1994) showed that CBLs are less dimorphic than head and body lengths. Consequently, canines appear to be responding to a form of selection different from that which has generated the large sexual dimorphism in body sizes of weasels. (Data from Dayan & Simberloff 1994.)

creasing size (Figure 14.3). They argued that this pattern could arise only if each morpho-species eats prey that differ in average size.

Unfortunately, sexual dimorphism in weasels gets even more complicated. While the perfect pattern appears almost to exist for canine diameters, it clearly does not exist for other measurements that should parallel such a pattern in canine diameter. Thor Holmes (Holmes 1987; Holmes & Powell 1994) analyzed over a dozen skeletal, skull, and dental measures from many mustelid species and found that the bodies of male and female weasels (and males and females of other mustelid species) were nearly perfect copies of each other at different sizes *except* that the heads were disproportionately more similar in size than expected. Those features most critically involved in catching and eating food, especially the jaws and teeth, showed *less* sexual dimorphism than did body size.

In addition, Holmes found that sexual dimorphism in North American weasels is smaller where their ranges do not overlap with competitors, such as other weasels or martens. Even the data presented by Dayan and Simberloff (1994) are consistent with those of Holmes: The canines of males and females in the same weasel species are more similar to each other than are their skulls, which Holmes showed to be more similar to each other than the bodies (Figure 14.3). We think that sexual selection drives the evolution of large size in males of each species, that efficiency of reproduction drives the evolution of small size in females, and that diets of males and females are more similar than expected from their differences in body size. It is true that diet studies have documented significant differences in the diets of males and females of sympatric weasel species, especially when corrected for seasonal effects (McDonald et al. 2000), but at some times and places diet overlap is substantial.

We consider that the best explanation of all these patterns is still that the differences in the diets of the two sexes arise as a *consequence* of their differences in body size, which itself is a consequence of sexual selection and reproductive constraints. In other words, sexual dimorphism came first, and different diets came second. Within each species, the members of the two sexes have similar diets because every weasel must catch a variety of prey to survive.

Males and females of each species are closer in size to each other than to members of other species and, therefore, the varieties of prey they catch must overlap extensively. In Dayan and Simberloff's data, the steps in canine diameter between females and males of the same species are smaller than those between species and, therefore, the stepwise increase in canine diameters is actually less even than expected from the animals' body sizes.

So, the stepwise increase in size from the smallest to the largest mustelids in any location and the relationship between body size and prey are still topics for discussion, awaiting a new set of data that might be able to support a conclusive explanation.

DELAYED IMPLANTATION

One of the keys to success for opportunists is that they can respond quickly to any local improvement in living conditions. For example, a sudden surge in the population density of voles presents adult weasels with the opportunity to produce more young. Also, while the surge lasts, it increases the young weasels' chances of surviving to become adults, and the adults' chances of surviving long enough to breed again. When this happens, common and least weasels can adjust their breeding efforts almost immediately; hence, the breeding success of common and least weasels is very closely linked to the density of rodents.

Small rodents comprise ideally convenient packages of meat to feed to all young weasels; they are easy to kill, light to carry, and wrapped in ready-made waterproof packets. Populations of common and least weasels have developed their opportunistic way of life to such a fine art that they can survive the most dangerous conditions, such as in the fierce cold and regular famines of the far north. Local extinction is common, but there are always a few patches of habitat with enough rodents to support at least a few individual weasels, which in turn can survive and disperse to recolonize other areas vacated when times were bad.

Most adult stoats and longtails are large enough to survive short periods of scarcity of small rodents by turning to other prey, but their chances of breeding success are still highest when rodents are abundant. Hence, the rate of survival of the juvenile stoats and longtails and the productivity of the adults are also linked to the density of rodents or rabbits, just as they are for the smaller least and common weasels.

Stoats and longtails are, therefore, also classed as opportunists, even if not quite such highly specialized ones. But stoats and longtails appear to have one huge disadvantage that common and least weasels do not have: a compulsory period of embryonic diapause, which introduces a long delay between fertilization and implantation.

Delayed implantation is obligatory for stoats and longtails, regardless of environmental conditions, and it prevents the females of both species from responding immediately to fluctuations in the numbers of voles available. Their birth season is fixed by day length, and their fertility (maximum potential litter size) in any given year is set by their fecundity (ovulation rate) of the previous year. Any increase in fecundity in response to abundant food cannot be translated into larger litters until the following season, by which time the rodents may well have disappeared (King 2002).

This handicap must be particularly serious for the stoats of the smallest races living in the far north, which depend on small rodents as much as do common or least weasels elsewhere. If the capacity to respond rapidly to variations in food supply is important to common and least weasels, then surely it must also be important to stoats and longtails. How, then, can stoats and longtails survive in the face of such a serious penalty? Indeed, how did such an apparently disadvantageous arrangement become established in the first place?

The Value of Delayed Implantation

Delayed implantation is more common in some groups of mammals, such as marsupials, mustelids, and cetaceans, than in others. It is controlled by complex physiological feedback interactions with environmental cues (Renfree & Shaw 2000). Explanations for the adaptive advantages of the ability to uncouple the fixed interval between mating and birth by keeping fertilized embryos in limbo for a while are easy to suggest for some species; for example, it allows female fur seals to mate again soon after producing their pups and spend the rest of the year at sea; it allows female wallabies living in arid regions to produce a new joey more rapidly after unexpected rain.

But for the small mustelids there are no such obvious explanations. Rather, there is the puzzling question: Why do stoats and longtails have delayed implantation while common and least weasels do not? Actually, this is really two questions: (1) Is delayed implantation useful to stoats and longtails? and (2) if it is useful, why do common weasels not have it too? Alternatively (looking at the question from the other perspective): (1) Are common and least weasels better off without delayed implantation? and (2) if so, why do stoats and longtails delay?

The ways in which modern animals make their livings are, by definition, ways that work. They are not necessarily the best possible ways, just the best of the ways that have been tried. Delayed implantation may be a useful adaptation original to stoats and longtails, or, alternatively, stoats and longtails may have retained the mechanism controlling delayed implantation, inherited from an ancestral species for which it was useful, because natural selection has not eliminated it from their lineage—even though they are much shorter-lived than all other species with a delay (Mead 1993; Thom et al. 2004).

The second explanation seems to us by far the most likely. Delayed implantation appears to have evolved first in an ancestor common to all the canoid carnivores (Lindenfors et al. 2003), and it has been lost only 11 times since. Stoats and longtails each belong to a lineage (not necessarily the same one) in which delayed implantation has been retained, conserving a mechanism so precise and consistent, there is practically no individual variation. All females of all ages and both species delay implantation every year, everywhere. Clearly, there is no profit in deviation. Surely, the entire suite of adaptations involved in delayed implantation must be maintained because the advantages of the total package outweigh the disadvantages of tinkering with inconvenient bits of it.

It is, in fact, easy to make a list of advantages of delayed implantation to stoats, and we have suggested some (King 1983a, 1984a; Powell 1985b). First, the adult female and her young are all ready for mating at about the same time. This means that any male that can get into a breeding den at the right moment stands to gain several mates all at once. Such extreme polygyny is rare among small carnivores, whose females are usually too widely dispersed for a male to monopolize access to a whole group of them. The precocity and immobility of the nestling females makes stoats the outstanding exception to this general rule, and the consequence for the successful male stoat is a huge payoff in future offspring. Only the old female can put up any resistance to his attentions, so, if she is at home when he calls, she makes the choice of a mate for all the females in the family.

Second, the young female stoats are fertilized before the family breaks up. This means that they are already assured of a litter next season, even if, by then, prospective mates are few. The blastocysts take very little energy to maintain, so pregnancy does not interfere with normal growth; come spring, the young female stoats are free to concentrate on hunting and on finding and establishing a den for giving birth. Young female longtails are also precocious, but not to quite the same extent. They do not mate in their natal dens, but shortly after independence. They still have the advantage of being free to concentrate next spring on hunting and finding a den, as do female stoats, but male longtails do not have the advantage of access to a ready-made harem of infant mates that male stoats have.

The trouble with explanations of evolutionary adaptations is that they are usually proposed in retrospect. As in Rudyard Kipling's famous *Just So Stories*, adaptation stories may be made to fit the observed outcome. Delayed implantation may well have many advantages for stoats and longtails, but for all we know, both species might have been just as successful without it. As in any theoretical problem, answers are easy; the trick is to find the right question. In this case, we need to know not only why stoats and longtails have retained delayed implantation, but also why (and how) common and least weasels have lost it.

Mikael Sandell (1984) proposed that stoats and common weasels have different patterns of reproduction because the keys to success in contemporary times are different for each. For stoats, he reasoned, the most important consideration is to produce the young as early in the season as possible. Sandell assumed that the young need plenty of time to grow before the next winter, yet the adults need to avoid mating too early in spring, while conditions may still be severe. Delayed implantation allows both. By contrast, King (1983a, 1984a) argued that the only thing that matters to common weasels is a high potential rate of increase, that is, the capacity to turn mouse meat into young weasels faster than anyone else. For this, small size, rapid maturation, and no delayed implantation are the best policies.

Nils Stenseth (1985) was stimulated by the exchange between Sandell (1984) and King (1984a) (in the columns of the journal *Oikos*) to produce a model incorporating the different reproductive capacities of the two species (stoat, six to

13 young once a year; common weasel, four to eight young once or twice a year). He then addressed Sandell's question on whether these differences could be maintained by the action of contemporary natural selection. He concluded that Sandell's idea is basically correct: The relative difference in litter size and frequency predicted by his model corresponds to the observed difference in nature.

Unfortunately, Stenseth's analysis included some known errors. For example, he assumed that common weasels always have two litters every year, whereas in nature they do this only when food is abundant (McDonald & Harris 2002). More controversially, Stenseth accepted Sandell's assumption that stoats must produce their young early in the year, whereas the advantage of early littering for stoats and longtails seems insufficiently consistent and compelling to account for the remarkable constancy of delayed implantation that both species show.

Stenseth's, Sandell's, and King's exchanges answered some questions but raised new ones, such as how do the tiny Canadian stoats manage to continue to behave like stoats when their body size and food resources are those of common and least weasels? And why must stoats avoid breeding too early in spring when the much smaller least weasels can breed under the snow all winter?

In the fully developed version of his hypothesis, Sandell (1990) proposed that delayed implantation is retained only in those contemporary mammals for whom the period required for active gestation is shorter than the period between the best time for mating and the best time to give birth. If the total gestation period could be lengthened, by inserting a delay in implantation, it might become possible both to breed and to give birth at the respective optimal times. Sandell reviewed the literature and found, as would be expected, that all mammals exhibiting delayed implantation can be interpreted to benefit from the delay.

The beauty of Sandell's hypothesis, however, is that it makes predictions beyond such *Just So Stories* interpretations. The hypothesis predicts that some mammals that might benefit from a large increase in the length of gestation do not delay because the assumed first step, a short increase, would actually decrease fitness. The hypothesis can also be used to predict delay for mammal species whose biology is poorly known. To date, unfortunately, no one has tested the hypothesis by making real predictions and assessing the outcomes. We hope to see such tests in the future.

With this background, we can search for possible answers to questions about the origins and uneven distribution of delayed implantation in small mustelids by looking more closely at the two best-studied contrasting species, the stoat and the common weasel.

The Origins of Delayed Implantation

Modern animals are descended by a continuous process of change from different animals living in the past. Ancestral characters, or modified remnants of them, are often still discernible in contemporary species, constituting what Stephen Jay Gould called "the footprints of history."

We suggested in Chapter 1 that weasels evolved just before and during the Pleistocene from larger ancestors somewhat similar to martens and polecats. These ancestors probably already had delayed implantation (Lindenfors et al. 2003; Thom et al. 2004), as martens still do. About 53 mammal species are known to have obligate delayed implantation—that is, fewer than 0.05% of all mammals (Thom et al. 2004). The life histories of only 37 of the 60+ species of mustelids are known in any respect; of these, 17 have delayed implantation, a vastly greater proportion (about 34% of all the mustelids) than in mammals in general (Mead 1989; Sandell 1990). How could delayed implantation have gotten started in that far distant ancestral line, and why was it worth keeping?

Some mustelids and their closely related kin, American minks and striped skunks, for example, have short, variable delays of implantation, and sometimes they do not delay at all. These species may hold the clue to the original appearance of delayed implantation. Minks continue to ovulate even when fertilized if more than 6 days elapse between matings—a habit known as superfetation (Shackelford 1952; Mead 1994). Different males can fertilize ova shed in different ovulations, so if a better male comes along after a female mink has bred, she can still mate with the new, better male. Meanwhile, the zygotes resulting from early fertilizations delay implantation so that all can implant together.

This arrangement allows female minks to increase the number of offspring they produce, often fathered by more than one male, and also to improve the quality of their mates (and offspring), yet have all offspring born at the same time at the same developmental stage. If this short delay in implantation of some offspring proved to be beneficial, it could be increased for those species that could gain a further advantage from delaying implantation of all offspring, so increasing the interval between mating and giving birth. The simplest interpretation of mustelid phylogeny suggests that the ancestral mustelid lineage did delay implantation, and that several lines of its descendants have since lost it (Lindenfors et al. 2003). This line of reasoning is plausible only if there were good reasons to abandon such a long-established and successful characteristic.

King (1989a) suggested that the weasels in general became smaller in response to the opportunities presented by the evolution of open grasslands inhabited by voles and lemmings. The characters most favored in predators specializing on these new prey included, of course, the ability to search through miles of rodent tunnels and small burrows (see Figure 2.2). But even more important was the capacity to breed rapidly in response to sudden increases in food supply and to disperse in response to the decreases.

Stoats and common weasels look much alike, and most phylogenetic analyses done so far support the view that they are closely related (Bininda-Edmonds et al. 1999). In the process of adapting their life history characteristics to the uncertainties of life as a specialist predator of voles, both stoats and common weasels have evolved characteristics that substantially increased their reproductive rates compared with their larger bodied ancestors. The most effective ways of doing this, especially for small, short-lived animals, are early maturity, large litter sizes, and multiple litters per year.

On this view, common and least weasels have achieved all these advantages simply by abandoning the delayed implantation inherited from their common ancestor. Analyses of the chromosomes of the two species in Japan confirm the idea that the living least weasel is derived from an ancestral form similar to the stoat (Obara 1985). The problem with this idea is, if this new arrangement was advantageous for *nivalis*, why didn't *erminea* do the same?

One possible reason is that stoats developed, instead, a different, *additional* character, the extraordinary precocity of the juvenile females (King 1984a). The combination of delayed implantation and juvenile precocity allowed simultaneous mating of adult and young females, which in turn meant enormous potential breeding success for dominant males and certain fertilization before dispersal for the young females.

The addition of juvenile precocity turned delayed implantation into an advantage rather than a handicap, which, allied with the larger litters and longer life expectancy of stoats relative to common weasels, made a set of reproductive characters that succeeded. Females inheriting the combination of juvenile precocity and delayed implantation were at an advantage compared with other females. Consider an imaginary population in which most females died at 2.5 years old and delayed implantation was obligatory. A single heritable change allowing a female to mate as a nestling as well as at a year old would double her lifetime fecundity compared with other females who could not mate until they were yearlings—and then could not produce the litter until they were 2 years old. The advantages of the new arrangement to the female, and perhaps even more important to her sons, would ensure that it spread through the population.

One could, of course, consider alternative scenarios. Perhaps stoats and common weasels descended from an ancestor that did not have delayed implantation, and stoats later acquired it. The trouble with this idea is that stoats would have had to acquire juvenile precocity first, since to add delayed implantation without it would have caused a drastic drop in reproductive rate—the opposite effect to that required. Yet juvenile precocity without delayed implantation would have been suicidal for the young females, which could not possibly bear their litters while still nestlings themselves.

One of us (King 1983a) first suggested this hypothesis at a meeting in Helsinki in 1982. Soon afterward, the other (Powell 1985b) published a comparison between real or reasonable life tables for modern stoats and common weasels and for a hypothetical ancestor, addressing the question of how the modern species evolved. He concluded that, as common weasels have such short life spans, they could not have evolved their present small size without the massive extra productivity made possible by direct implantation and opportunistic late summer breeding, first by adults and later, as size decreased, by the rapidly maturing spring-born young. He added that, if stoats are descended from a larger ancestor with good adult survival rates and delayed implantation, the reduction in life expectancy that accompanied their decrease in size "would have required the evolution of juvenile breeding." In other words, these calculations confirmed that the modern stoats could have evolved from an ancestor with delayed implantation, provided they added juvenile precocity later.

The only other logical possibility was that stoats and common weasels are not as closely related as they look; perhaps stoats descended from marten-like ancestors that had delayed implantation, whereas common weasels came from polecat-like ones that did not (King 1983a). At the time it was suggested, this idea had something to recommend it; stoats are more similar to martens than to weasels in several curious characters, for example, in the shape of the baculum and in delayed implantation and the whole physiological mechanism that controls it. This apparently plausible suggestion has since been supported by at least one phylogenetic analysis, which considers *nivalis* and *erminea* different enough to be placed in separate subgenera (Abramov 2000), although the general view is usually that stoats and common/least weasels are descended from a fairly recent common ancestor (Chapter 1).

The evolutionary history of common weasels and stoats is therefore equivalent to a natural experiment conducted over several million years. That was plenty of time to test the effect of a single inherited variable—the presence or absence of delayed implantation—on a pair of sympatric species of common origin, adapted to and still living in similar ecological conditions. The results are unusually clear, and their implications have a particular fascination for speculations on the evolution of reproductive strategies: They show how similar, closely related species can evolve different solutions to a common problem.

How do longtails fit into this story? As we saw in Chapter 1, independent lineages of mustelids have lived in North America and Eurasia since the late Miocene, and *Mustela frenata* has survived in its recognizably modern form much longer than has either *erminea* or *nivalis*. Both longtails and stoats have lengthy and obligate delayed implantation and large litters, but while both also have precocial maturation of young females, precocity in longtails is much less extreme than in stoats. Young female longtails breed, not as blind infants, but when they are older, well grown, and independent.

The long evolutionary separation of stoats and longtails suggests that the *frenata/erminea* story fits a scenario slightly different from the *erminea/nivalis* comparison. Perhaps longtails inherited delayed implantation from an early common ancestor with stoats and still control it by a similar, invariable genetic code, and later longtails independently added juvenile precocity, but by a slightly different genetic code. If this hypothesis is true, longtails and stoats demonstrate a remarkable degree of convergence—the tendency for similar-looking animals to acquire, regardless of background, adaptations advantageous in competition for particularly abundant resources.

COMPETITION AND COEXISTENCE AMONG WEASEL SPECIES

The three species of northern weasels all have a strong family resemblance and all depend on more or less the same prey. At least two of them may be found in almost all countries north of 40°N, and in some places in North America all three species may be found together. Yet, it is one of the basic ideas of ecology that two or more similar species cannot coexist indefinitely.

The theory starts from the assumption that similar species tend to depend on the same resources, harvested in the same way, and then it predicts that, unless those resources are so abundant that there is always enough for both (in which case there is no competition), one species will eventually displace the other. Either one will be better at harvesting, so that the other starves, or one will aggressively drive the other away. The two forms of interaction are known as exploitation, or "scramble," competition and interference, or "contest," competition. The expectation is that, unless one of a pair of similar species develops some means of evasion (such as hunting at a different time of day or in different habitats), the other will exclude it from a habitat by one means or the other, or by a combination of both.

Where two or more weasel species coexist, they occupy slightly different niches: Members of the small species concentrate on small rodents, while members of the large species favor large rodents and rabbits (Chapter 5). In a diverse habitat with a variety of prey of different sizes, the overlap between their niches is minimal and the two can probably ignore each other for much of the year. But, whenever or wherever there is less choice of prey, both must search out anything available, and then competition is inevitable.

Because populations of small rodents are so unreliable, this must happen often enough to have a real effect on the distribution and population density of whichever weasel species is less well equipped to meet the conditions of the moment. Sooner or later one is likely to be eliminated, at least temporarily.

Mick Southern (a perceptive observer both of nature and of graduate students) always used to say that, although it is fun to discuss theoretical questions in the library at tea time (such as, what controls niche overlap?), if you really want to know, it is quicker and easier to go out and ask the animals themselves. This is true, provided first that the question is carefully framed, and second that the observations to be made are defined in terms that could provide a definite answer, one way or the other. This process has given us a preliminary idea of how stoats and common weasels coexist in

Britain, and how stoats and longtails or stoats and least weasels coexist in North America.

A Hypothesis

In Aberdeen in 1977, Chris Pounds chose to study the coexistence of common weasels and stoats for his doctoral thesis. In Ontario in 1973–1976, another graduate student, Dave Simms, was doing field research on stoats and longtails and thinking along the same lines. Unaware of either, King and Moors (1979a) proposed a verbal hypothesis on how stoats and common weasels might coexist, hoping that someone would undertake the fieldwork necessary to see whether or not it was right. Pounds (1981) was able to test parts of the hypothesis before he finished fieldwork, and the rest of it was tested by Erlinge and Sandell (1988).

The original hypothesis was developed with common weasels and stoats in mind, but is easily generalized for any pair of weasel species. It was based on the idea of an unstable balance of opposing advantages. Because common weasels are smaller than stoats, they are better able to reach small rodents in their burrows and nests; they can survive in a smaller area, and for long after the rodents have become too scarce to support stoats. They can also respond more rapidly to a glut of rodents. Common weasels therefore have an advantage over stoats in exploitation competition, although their specialization also makes them vulnerable to local extinction if the rodents disappear altogether. This certainly happens occasionally, but the vacated area can be recolonized from elsewhere in the next few years (Chapter 9).

On the other hand, because stoats are larger than common weasels, they can turn to larger prey when rodents become scarce (Chapter 5), and they will also always win in direct confrontations with common weasels, so are better able to evict common weasels from a choice area or to steal their catch. Stoats, therefore, have an advantage in interference competition, but only so long as they have access to larger prey. Delayed implantation means they cannot take immediate advantage of a rodent peak, but their wider choice of prey and longer average life span means that individual stoats have more chance of surviving to the next breeding season.

Hence, each species has a different combination of advantages in foraging and reproduction, and each is best adapted to exploit slightly different conditions. In any one place and time, one may be present and the other not, but over the long term the two coexist because in a patchy environment there will always be places and times where common weasels can avoid confrontations with stoats and stoats can avoid overdependence on rodents. The hypothesis predicts that, when they are forced to face up to each other, one or the other always wins, depending on the circumstances. A variable environment permits coexistence by constantly changing the balance of the different advantages of each.

As examples of how these ideas might work, King and Moors (1979a) cited two cases in which coexistence broke down. In one the outcome favored common weasels, and in the other it favored stoats.

The Consequences of Myxomatosis

The arrival of myxomatosis in Britain in 1953–1955 constituted a massive, far-reaching, unplanned field experiment. Myxomatosis is a lethal disease of European rabbits, and within a couple of years the national rabbit population crashed to less than 1% of its former size (Chapter 10). This sudden removal of the formerly abundant rabbits had broad environmental effects.

For example, the lifting of the enormous grazing pressure once exerted by rabbits permitted unprecedented, spectacular flowering of the countryside in the springs of 1955 and 1956 (Sumption & Flowerdew 1985). The consequences of these events for common weasels and stoats were not observed directly, but were strong enough to make unmistakable ripples in the vermin books of many game estates (see Figures 10.2 and 10.6). The numbers of each species caught can be compared directly, since both are collected in the same traps by identical methods. The records clearly show that, for at least 15 or 20 years after the epizootic, common weasels flourished, while stoats virtually disappeared.

King and Moors (1979a) suggested that one possible reason why myxomatosis had such different consequences for the two small mustelids was that it changed the balance of their relative advantages. When rabbits disappeared, the broad range of sizes of prey animals available to stoats suddenly narrowed. Stoats lost their main advantage over common weasels, the freedom of choice between large and small prey, and instead found themselves in fierce competition with common weasels for small rodents and with larger predators, such as foxes, feral cats, and raptors, for the remaining rabbits. Stoats lost against both the common weasels, which can reach rodents in their burrows, and the larger predators, which would not hesitate to attack a stoat. The balance of advantages was temporarily tipped in favor of common weasels.

The crash in the rabbit population resulted in a general shortage of prey at first, because all surviving predators had to concentrate on small rodents. This shortage did not affect common weasels nearly as much as it affected the larger predators, which can hunt only those rodents that show themselves out of their nests and burrows. Then, in 1956, the populations of small rodents soared, and the common weasel population followed them.

For the next few years over most of the country, the average ratio of common weasels to stoats killed was reversed, from about 1:2 before myxomatosis to about 2:1 after, depending on the habitat (Craster 1970; Hewson 1972). On one estate in eastern England, common weasels never exceeded stoats in the annual bag, but the average number killed more than doubled, from 15 per year in the 7 years 1947 to 1953, when stoats averaged 650 a year (1:40), to 38 per year in the 7 years 1957 to 1963, when stoats were at their lowest ebb, 69 a year

(1:2) (King 1980c). The main reason for the extraordinary increase in numbers of common weasels was certainly the sudden glut of small rodents, although the removal of interference from stoats (very scarce by then) probably helped.

Introductions and Distributions on Islands

Rabbits were also the key players in the second example cited by King and Moors (1979a), but which had the opposite result. The common weasels and stoats introduced into New Zealand both survived and spread, and now, over a hundred years later, stoats are common in virtually all forested areas. Common weasels, originally imported in much greater numbers than stoats, are now among the rarest of all New Zealand's mammals.

Why did the pioneering stocks of the two small mustelids respond so differently to their new environment? The hypothesis suggests that it was because the radically unfamiliar conditions in New Zealand changed the balance of their relative advantages. Although rabbits and rats are abundant in New Zealand, voles are absent and other smaller prey are scarce; feral house mice are the only rodents under 50 g in body weight (Chapter 5).

Because common weasels are so strongly specialized to prey on small rodents, especially voles, they are handicapped in New Zealand. Their main advantage over stoats, their greater efficiency as specialist hunters of small rodents in tight spaces, is of little use to them where small rodents are always hard to find. The only alternative small prey, mostly large native insects and lizards, were insufficient substitutes for voles. Stoats, on the other hand, have the range of large and small prey that they prefer (see Figure 5.4), and they have hardly any larger predators to avoid. The consequence was that the balance of advantages was permanently tipped in favor of stoats.

Common weasels are never likely to be abundant in New Zealand but, even after decades of disadvantage, they are not yet extinct either. The reason they have survived despite the lack of voles is probably that the two main islands of New Zealand are very large and diverse, giving them plenty of room to disperse to find habitat patches with sufficient mice, and to avoid stoats. Where stoats are being effectively removed, weasels seem to be becoming more common.

Matters were different, however, on Terschelling, a much smaller island (680 km²) off the Dutch coast. As in New Zealand, many more common weasels were released than stoats, and like New Zealand, Terschelling has no voles. By 1934 the common weasels had disappeared, while the stoats were well established. The size distribution of prey available on Terschelling was much better for common weasels than was that in New Zealand (King & Moors 1979a), which suggests that common weasels might have survived on Terschelling if there had been no interference from stoats, but on such a small island, a weasel could not always avoid meeting a stoat in a tight corner.

The same hypothesis also offers an explanation for the puzzling distribution of stoats and common weasels on the offshore islands of Britain. Eleven of these islands, including relatively small ones (<60 km²) are, or have been, oc-cupied by stoats, but only four by common weasels—all large (>380 km²) or connected to the mainland by bridges.

This is odd, because one might expected the small common weasels to sur-vive better than stoats in a restricted area. But in fact, for common weasels, the size of an area is less important than its accessibility. Almost all populations of common (or least) weasels and stoats become locally extinct periodically, and depopulated habitats have to be recolonized from further afield. On small is-lands with voles but without stoats, common weasels are too vulnerable to local extinction to last long. On larger islands they might be better off, except that they are vulnerable to interference from stoats.

Only if the island is large enough to support diverse habitats, or is easily recolonized, can populations of common weasels survive. Terschelling, one of the few islands on which we know that both arrived in numbers sufficient to found a population, is twice the size of the Isle of Wight (380 km²), the smallest British island on which both may be found together, but like New Zealand, Terschelling has no voles.

Ireland is much larger than Wight, and the absence of common weasels there is a mystery. One suggestion (King & Moors 1979a) is that common weasels were there in earliest postglacial times, along with a whole fauna of hardy north-ern species including stoats and lemmings (Yalden 1999), but that they disap-peared later when the lemmings became extinct and were not replaced by voles. Yalden rightly points out that there is no evidence for this idea, which is true, but in the fossil record, more than in most places, absence of evidence is not evidence of absence.

Testing the Hypothesis

Pounds (1981) radio tracked both stoats and common weasels in the farm-lands and sand dunes near Aberdeen, Scotland (see Table 8.1). He found that both preferred to hunt in the same habitats, the field margins and rough grass-lands where small mammals were most abundant. Both stoats and common weasels could be found in such places at any time. Overall, there was a constant ratio of five stoats to 10 common weasels over an area of 54 km², which implies, since stoats are twice the size of common weasels, that both species were mak-ing roughly equal demands on the local prey resources. Competition seemed to be unavoidable, especially in spring when the numbers of small mammals were at their annual low after 7 months (October to April) of unreplaced losses.

Pounds watched both species at a distance (30 to 50 m) with strong binocu-lars, using infrared lights after dark, and followed them on hunting expeditions. Both hunted at any time of day or night. The diameters of the burrows of the local field voles averaged 23 mm, and Pounds estimated that female common weasels could get into any but the narrowest ones. Males could get into the larger ones with a squeeze, but stoats of both sexes were excluded from all burrows.

By scat analysis, Pounds showed that field voles were the most important prey of common weasels all the year round, and also of stoats in autumn; for the rest of the year stoats concentrated on rabbits. There were differences in emphasis but, still, the overlap in their diets was substantial, especially between male weasels and female stoats. Members of the two species had no obvious way to partition their common prey resources, either in space or time. Pounds reckoned that competition for food between the two must be serious, and that no other consideration (e.g., shortage of den sites, predation by human or other large predators) was anywhere near as significant to them. Nevertheless, he concluded that the exploitation advantage of the common weasels' ability to hunt in tunnels was, most of the time, sufficient to ensure their survival and that the stoats' ability to hunt alternative prey over a larger area was sufficient to compensate for their restricted access to rodents.

The other side of the hypothesis suggests that stoats should be able to evict common weasels from choice hunting areas. In fact, of course, no open aggression is needed. The parties need not even meet face to face. The effect is the same so long as a common weasel always knows when a stoat is about and is scared enough of meeting it to move elsewhere rather than risk an encounter. Pounds noted that both species routinely marked their home ranges with scent signals, and these could be quite enough on their own to have the required effect, perhaps reinforced occasionally by an actual meeting.

Pounds carefully refrained from speculating any further, but he reported that, in the 1,300 hours of radio tracking he logged, he recorded no direct encounters between free-living stoats and common weasels. His experiments with captive animals in enclosures and in cages in the field strongly suggested that common weasels certainly could detect the presence of a stoat, and avoided it if they could. He reckoned the effect in the wild would be very limited in time and extent. Female weasels, which might be considered most vulnerable to interference from stoats, were actually least concerned by them, because they could always escape to a rodent tunnel.

Erlinge and Sandell (1988) made much use of enclosures, in which animals cannot be guaranteed to behave naturally, yet their observations complement Pounds' fieldwork in two of three respects. First, they showed that common weasels really are scared to death of face-to-face encounters with stoats. When one of each was released into a 30-m^2 enclosure, the common weasel always fled to a refuge box, while the stoat took over the open area, moving about confidently and ignoring the weasel with lordly disdain. The weasel remained hidden but alert, watching the stoat and ready to react with threats if ever the stoat approached its refuge. If the stoat settled down somewhere quietly, the weasel cautiously emerged and began to make desperate attempts to escape.

Erlinge and Sandell also showed that common weasels tended to avoid traps that had previously been occupied by stoats and still reeked of stoat scent, whereas stoats treated weasel-scented traps the same as any others.

Finally, Erlinge and Sandell looked for field evidence to support their observations in the enclosure. They searched back through years of trapping records (1973 to 1984), checking for any evidence of reciprocal distributions of the two species in their study area. The two species were considered to be in potential contact if one was caught within 200 m of a site, and within 2 months before or after the date on which the other had been caught. In habitats that were potentially suitable for both, they recorded 21 weasels caught in places without stoats, but only six in places with them. Erlinge and Sandell concluded that common weasels avoid areas occupied by stoats, although they use them when the stoats are absent.

This result disagreed quite specifically with Pounds' radio-tracking observations. Consequently, King (unpublished) decided to repeat the test for reciprocal distribution, using the detailed trapping records kept by three gamekeepers at North Farm, Sussex, in 1974 and 1975, and kindly lent by the Game Conservancy. The keepers had a network of about 300 permanent Fenn steel trap sites spread evenly over about 1,500 ha, mostly along field boundaries and rough vehicle tracks, as mapped by Tapper (1982). These were kill traps, so each individual was removed as soon as it was caught, but the scent marks left by residents continued to proclaim home range boundaries for some time after the owners were gone.

The entire area, except in the centers of open fields, was suitable habitat for both species. The first step was to mark the trap sites where stoats were caught, then to mark all captures of common weasels, distinguishing those that could have been in contact with a stoat by Erlinge and Sandell's definition. The common weasels clearly did not avoid the areas occupied by stoats; over a quarter of them (64 of 246) were potentially "in contact," and 15 were caught in the same trap as a stoat within 30 days.

This is, of course, a very rough method of estimating the relative distributions of the two species over such a large area, and even if the results had shown the mutual avoidance claimed, there would be no way to tell why. "Unfavored" areas could be unoccupied by common weasels not only because stoats were there but also because hunting conditions were better elsewhere, or simply because the density of common weasels was low in that year.

The only reliable method of tackling the problem is the way Pounds did it, by direct observation, and his conclusion was that, if the habitat is favorable to both, both will use it. There may well be mutual avoidance, but it must be on a rather fine scale under normal conditions, and becomes a critical factor only in places (such as on small islands) where food is severely limited and confrontations inescapable.

Recent studies have documented in great detail the differences in habitat selection, activity, and diet between sympatric populations of two weasel species (McDonald 2002; Aunapuu & Oksanen 2003; St.-Pierre et al. in press-a; Sidorovich et al. unpublished). The larger species usually chooses larger prey and the most productive habitats, capable of supplying their higher demands for food; the smaller species is more affected by changes in the abundance of small rodents, because it has fewer alternatives. But the question of whether the larger one can actively exclude the smaller, or whether the smaller simply avoids the larger, remains very difficult to answer. As Aunapuu and Oksanen (2003) put it, "The mechanisms of competitive coexistence cannot be derived from the preferences or from the food habits in times of prey abundance. The theory of exploitation competition . . . does not deal with preferences but with the ability of organisms to persist in times of shortage." Weasels are very scarce at such times, so the chances of getting field data on their interactions then depend on the development of new technology for field observation.

Relationships among the North American species might be different, or at least more complicated, because both stoats and longtails can be the large weasel of a pair. Stoats may be either the larger weasel, when stoats are sympatric with least weasels, or the small weasel, when sympatric with longtails. Inside a narrow zone across North America (Chapter 1), all three weasels may be found, although they seldom all live in the same place at the same time. Thus, the scenario of conditional coexistence, followed by local extinction of one or both (or all three!) and later recolonization, seems to be generally true. Indeed, several researchers have reported locally reciprocal distributions among sympatric weasels, such as stoats moving out when longtails arrived (Fitzgerald 1977; Simms 1979b; Gamble 1980).

The distribution of North American weasels has one curious feature. Stoats and least or common weasels are found together from the Arctic right down to around 40°N on both sides of the Atlantic, but longtails live only south of about 55°N. If longtails are capable of displacing the smaller species by aggressive interference, why do they not extend into the forests and grasslands further north?

The answer may lie in a thoughtful hypothesis by Simms (1979a), which he suggested quite independently but has intriguing similarities with the one discussed above. Simms noted that the small stoats of the snowy north feed primarily on voles. They are well adapted to hunt in rodent tunnels and are strongly specialized as underground or under-snow predators. Long-tailed weasels are larger and have more general food habits. Simms suggested that longtails are confined to the south by the prolonged snow cover of the north, because they are less capable of hunting through the confined under-snow spaces than smaller weasels, and because the only alternative in the far north is to compete with the larger predators such as foxes and martens, which hunt above the snow. Conversely, stoats are confined to the north by interference competition from longtails.

Longtails can and do, in fact, hunt under the snow when it is deep enough and when coarse woody debris and sugar snow allow access to subnivean spaces (see Figure 6.2). From the Upper Midwest and across the Rockies to the western, coastal mountains, longtails and American martens, their larger cousins, spend a lot of time under the snow (Francis & Stephenson 1972; Fitzgerald 1977; Buskirk et al. 1989; Buskirk & Powell 1994). The key to the longtail's success is its ability to switch to alternative prey when rodents are scarce, but boreal forests have few prey for weasels other than small rodents. Perhaps it is this poor diversity of prey that keeps longtails mainly in the south, regardless of snow or other weasels (Fitzgerald 1977; Gamble 1981). If so, we are left with another question: If boreal forests lack diverse prey, how can stoats and least weasels coexist there?

Another Approach

The King and Moors hypothesis on the coexistence of two or more weasel species was formulated at a time when competitive exclusion was considered to be a dominant ecological process. For example, most biologists generally accepted that the reason why the members of a group of similar-looking animals, such as the weasels, are graded by size is that size differences help to reduce competition.

It is true that small weasels specialize on small voles, whereas large ones kill prey the size of cotton rats, chipmunks, water voles, and rabbits (McDonald 2002; Sidorovich et al. unpublished). The trouble is, these are differences of degree, not of kind. The large weasels do eat a great many small voles when they are profitable, and the small weasels are still able, at times, to kill rabbits. Differences in size alone do not, in fact, ensure that the weasel species exploit different resources. There must be more to it than that.

Powell and Zielinski (1983) took a different approach to the question and came up with some suggestions as to what that extra factor might be. They added two twists to the argument put up by the simple basic hypothesis, derived from modeling a community with two weasel species, a range of prey from small to large, and other predators that compete with weasels for prey but also eat weasels on occasion (Figure 14.4).

The first twist is that, when both weasel species become locally extinct, the species that recolonizes first gains an advantage by building up a large population first. That large population increases its probability of survival locally during the next crash of the prey population. The second twist is that predation on weasels by other predators reduces the chances that their populations will grow to sizes large enough to lead to intense competition. Some field observations seem to support the hypothesis that predation facilitates coexistence, at least at landscape scale when populations of both species are present. In northeastern Canada, where longtails are extending their distribution into forests formerly occupied only by stoats and least weasels, the presence of larger fur-bearers seems beneficial to coexistence (St.-Pierre et al. in press-a).

Computer simulation models constructed to imitate the competitive interactions of weasels and analyses of the dynamics of communities including weasels confirm the core of the argument: Local coexistence of two weasel species depends on geographic scale. At the continental scale, two or more species of

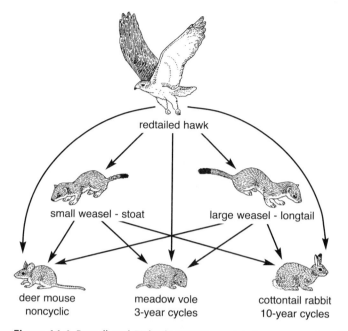

Figure 14.4 Powell and Zielinski (1983) modeled a community with two weasel species, a range of prey from small to large, and other predators that compete with weasels for prey but also eat weasels on occasion. The presence of large prey allowed the model population of large weasels to avoid extinction when small mammals disappeared. The ability to enter vole holes gave the small weasels an advantage in catching voles when they were rare. When both weasel species went extinct locally in the model, the species that recolonized first gained an advantage by building a large population first. Predation on weasels by other predators reduced both weasel populations and reduced competition between them. Thus, predation facilitated coexistence when populations of both weasels were present.

weasels have lived together in Eurasia and North America for thousands of years; at the local scale, living together is a dynamic matter of repeated local extinctions and recolonizations.

CONCLUSION

Some ecologists now see the theory of competitive exclusion as not very useful in explaining the real world, because its assumptions are too restrictive to be realistic outside the laboratory and its reasoning is suspected of being circular. After all, every species is by definition slightly different from every other, every patch of habitat is in practice unique in some respect, and very few environments are truly in stable equilibrium.

On the other hand, opportunistic species such as the weasels never do live in equilibrium conditions, so any theory based on that assumption simply cannot be applied to them. Instead, we emphasize that the coexistence of weasel species actually *depends* on lack of equilibrium. Weasels simply use the fluctuations in rodent populations as a resource (Levins 1979; Powell & Zielinski 1983).

Fluctuations in food supply and the patchy distribution of habitats orchestrate the frequent local extinctions and recolonizations of weasel populations. Throughout this book we have discovered very many costs attached to being a small, thin specialist rodent hunter, and the problem of unstable numbers is one of them. Our guess is that, given enough space and time, these in turn permit the long-term survival in one place of more than one variation on the weasel way of life. Confirmation of this idea awaits some ingenious future project. We are glad that there are still plenty of puzzles left for future weasel biologists to study.

Appendix | Latin Names of Taxa Mentioned Only by Common Name in the Text and Tables

aspen, *Populus* sp.
badger, European, *Meles meles*
bat, big brown, *Eptesicus fuscus*
bat, NZ short-tailed, *Mystacina tuberculata*
bear, black, *Ursus americanus*
bear, brown, *Ursus arctos*
bear, polar, *Ursus maritimus*
beaver, mountain, *Aplodontia rufa*
beech, southern, *Nothofagus* sp.
beetle, vedalia, *Rodolia cardinalis*
bellbirds, *Anthornis melanura*
birch, *Betula* sp.
buzzard, common, *Buteo buteo*
cat, big, *Panthera* sp.
cat, house, *Felis sylvestris catus*
cat, wild, *Felis sylvestris*
cats, Family Felidae
cetaceans, Order Cetacea
chaffinch, *Fringilla coelebs*
chipmunks, *Tamias* sp.
corvids, Family Corvidae
cottonwood, *Populus deltoides*
coyote, *Canis latrans*
crow, carrion, *Corvus corone*
curlew, *Numenius* sp.
deer, Family Cervidae
dormouse, garden, *Eliomys* sp.
dotterel, NZ, *Charadrius obscurus*
earthworm, *Lumbricus* sp.
elephants, Family Elephantidae
ermine, *Mustela erminea*

falcon, NZ bush, *Falco novaeseelandiae*
ferret, *Mustela furo*
flycatcher, willow, *Empidonax traillii*
fox, arctic, *Vulpes lagopus*
fox, red, *Vulpes vulpes*
foxes, Family Vulpidae
frogs, Order Anura
goat, *Capra hircus*
goldcrest, *Regulus regulus*
gophers, pocket, Family Geomyidae
grouse, Family Tetraonidae
gulls, Family Laridae
hare, brown, *Lepus europaeus*
hare, snowshoe, *Lepus americanus*
harrier, hen, *Circus cyaneus*
hawk, Australasian harrier, *Circus approximans*
hawk, red-tailed, *Buteo jamaicensis*
hedgehog, *Erinaceus europaeus*
hornbeam, European, *Carpinus betulus*
horse, *Equus caballus*
hyena, spotted, *Crocuta crocuta*
kaka, *Nestor meridionalis*
kakapo, *Strigops habroptilus*
kestrel, European, *Falco tinnunculus*
kiwi, *Apteryx* sp.
kokako, *Callaeas cinerea cinerea*, *Callaeas c wilsoni*
lagomorphs, Order Lagomorpha
larch, *Larix laricina*

squirrel, Columbian ground,
Spermophilus columbianus
squirrel, golden-mantled ground,
Spermophilus lateralis
squirrel, grey, *Sciurus carolinensis*
squirrels, ground, Family Sciuridae
squirrel, red (American),
Tamiasciurus hudsonicus
squirrel, red (European), *Sciurus vulgaris*
squirrels, tree, Family Sciuridae
stilt, black, *Himantopus novaezelandiae*
stilt, pied, *Himantopus himantopus*
stoat, Irish, *Mustela erminea hibernica*
stoat, *Mustela erminea*
takahe, *Porphyrio mantelli*
terns, Family Laridae
thrush, NZ, *Turnagra c. capensis*
thrush, wood, *Hylo cichla mustelina*
tit, blue, *Parus caeruleus*
tit, great, *Parus major*
tomtit, NZ, *Petroica macrocephala*
vetch, *Vica* sp.
vireo, black-capped, *Vireo atricapillus*
vole, bank, *Clethrionomys glareolus*
vole, common, *Microtus arvalis*
vole, field, *Microtus agrestis*
vole, long-tailed, *Microtus longicaudus*
vole, meadow, *Microtus pennsylvanicus*
vole, montane, *Microtus montanus*
vole, Oregon, *Microtus oregoni*
vole, red-backed, *Clethrionomys gapperi*
vole, root, *Microtus oeconomus*
vole, sibling, *Microtus rossiaemeridionalis*
vole, water, *Arvicola terrestris*
vole, woodland (pine vole), *Microtus pinetorum*
wallabies, Family Macropodidae
waterhen, *Gallinula chloropus*
weasel, back-striped, *Mustela strigidorsa*
weasel, barefoot, *Mustela nudipes*
weasel, Columbian, *Mustela felipei*
weasel, common, *Mustela nivalis vulgaris*
weasel, Egyptian, *Mustela nivalis subpalmata*
weasel, least, *Mustela nivalis nivalis*
weasel, long-tailed, *Mustela frenata*
weasel, mountain, *Mustela altaica*
weasel, tropical, *Mustela africana*
weasel, yellow-bellied, *Mustela kathiah*
weta, Order Orthopera
willows, *Salix* sp.
wolf, *Canis lupus*
wolverine, *Gulo gulo*
woodpeckers, Family Picidae
wren, bush *Xenicus l. longipes*
yellowhead (mohua), *Mohoua ochrocephala*

References

Abramov, A. V. 2000. A taxonomic review of the genus *Mustela. Zoosystematica Rossica* 8: 357–364.

Abramov, A. V., and Baryshnikov, G. F. 2000. Geographic variation and intraspecific taxonomy of weasel *Mustela nivalis* (Carnivora, Mustelidae). *Zoosystematica Rossica* 8: 365–402.

Allen, D. L. 1938a. Ecological studies on the vertebrate fauna of a 500-acre farm in Kalamazoo county, Michigan. *Ecological Monographs* 8: 347–436.

———. 1938b. Notes on the killing technique of the New York weasel. *Journal of Mammalogy* 19: 225–229.

Alley, J. C., Berben, P. H., Dugdale, J. S., Fitzgerald, B. M., Knightbridge, P. I., Meads, M. J., and Webster, R. A. 2001. Responses of litter-dwelling arthropods and house mice to beech seeding in the Orongorongo Valley, New Zealand. *Journal of the Royal Society of New Zealand* 31: 425–452.

Alterio, N. 1996. Secondary poisoning of stoats (*Mustela erminea*), feral ferrets (*Mustela furo*) and feral house cats (*Felis catus*) by the anticoagulant poison, brodifacoum. *New Zealand Journal of Zoology* 23: 331–338.

———. 1998. Spring home range, spatial organisation and activity of stoats *Mustela erminea* in a South Island *Nothofagus* forest, New Zealand. *Ecography* 21: 18–24.

———. 2000. Controlling small mammal predators using sodium monofluoroacetate (1080) in bait stations along forestry roads in a New Zealand beech forest. *New Zealand Journal of Ecology* 24: 3–9.

Alterio, N., and Moller, H. 1997a. Daily activity of stoats (*Mustela erminea*), feral ferrets (*Mustela furo*) and feral house cats (*Felis catus*) in coastal grassland, Otago Peninsula, New Zealand. *New Zealand Journal of Ecology* 21: 89–95.

———. 1997b. Diet of feral house cats *Felis catus*, ferrets *Mustela furo* and stoats *M. erminea* in grassland surrounding yellow-eyed penguin *Megadyptes antipodes* breeding areas, South Island, New Zealand. *Journal of Zoology (London)* 243: 869–877.

Alterio, N., Moller, H., and Brown, K. 1999. Trappability and densities of stoats (*Mustela erminea*) and ship rats (*Rattus rattus*) in a South Island *Nothofagus* forest, New Zealand. *New Zealand Journal of Ecology* 23: 95–100.

Alterio, N., Moller, H., and Ratz, H. 1998. Movements and habitat use of feral house cats *Felis catus*, stoats *Mustela erminea* and ferrets *Mustela furo*, in grassland surrounding yellow-eyed penguin *Megadyptes antipodes* breeding areas in spring. *Biological Conservation* 83: 187–194.

Amstislavsky, S., and Ternovskaya, Y. 2000. Reproduction in mustelids. *Animal Reproduction Science* 60–61: 571–581.

Anderson, D. R. 2001. The need to get the basics right in wildlife field studies. *Wildlife Society Bulletin* 29: 1294–1297.

Anderson, E. 1966. Weasel kills buzzard. *The Countryman* Spring 1966: 122.

Anon. 1960. Vermin bag records. *Annual Report of ICI Game Services for 1959*: 56.

———. 1981. *Predator and squirrel control.* Fordingbridge, Hampshire, England: The Game Conservency.

———. 1987. *Bovine Tuberculosis in Badgers.* London: UK Ministry of Agriculture, Fisheries and Food. Report 11.

———. 1998. Rail tracks. *Mammal News (UK)* 113: 11.

Asdell, S. A. 1964. *Patterns of mammalian reproduction.* Ithaca, New York: Cornell University Press.

Aspisov, D. I., and Popov, V. A. 1940. Factors determining fluctuations in the numbers of ermines. In: King, C. M., editor. *Biology of mustelids. Some Soviet research.* Volume 2. Wellington, New Zealand: New Zealand Department of Scientific and Industrial Research, Bulletin 227 (1980). pp. 109–131.

Atkinson, I. A. E. 2001. Introduced mammals and models for restoration. *Biological Conservation* 99: 81–96.

Aunapuu, M., and Oksanen, T. 2003. Habitat selection of coexisting competitors: a study of small mustelids in northern Norway. *Evolutionary Ecology* 17: 371–392.

Barlow, N. D., and Barron, M. C. 2005. Modelling the dynamics and control of stoats in New Zealand forests. *Science for Conservation* 252: 1–40.

Barlow, N. D., and Choquenot, D. 2002. Predicting the impact and control of stoats: a review of modelling approaches. *Science for Conservation* 191: 1–46.

Barrow, C. 1953. Weasel and waterhen. *The Countryman* Spring 1953: 115.

Baryshnikov, G. F., Bininda-Emonds, O. R. P., and Abramov, A. V. 2003. Morphological variability and evolution of the baculum (os penis) in Mustelidae (Carnivora). *Journal of Mammalogy* 84: 673–690.

Basse, B., Flux, I., and Innes, J. 2003. Recovery and maintenance of North Island kokako (*Callaeas cinerea wilsoni*) populations through pulsed pest control. *Biological Conservation* 109: 259–270.

Basse, B., and McLennan, J. A. 2003. Protected areas for kiwi in mainland forests of New Zealand: how large should they be? *New Zealand Journal of Ecology* 27: 95–105.

Basse, B., McLennan, J. A., and Wake, G. C. 1999. Analysis of the impact of stoats, *Mustela erminea*, on northern brown kiwi, *Apteryx mantelli*, in New Zealand. *Wildlife Research* 26: 227–237.

Belant, J. L. 1992. Efficacy of three types of live traps for capturing weasels, *Mustela* spp. *The Canadian Field-Naturalist* 106: 394–397.

Biknevicius, A. R., and van Valkenburgh, B. 1996. Design for killing: craniodental adaptations of predators. In: Gittleman, J. L., editor. *Carnivore behavior, ecology, and evolution.* Ithaca, NY: Comstock/Cornell University Press. pp. 393–428.

Bininda-Emonds, O. R. P., Gittleman, J. L., and Purvis, A. 1999. Building large trees by combining phylogenetic information: a complete phylogeny of the extant Carnivora (Mammalia). *Biological Reviews* 74: 143–175.

Bishop, S. C. 1923. Note of the nest and young of the small brown weasel. *Journal of Mammalogy* 4: 26–27.

Blackwell, G. L., Potter, M. A., McLennan, J. A., and Minot, E. O. 2003. The role of predators in ship rat and house mouse population eruptions: drivers or passengers? *Oikos* 100: 601–613.

Blackwell, G. L., Potter, M. A., and Minot, E. O. 2001. Rodent and predator population dynamics in an eruptive system. *Ecological Modelling* 142: 227–245.

Blomquist, L., Muuronen, P., and Rantanen, V. 1981. Breeding the least weasel (*Mustela rixosa*) in Helsinki Zoo. *Zoologische Garten Jena* 51: 363–368.

Bolbroe, T., Jeppesen, L. L., and Leirs, H. 2000. Behavioural response of field voles under mustelid predation risk in the laboratory: more than neophobia. *Annales Zoologici Fennici* 37: 169–178.

Borowski, Z. 1998. Influence of weasel (*Mustela nivalis* Linnaeus, 1766) odour on spatial behaviour of root voles (*Microtus oeconomus* Pallas, 1776). *Canadian Journal of Zoology–Revue Canadienne de Zoologie* 76: 1799–1804.

———. 2000. The odour of a terrestrial predator as an olfactory risk signal detected by rodents, and its consequences. *Biological Bulletin of Poznan* 37: 159–169.

Boxall, P. C. 1979. Interaction between a longtailed weasel and a snowy owl. *Canadian Field-Naturalist* 93: 67–68.

Brander, R. B. 1971. Longevity of wild porcupines. *Journal of Mammalogy* 50: 835.

Brandt, M. J., and Lambin, X. 2005. Summertime activity patterns of common weasels *Mustela nivalis vulgaris* under differing prey abundances in grassland habitats. *Acta Theriologica* 50: 67–79.

Brinck, C., Erlinge, S., and Sandell, M. 1983. Anal sac secretion in mustelids: a comparison. *Journal of Chemical Ecology* 9: 727–745.

Brown, G. H. 1952. Illustrated skull key to the recent land mammals of Virginia. Virginia Cooperative Wildlife Research Unit, USF&WS, USDI. Report 52–2.

Brown, J. A., and Miller, C. J. 1998. Monitoring stoat (*Mustela erminea*) control operations: power analysis and design. *Science for Conservation* 96: 1–21.

Brown, J. H., and Lasiewski, R. C. 1972. Metabolism of weasels: the cost of being long and thin. *Ecology* 53: 939–943.

Brown, K. P. 1997. Predation at nests of two New Zealand endemic passerines; implications for bird community restoration. *Pacific Conservation Biology* 3: 91–98.

Brown, K. P., Moller, H., Innes, J., and Alterio, N. 1996. Calibration of tunnel tracking rates to estimate relative abundance of ship rats (*Rattus rattus*) and mice (*Mus musculus*) in a New Zealand forest. *New Zealand Journal of Ecology* 20: 271–275.

Brugge, T. 1977. Prey selection of weasel, stoat and polecat in relation to sex and size. *Lutra* 19: 39–49.

Buchanan, J. B. 1987. Seasonality in the occurrence of long-tailed weasel road-kills. *Murrelet* 68: 67–68.

Buckingham, C. J. 1979. The activity and exploratory behaviour of the weasel *Mustela nivalis*. Ph.D. thesis. Exeter: Exeter University.

Bullock, D., and Pickering, S. 1982. Weasels (*Mustela nivalis*) attacking a young and an adult brown hare (*Lepus capensis*). *Journal of Zoology, London* 197: 307–308.

Bumann, G. B., and Stauffer, D. F. 2002. Scavenging of ruffed grouse in the Appalachians: influences and implications. *Wildlife Society Bulletin* 30: 853–860.

Bunin, J. S., and Jamieson, I. G. 1996. Responses to a model predator of New Zealand's endangered takahe and its closest relative, the pukeko. *Conservation Biology* 10: 1463–1466.

Burns, J. J. 1964. Movements of a tagged weasel in Alaska. *The Murrelet* 45: 10.

Burt, W. H. 1943. Territoriality and home range concepts as applied to mammals. *Journal of Mammalogy* 24: 346–352.

Burt, W. H. 1960. Bacula of North American mammals. Miscellaneous Publications 113. Ann Arbor, MI: Museum of Zoology, University of Michigan.

Buskirk, S. W., and Lindstedt, S. L. 1989. Sex biases in trapped samples of Mustelidae. *Journal of Mammalogy* 70: 88–97.

Buskirk, S. W., Forrest, S. C., Raphael, M. G., and Harlow, H. J. 1989. Winter resting site ecology of martens in the central Rocky Mountains. *Journal of Wildlife Management* 53: 191–196.

Buskirk, S. W., and Powell, R. A. 1994. Habitat ecology of fishers and American martens. In: Buskirk, S. W., Harestad, A. S., Raphael, M. G., and Powell, R. A., editors. *Martens, sables and fishers: biology and conservation.* Ithaca, NY: Cornell University Press. pp. 283–296.

Byers, J. A. 1998. Biological effect of locomotor play: getting in shape or something more? In: Bekoff, M., and Byers, J. A., editors. *Animal play.* Cambridge: Cambridge University Press. pp. 205–220.

Byers, J. A., and Walker, C. 1995. Refining the motor training hypothesis for the evolution of play. *American Naturalist* 146: 1–24.

Cahn, A. R. 1936. A weasel learns by experience. *Journal of Mammalogy* 17: 386.

Cain III, J. W., Morrison, M. L., and Bombay, H. L. 2003. Predator activity and nest success of willow flycatchers and yellow warblers. *Journal of Wildlife Management* 67: 600–610.

Calder, C. J., and Gorman, M. L. 1991. The effects of red fox *Vulpes vulpes* fecal odors on the feeding behavior of Orkney voles *Microtus arvalis*. *Journal of Zoology (London)* 224: 599–606.

Carbone, C., Mace, G. M., Roberts, S. C., and Macdonald, D. W. 1999. Energetic constraints on the diet of large carnivores. *Nature* 402: 286–288.

Carpenter, F. L., and MacMillen, R. E. 1976. Threshold model of feeding territoriality and test with a Hawaiian honeycreeper. *Science* 194: 634–642.

Casey, T. M., and Casey, K. K. 1979. Thermoregulation of arctic weasels. *Physiological Zoology* 52: 153–164.

Caughley, G. 1977. *Analysis of vertebrate populations.* London: John Wiley & Sons.

Caughley, G. C., and Sinclair, A. R. E. 1994. *Wildlife ecology and management.* Cambridge, MA: Blackwell Science.

Chernov, Y. I. 1985. *The living tundra.* Cambridge: Cambridge University Press.

Choate, J. R., Engstrom, M. D., and Wilhelm, R. B. 1979. Historical biogeography of the least weasel (*Mustela nivalis campestris*) in Kansas, USA. *Tranactions of the Kansas Academy of Sciences* 82: 231–234.

Choquenot, D., and Ruscoe, W. A. 2000. Mouse population eruptions in New Zealand forests: the role of population density and seedfall. *Journal of Animal Ecology* 69: 1058–1070.

Choquenot, D., Ruscoe, W. A., and Murphy, E. 2001. Colonisation of new areas by stoats: time to establishment and requirements for detection. *New Zealand Journal of Ecology* 25: 83–88.

Clapperton, B. K., and Day, T. D. 2001. Cost-effectiveness of exclusion fencing for stoat and other pest control compared with conventional control. *DoC Science Internal Series* 14: 1–19.

Clapperton, B. K., McLennan, J. A., and Woolhouse, A. D. 1999. Responses of stoats to scent lures in tracking tunnels. *New Zealand Journal of Zoology* 26: 175–178.

Clapperton, B. K., Phillipson, S. M., and Woolhouse, A. D. 1994. Field trials of slow-release synthetic lures for stoats (*Mustela erminea*) and ferrets (*M. furo*). *New Zealand Journal of Zoology* 21: 279–284.

Clark, B. S., and Clark, B. K. 1988. First record of the least weasel in Oklahoma. *Prairie Naturalist* 20: 134.

Clout, M. N. 2001. Where protection is not enough: active conservation in New Zealand. *Trends in Ecology and Evolution* 16: 415–416.

Clout, M. N., and Hay, J. R. 1981. South Island kokako (*Callaeas cinerea cinerea*) in *Nothofagus* forest. *Notornis* 28: 256–259.

Clowes, E. F. 1933. Stoat feeding young on rock pipits. *Irish Naturalist's Journal* 4: 217–218.

Clutton-Brock, T. 1989. Mammalian mating systems. *Proceedings of the Royal Society of London* 236B: 339–372.

Conroy, C. J., and Cook, J. A. 2000. Phylogeny of a post-glacial colonizer: *Microtus longicaudus* (Rodentia: Muridae). *Molecular Ecology* 9: 165–175.

Cook, J. A., Bidlack, A. L., Conroy, C. J., Demboski, J. R., Fleming, M. A., Runck, A. M., Stone, K. D., and MacDonald, S. O. 2001. A phylogeographic perspective on endemism in the Alexander Archipelago of southeast Alaska. *Biological Conservation* 97: 215–227.

Cooper, D. 1999. Evolution of resistance to biocontrol agents. *Royal Society of New Zealand, Miscellaneous Series* 56: 96–99.

Corbet, G. B., and Harris, S., editors. 1991. *The Handbook of British Mammals*, 3rd edition. Oxford: Blackwell Scientific Publications.

Coues, E. 1877. *Fur-bearing animals: a monograph of the North American Mustelidae.* Washington: Government Printing Office.

Cowan, P. E., and Tyndale-Biscoe, C. H. 1997. Australian and New Zealand mammal species considered to be pests or problems. *Reproduction, Fertility and Development* 9: 27–36.

Craig, J. L., Anderson, S., Clout, M. N., Creese, B., Mitchell, N., Ogden, J., Roberts, M., and Ussher, G. 2000. Conservation issues in New Zealand. *Annual Review of Ecology and Systematics* 31: 61–78.

Craighead, J. J., and Craighead Jr, F. C. 1956. *Hawks, owls and wildlife.* Washington DC: Stackpole Co. and Wildlife Management Institute.

Craik, C. 1997. Long-term effects of North American mink *Mustela vison* on seabirds in western Scotland. *Bird Study* 44: 303–309.

Craster, J. 1970. Stoats and weasels: a new contrast. *The Field* 236: 786–787.

Criddle, N., and Criddle, S. 1925. The weasels of southern Manitoba. *The Canadian Field-Naturalist* 39: 142–148.

Cringle, P. 1968. Marshland incidents. Shooting Times: 9 March 1968.

Crouchley, D. 1994. Stoat control on Maud Island 1982–1993. *Ecological Management* 2: 39–45.

Cushing, B. S. 1985. Estrous mice and vulnerability to weasel predation. *Ecology* 66: 1976–1978.

Cuthbert, R. 2002. The role of introduced mammals and inverse density-dependent predation in the conservation of Hutton's shearwater. *Biological Conservation* 108: 69–78.

Cuthbert, R., and Davis, L. S. 2002. The impact of predation by introduced stoats on Hutton's shearwaters, New Zealand. *Biological Conservation* 108: 79–92.

Cuthbert, R., and Sommer, E. 2002. Home range, territorial behaviour and habitat use of stoats (*Mustela erminea*) in a colony of Hutton's shearwater (*Puffinus huttoni*), New Zealand. *New Zealand Journal of Zoology* 29: 149–160.

Cuthbert, R., Sommer, E., and Davis, L. S. 2000. Seasonal variation in the diet of stoats in a breeding colony of Hutton's shearwaters. *New Zealand Journal of Zoology* 27: 367–373.

Danilov, P. I., and Tumanov, I. L. 1975. Female reproductive cycles in the mustelidae. In: King, C. M., editor. *Biology of mustelids. Some Soviet research.* Volume 2. Bulletin 227 (1980) ed. Wellington, New Zealand: New Zealand Department of Scientific and Industrial Research. pp. 81–92.

Day, M. G. 1963. An ecological study of the stoat (*Mustela erminea* L.) and the weasel (*M. nivalis* L.) with particular references to their food and feeding habits. Ph.D. thesis: Exeter University, UK.

———. 1966. Identification of hair and feather remains in the gut and faeces of stoats and weasels. *Journal of Zoology* 148: 201–217.

———. 1968. Food habits of British stoats (*Mustela erminea*) and weasels (*Mustela nivalis*). *Journal of Zoology (London)* 155: 485–497.

Dayan, T., and Simberloff, D. 1994. Character displacement, sexual dimorphism, and morphological variation among British and Irish mustelids. *Ecology* 75: 1063–1071.

———. 1996. Patterns of size separation in carnivore communities. In: Gittleman, J. L., editor. *Carnivore behavior, ecology and evolution.* Volume 2. Ithaca, New York, USA, and London, England: Comstock Publishing Associates, Cornell University Press. pp. 243–266.

———. 1998. Size patterns among competitors: ecological character displacement and character release in mammals, with special reference to island populations. *Mammal Review* 28: 99–124.

Dayan, T., and Tchernov, E. 1988. On the first occurrence of the common weasel (*Mustela nivalis*) in the fossil record of Israel. *Mammalia* 52: 165–168.

Deanesly, R. 1935. The reproductive processes of certain mammals. Part IX–Growth and reproduction in the stoat (*Mustela erminea*). *Philosophical Transactions of the Royal Society of London* 225B: 459–492.

———. 1943. Delayed implantation in the stoat (*Mustela mustela*) (sic). *Nature* 151: 365–366.

———. 1944. The reproductive cycle of the female weasel (*Mustela nivalis*). *Proceedings of the Zoological Society of London* 114: 339–349.

Debrot, S. 1983. Fluctuations de populations chez l'hermine (*Mustela erminea* L.). *Mammalia* 47: 323–332.

———. 1984. Dynamique du renouvellement et structure d'age d'une population d'hermines (*Mustela erminea*). *Revue Ecologique (La Terre et la Vie)* 39: 77–88.

Debrot, S., and Mermod, C. 1981. Cranial helminth parasites of the stoat and other mustelids in Switzerland. In: Chapman, J. A., and Pursley, D., editors. *Proceedings of the Worldwide Furbearer Conference.* Frostburg, MA: Wordlwide Furbearer Conference Inc. pp. 690–705.

———. 1982. Quelques siphonaptères de Mustélidés, dont *Rhadinopsylla pentacantha* (Rothschild, 1897), nouvelle espèce pour la Suisse. *Revue Suisse de Zoologie* 89: 27–32.

———. 1983. The spatial and temporal distribution pattern of the stoat (*Mustela erminea* L.). *Oecologia (Berl.)* 59: 69–73.

Delattre, P. 1983. Density of weasel (*Mustela nivalis* L.) and stoat (*Mustela erminea* L.) in relation to water vole abundance. *Acta Zoologica Fennica* 174: 221–222.

———. 1984. Influence de la pression de prédation exercée par une population de belettes (*Mustela nivalis* L.) sur un peuplement de microtidae. *Acta Œcologica/ Œcologia Generalis* 5: 285–300.

Delattre, P., Angibault, J.-M., and Poser, S. 1985. [Effects of radio-collars on movements and activity patterns in the weasel (*Mustela nivalis*)]. *Gibier Faune Sauvage* 2: 5–13.

Derting, T. L. 1989. Prey selection and foraging characteristics of least weasels (*Mustela nivalis*) in the laboratory. *American Midland Naturalist* 122: 394–400.

DeVan, R. 1982. The ecology and life history of the long-tailed weasel (*Mustela frenata*). Ph.D. thesis: University of Cincinnati.

Devos, A. 1960. *Mustela frenata* climbing trees. *Journal of Mammalogy* 41: 520.

Dilks, P. 1999. Recovery of a mohua (*Mohoua ochrocephala*) population following predator control in the Eglinton Valley, Fiordland, New Zealand. *Notornis* 46: 323–332.

Dilks, P., Willans, M., Pryde, M., and Fraser, I. 2003. Large scale stoat control to protect mohua (*Mohoua ochrocephala*) and kaka (*Nestor meridionalis*) in the Eglinton Valley, Fiordland, New Zealand. *New Zealand Journal of Ecology* 27: 1–9.

Dilks, P. J., and Lawrence, B. 2000. The use of poison eggs for the control of stoats. *New Zealand Journal of Zoology* 27: 173–182.

Dilks, P. J., O'Donnell, C. F. J., Elliott, G. P., and Phillipson, S. M. 1996. The effect of bait type, tunnel design, and trap position on stoat control operations for conservation management. *New Zealand Journal of Zoology* 23: 295–306.

Dixon, T. 1931. Pika versus weasel. *Journal of Mammalogy* 12: 72.

Dobson, M. 1998. Mammal distributions in the western Mediterranean: the role of human intervention. *Mammal Review* 28: 77–88.

DonCarlos, M. W., Peterson, J. S., and Tilson, R. L. 1986. Captive biology of an asocial mustelid; *Mustela erminea. Zoo Biology* 5: 363–370.

Dougherty, E. C., and Hall, E. R. 1955. The biological relationships between American weasels (genus *Mustela*) and nematodes of the genus *Skrjabingylus* Petrov, 1927 (Nematoda: Metastrongylidae) the causative organisms of certain lesions in weasel skulls. *Revista Ibérica de Parasitologia (Granada, España)* Tomo Extraordinario: 531–576.

Dowding, J. E., and Elliott, M. J. 2003. *Ecology of stoats in a South Island braided river valley.* Wellington: New Zealand Department of Conservation.

Dowding, J. E., and Murphy, E. C. 2001. The impact of predation by introduced mammals on endemic shorebirds in New Zealand: a conservation perspective. *Biological Conservation* 99: 47–64.

Drabble, P. 1973. Life with a stoat, and a weasel. *The Field*: 6 December 1973: 1508–1509.

———. 1977. *A weasel in my meatsafe.* London: Michael Joseph.

Dubnitskii, A. A. 1956. A study of the development of the nematode *Skrjabingylus nasicola*, a parasite of the frontal sinuses of mustelids. In: King, C. M., editor. *Biology of mustelids. Some Soviet research.* Boston Spa, Yorkshire, England: British Library Lending Division. pp. 235–241.

Dunlap, T. R. 1999. *Nature and the English diaspora: environment and history in the United States, Canada, Australia and New Zealand.* Cambridge: Cambridge University Press.

Dunn, E. 1977. Predation by weasels (*Mustela nivalis*) on breeding tits (*Parus* spp.) in relation to the density of tits and rodents. *Journal of Animal Ecology* 46: 633–652.

Dyczkowski, J., and Yalden, D. W. 1998. An estimate of the impact of predators on the British field vole *Microtus agrestis* population. *Mammal Review* 28: 165–184.

East, K., and Lockie, J. D. 1964. Observations on a family of weasels (*Mustela nivalis*) bred in captivity. *Proceedings of the Zoological Society of London* 143: 359–363.

———. 1965. Further observations on weasels (*Mustela nivalis*) and stoats (*Mustela erminea*) born in captivity. *Journal of Zoology (London)* 147: 234–238.

Easterla, D. A. 1970. First records of the least weasel, *Mustela nivalis*, from Missouri and southwestern Iowa. *Journal of Mammalogy* 51: 333–340.

Edson, J. M. 1933. A visitation of weasels. *Murrelet* 14: 76–77.

Edwards, M. A., Forbes, G. J., and Bowman, J. 2001. Fractal dimension of ermine, *Mustela erminea* (Carnivora: Mustelidae), movement patterns as an indicator of resource use. *Mammalia* 65: 220–225.

Efford, M. 2004. Density estimation in live-trapping studies. *Oikos* 106: 598–610.

Eger, J. L. 1990. Patterns of geographic variation in the skull of Nearctic ermine (*Mustela erminea*). *Canadian Journal of Zoology* 68: 1241–1249.

Eggleston, J. E., Rixecker, S. S., and Hickling, G. J. 2003. The role of ethics in the management of New Zealand's wild mammals. *New Zealand Journal of Zoology* 30: 361–376.

Ehlers, J., and Gibbard, P. 2004. *Quaternary glaciations–extent and chronology.* New York: Elsevier.

Ekerholm, P., Oksanen, L., Oksanen, T., and Schneider, M. 2004. The impact of short-term predator removal on vole dynamics in an arctic-alpine landscape. *Oikos* 106: 457–468.

Elliott, G. P. 1996a. Productivity and mortality of mohua (*Mohoua ochrocephala*). *New Zealand Journal of Zoology* 23: 229–237.

———. 1996b. Mohua and stoats: a population viability analysis. *New Zealand Journal of Zoology* 23: 239–247.

Elliott, G. P., Dilks, P. J., and O'Donnell, C. F. J. 1996. The ecology of yellow-crowned parakeets (*Cyanoramphus auriceps*) in *Nothofagus* forest in Fiordland, New Zealand. *New Zealand Journal of Zoology* 23: 249–265.

Elton, C. 1942. *Voles, mice and lemmings: problems in population dynamics*. Oxford: Oxford University Press.

Erlinge, S. 1974. Distribution, territoriality and numbers of the weasel *Mustela nivalis* in relation to prey abundance. *Oikos* 25: 308–314.

———. 1975. Feeding habits of the weasel *Mustela nivalis* in relation to prey abundance. *Oikos* 26: 378–384.

———. 1977a. Agonistic behaviour and dominance in stoats (*Mustela erminea* L.). *Zeitschrift für Tierpsychologie* 44: 375–388.

———. 1977b. Spacing strategy in stoat *Mustela erminea*. *Oikos* 28: 32–42.

———. 1977c. Home range utilization and movements of the stoat, *Mustela erminea*. *International Congress of Game Biologists* 13: 31–42.

———. 1979a. Adaptive significance of sexual dimorphism in weasels. *Oikos* 33: 233–245.

———. 1979b. Movements and daily activity pattern of radio tracked male stoats, *Mustela erminea*. In: Amlaner, C. J. J., and Macdonald, D. W., editors. *A handbook on biotelemetry and radiotracking*. Oxford and New York: Pergamon Press. pp. 703–710.

———. 1981. Food preference, optimal diet and reproductive output in stoats *Mustela erminea* in Sweden. *Oikos* 36: 303–315.

———. 1983. Demography and dynamics of a stoat *Mustela erminea* population in a diverse community of vertebrates. *Journal of Animal Ecology* 52: 705–726.

———. 1987. Why do European stoats *Mustela erminea* not follow Bergmann's rule? *Holarctic Ecology* 10: 33–39.

Erlinge, S., Göransson, G., Hansson, L., Högstedt, G., Liberg, O., Nilsson, I. N., Nilsson, T., von Schantz, T., and Sylvén, M. 1983. Predation as a regulating factor on small rodent populations in southern Sweden. *Oikos* 40: 36–52.

Erlinge, S., Göransson, G., Högstedt, G., Jansson, G., Liberg, O., Loman, J., Nilsson, I. N., Schantz, T. V., and Sylvén, M. 1984. Can vertebrate predators regulate their prey? *The American Naturalist* 123: 125–133.

Erlinge, S., Jonsson, B., and Willstedt, H. 1974. Jaktbeteende och bytesval hos småvesslan. *Fauna och flora* 69: 95–101.

Erlinge, S., and Sandell, M. 1986. Seasonal changes in the social organization of male stoats, *Mustela erminea*: an effect of shifts between two decisive resources. *Oikos* 47: 57–62.

———. 1988. Coexistence of stoat, *Mustela erminea*, and weasel, *M. nivalis*: social dominance, scent communication, and reciprocal distribution. *Oikos* 53: 242–246.

Erlinge, S., Sandell, M., and Brinck, C. 1982. Scent-marking and its territorial significance in stoats, *Mustela erminea*. *Animal Behaviour* 30: 811–818.

Errington, P. L. 1963. The phenomenon of predation. *Smithsonian Report* 1964: 507–519.

Etheridge, N., and Powlesland, R. G. 2001. High productivity and nesting success of South Island robins (*Petroica australis australis*) following predator control at St Arnaud, Nelson Lakes, South Island. *Notornis* 48: 179–180.

Ewer, R. F. 1973. *The carnivores.* Ithaca, New York, USA: Cornell University Press.

Fagen, R. 1981. *Animal play.* New York: Oxford University Press.

Fagerstone, K. A. 1987. Black-footed ferret, long-tailed weasel, short-tailed weasel, and least weasel. In: Nowak, R., Baker, J. A., Obbard, M. E., and Malloch, B., editors. *Wild furbearer management and conservation in North America.* Toronto, Canada: Ontario Ministry of Natural Resources. pp. 548–573.

Fairley, J. S. 1971. New data on the Irish stoat. *Irish Naturalists' Journal* 17: 49–57.

———. 1981. A north-south cline in the size of the Irish stoat. *Proceedings of the Royal Irish Academy* 81B: 5–10.

———. 1984. *An Irish beast book,* 2nd edition. Belfast: Blackstaff Press.

Feder, S. 1990. Environmental determinants of seasonal coat color change in weasels (*Mustela erminea*) from two populations. MS thesis. Fairbanks: University of Alaska.

Fitzgerald, B. M. 1964. *The ecology of mustelids in New Zealand.* M. Sc. thesis thesis. Christchurch: University of Canterbury.

———. 1977. Weasel predation on a cyclic population of the montane vole (*Microtus montanus*) in California. *Journal of Animal Ecology* 46: 367–397.

———. 1981. Predatory birds and mammals. In: Bliss, L. C., Cragg, J. B., Heal, D. W., and Moore, J. J., editors. *Tundra ecosystems: a comparative analysis.* Cambridge: Cambridge University Press. pp. 485–508.

Fitzgerald, B. M., and Gibb, J. A. 2001. Introduced mammals in a New Zealand forest: long-term research in the Orongorongo Valley. *Biological Conservation* 99: 97–108.

Fitzgerald, G., Fitzgerald, N., and Wilkinson, R. 2002. Social acceptability of stoats and stoat control methods: focus group findings. *Science for Conservation* 207: 1–45.

Flack, J. A. D., and Lloyd, B. D. 1978. The effect of rodents on the breeding success of the South Island robin. In: Dingwall, P. R., Atkinson, I. A. E., and Hay, C., editors. *The ecology and control of rodents in New Zealand nature reserves.* Wellington: Department of Lands and Survey, Information Series 4. pp. 59–66.

Fleskes, J. P. 1988. Predation by ermine and longtailed weasels on duck eggs. *Journal of the Iowa Academy of Sciences* 95: 14–17.

Flintoff, R. J. 1933. Stoats and weasels, brown and white. *Northwestern Naturalist* 8: 36–45.

———. 1935. Stoats and weasels, brown and white. *Northwestern Naturalist* 10: 214–229.

Florine, C. 1942. Weasel in a pocket gopher burrow. *Journal of Mammalogy* 23: 213.

Flowerdew, J. R. 1972. The effect of supplementary food on a population of wood mice (*Apodemus sylvaticus*). *Journal of Animal Ecology* 41: 553–566.

Flynn, J. J., and Wesley-Hunt, G. D. 2005. Carnivora. In: Rose, K. D., and Archibald, J. D., editors. *The rise of placental mammals.* Baltimore: Johns Hopkins University Press. pp. 175–198.

Forester, N. T., Lumsden, J. S., Parton, K., Cowan, P. E., and O'Toole, P. W. 2003. Detection and isolation of Helicobacter mustelae from stoats in New Zealand. *New Zealand Veterinary Journal* 51: 142–145.

Francis, G. R., and Stephenson, A. B. 1972. Marten ranges and food habits in Algonquin Provincial Park, Ontario. *Ontario Minstry of Natural Resources, Research Report* 91: 1–53.

Frank, F. 1974. Wurfzahl und wurffolge biem nordischen wiesel (*Mustela nivalis rixosa* Bangs, 1896). *Zeitschrift für Säugetierkunde* 39: 248–250.

———. 1985. Zur evolution und systematik der kleinen wiesel (*Mustela nivalis* Linnaeus, 1766). *Zeitschrift für Säugetierkunde* 50: 208–225.

Fryxell, J., Falls, J. B., Falls, E. A., Brooks, R. J., Dix, L., and Strickland, M. 2001. Harvest dynamics of mustelid carnivores in Ontario, Canada. *Wildlife Biology* 7: 151–159.

Fulton, G. R., and Ford, G. R. 2001. The conflict between animal welfare and conservation. *Pacific Conservation Biology* 7: 152–153.

Gaiduk, V. E. 1977. Control of moult and winter whitening in the ermine (*Mustela erminea*). In: King, C. M., editor. *Biology of mustelids. Some Soviet research.* Volume 2. Wellington, New Zealand: New Zealand Department of Scientific and Industrial Research, Bulletin 227 (1980). pp. 56–61.

Galbreath, R. 1989. *Walter Buller: the reluctant conservationist.* Wellington: GP Books.

Gamberg, M., and Atkinson, J. L. 1988. Prey hair and bone recovery in ermine scats. *Journal of Wildlife Management* 52: 657–660.

Gamble, R. L. 1980. The ecology and distribution of *Mustela frenata longicauda* Bonaparte and its relationships to other *Mustela* spp. in sympatry. MS thesis. Winnipeg, Manitoba, Canada: The University of Manitoba.

———. 1981. Distribution in Manitoba of *Mustela frenata longicauda* Bonaparte, the long-tailed weasel, and the interrelation of distribution and habitat selection in Manitoba, Saskatchewan, and Alberta. *Canadian Journal of Zoology* 59: 1036–1039.

Gamble, R. L., and Riewe, R. R. 1982. Infestations of the nematode *Skrjabingylus nasicola* (Leukart 1842) in *Mustela frenata* (Lichtenstein) and *M. erminea* (L.) and some evidence of a paratenic host in the life cycle of this nematode. *Canadian Journal of Zoology* 60: 45–52.

Gehring, T. M., and Swihart, R. K. 2000. Field immobilization and use of radio collars on long-tailed weasels. *Wildlife Society Bulletin* 28: 579–585.

———. 2003. Body size, niche breadth, and ecologically scaled responses to habitat fragmentation: mammalian predators in an agricultural landscape. *Biological Conservation* 109: 283–295.

———. 2004. Home range and movements of long-tailed weasels in a landscape fragmented by agriculture. *Journal of Mammalogy* 85: 79–86.

Geist, V. 1975. *Mountain sheep and man in the northern wilds.* Ithaca, New York: Cornell University Press.

Gewalt, W. 1959. [A contribution to knowledge of the optical differentiation ability of some mustelids with special reference to colour vision]. *Zoologische Beiträge* 5: 117–175.

Gibbs, S. J., and Clout, M. N. 2003. Behavioural vulnerability of juvenile brown kiwi: habitat use and overlap with predators. *DoC Science Internal Series* 102: 1–12.

Gilg, O., Hanski, I., and Sittler, B. 2003. Cyclic dynamics in a simple vertebrate predator-prey community. *Science* 302: 866–868.

Gillies, C. A., Leach, M. R., Coad, N. B., Theobald, S. W., Campbell, J., Herbert, T., Graham, P. J., and Pierce, R. J. 2003. Six years of intensive pest mammal control at Trounson Kauri Park, a Department of Conservation "mainland island," June 1996–July 2002. *New Zealand Journal of Zoology* 30: 399–420.

Gillies, C. A., and Pierce, R. J. 1999. Secondary poisoning of mammalian predators during possum and rodent control operations at Trounson Kauri Park, Northland, New Zealand. *New Zealand Journal of Ecology* 23: 183–192.

Gillingham, B. J. 1984. Meal size and feeding rate in the least weasel (*Mustela nivalis*). *Journal of Mammalogy* 65: 517–519.

———. 1986. Sensory aspects of foraging behavior in the least weasel (*Mustela nivalis*). Ph.D. thesis. Urbana-Champaign: University of Illinois.

Gleeson, D., Byrom, A., Howitt, R., and Nugent, G. 2003. Monitoring vertebrate pests by using their DNA. Vertebrate Pest Research (Newsletter of Landcare Research, New Zealand): June 2003. pp. 13–14.

Gliwicz, J. 1988. Sexual dimorphism in small mustelids: body diameter limitation. *Oikos* 53: 411–414.

Glover, F. A. 1942a. Spring color change of the New York weasel. *Pennsylvania Game News* 13: 18 and 33.

———. 1942b. A population study of weasels in Pennsylvania. MS thesis: The Pennsylvania State College.

———. 1943. A study of the winter activities of the New York weasel. *Pennsylvania Game News* 14: 8–9.

Golley, F. B. 1960. Energy dynamics of a food chain of an old-field community. *Ecological Monographs* 30: 187–206.

Gorman, M. L. 1976. A mechanism for individual recognition by odour in *Herpestes auropunctatus* (Carnivora: Viverridae). *Animal Behaviour* 24: 141–145.

———. 1984. The response of prey to stoat (*Mustela erminea*) scent. *Journal of Zoology (London)* 202: 419–423.

Goszczynski, J. 1977. Connections between predatory birds and mammals and their prey. *Acta Theriologica* 22: 399–430.

Graham, I. M. 2002. Estimating weasel *Mustela nivalis* abundance from tunnel tracking indices at fluctuating field vole *Microtus agrestis* density. *Wildlife Biology* 8: 279–287.

Grue, H. E., and King, C. M. 1984. Evaluation of age criteria in New Zealand stoats (*Mustela erminea*) of known age. *New Zealand Journal of Zoology* 11: 437–443.

Gulamhusein, A. P., and Tam, W. H. 1974. Reproduction in the male stoat, *Mustela erminea*. *Journal of Reproduction and Fertility* 41: 303–312.

Gulamhusein, A. P., and Thawley, A. R. 1974. Plasma progesterone levels in the stoat. *Journal of Reproduction and Fertility* 36: 405–408.

Güttinger, R., and Müller, J. P. 1988. Zur verbreitung von Zwergwiesel und Mauswiesel im Kanton Graubünden (Schweiz). *Jahresbericht der Naturforschenden Gesellschaft Graubünden* 105: 103–114.

Hackwell, K., and Bertram, G. 1999. *Pests and weeds: a blueprint for action*. Wellington: New Zealand Conservation Authority.

Haley, D. 1975. *Sleek and savage. North America's weasel family*. Seattle, Washington, USA: Pacific Search Books.

Hall, E. R. 1951. *American weasels*. Lawrence, Kansas, USA: University of Kansas.

Hamilton Jr., W. J. 1933. The weasels of New York: their natural history and economic status. *The American Midland Naturalist* 14: 289–344.

Hammel, H. T. 1956. Infrared emmissivities of some arctic fauna. *Journal of Mammalogy* 37: 375–378.

Hanski, I., and Henttonen, H. 1996. Predation on competing rodent species–a simple explanation of complex patterns. *Journal of Animal Ecology* 65: 220–232.

Hanski, I., Henttonen, H., Korpimaki, E., Oksanen, L., and Turchin, P. 2001. Small-rodent dynamics and predation. *Ecology* 82: 1505–1520.

Hansson, I. 1967. Transmission of the parasitic nematode *Skrjabingylus nasicola* (Leuckart 1842) to species of *Mustela* (Mammalia). *Oikos* 18: 247–252.

———. 1970. Cranial helminth parasites in species of Mustelidae. II. Regional frequencies of damage in preserved crania from Denmark, Finland, Sweden, Greenland and the northeast of Canada compared with the helminth invasion in fresh mustelid skulls from Sweden. *Arkiv För Zoologi (Series 2)* 22: 571–594.

———. 1974. Seasonal and environmental conditions affecting the invasion of mustelids by larvae of the nematode *Skrjabingylus nasicola. Oikos* 25: 61–70.

Hansson, L. 1995. Size dimorphism in microtine populations: characteristics of growth and selection against large-sized individuals. *Journal of Mammalogy* 76: 867–872.

Hansson, L., and Henttonen, H. 1985. Gradients in density variations of small rodents: the importance of latitude and snow cover. *Oecologia (Berl.)* 67: 394–402.

Harestad, A. S. 1990. Mobbing of a long-tailed weasel, *Mustela frenata*, by Columbian ground squirrels, *Spermophilius columbianus. The Canadian Field-Naturalist* 104: 483–484.

Harlow, H. J. 1994. Trade-offs associated with the size and shape of American martens. In: Buskirk, S. W., Harestad, A. S., Raphael, M. G., and Powell, R. A., editors. *Martens, sables and fishers: biology and conservation.* Ithaca, New York: Cornell University Press. pp. 391–403.

Harris, A. H. 1993. A Late-Pleistocene occurrence of ermine (*Mustela erminea*) in southeastern New Mexico. *The Southwestern Naturalist* 38: 279–280.

Harris, S., Morris, P., Wray, S., and Yalden, D. 1995. *A review of British mammals: population estimates and conservation status of British mammals other than cetaceans.* Peterborough, England: Joint Nature Conservation Committee.

Hartman, L. 1964. The behaviour and breeding of captive weasels (*Mustela nivalis* L.). *New Zealand Journal of Science* 7: 147–156.

Hattar, S., Liao, H.-W., Takao, M., Berson, D. M., and Yau, K.-W. 2002. Melanopsin-containing retinal ganglion cells: architecture, projections, and intrinsic sensitivity. *Science* 295: 1065–1070.

Hayne, D. W. 1949. Calculation of size of home range. *Journal of Mammalogy* 30: 1–18.

Hayward, G. F. 1983. The bioenergetics of the weasel, *Mustela nivalis* L. D. Phil. thesis. Oxford: Oxford University.

Hayward, G. F., and Phillipson, J. 1979. Community structure and functional role of small mammals in ecosystems. In: Stoddart, D. M., editor. *Ecology of small mammals.* London: Chapman & Hall. pp. 135–211.

Heffner, R. S., and Heffner, H. E. 1985. Hearing in mammals: the least weasel. *Journal of Mammalogy* 66: 745–755.

Heidt, G. A. 1970. The least weasel *Mustela nivalis* Linnaeus. Developmental biology in comparison with other North American *Mustela. Publications of the Museum, Michigan State University. Biological Series* 4: 227–282.

———. 1972. Anatomical and behavioral aspects of killing and feeding by the least weasel, *Mustela nivalis* L. *Proceedings of the Arkansas Academy of Science* 26: 53–54.

Heidt, G. A., Petersen, M. K., and Kirkland, G. L., Jr. 1968. Mating behavior and development of least weasels (*Mustela nivalis*) in captivity. *Journal of Mammalogy* 49: 413–419.

Heikkila, J., Below, A., and Hanski, I. 1994. Synchronous dynamics of microtine rodent populations on islands in Lake Inari in northern Fennoscandia: evidence for regulation by mustelid predators. *Oikos* 70: 245–252.

Heitkamp, P. H. A. M., and van der Schoot, P. J. C. M. 1966. [The stoats, *Mustela erminea* L. in the dune reserve in north Holland]. Amsterdam: Koninklijke Nederlandse Jagers Vereniging, Unpublished report.

Hellstedt, P., and Henttonen, H. In press. Home ranges, habitat choice and activity of stoats (*Mustela erminea*) in relation to sex and dominance status during the snowless seasons in a subarctic area. *Journal of Zoology (London)*.

Hellstedt, P., and Kallio, E. R. 2005. Survival and behaviour of captive-born weasels (*Mustela nivalis nivalis*) released in nature. *Journal of Zoology (London)* 266: 37–44.

Henttonen, H., Oksanen, T., Jortikka, A., and Haukisalmi, V. 1987. How much do weasels shape microtine cycles in the northern Fennoscandian taiga? *Oikos* 50: 353–365.

Heptner, V. G., Naumov, N. P., Yurgenson, P. B., Sludski, A. A., Chirkova, A. F., and Bannikov, A. G. 1967. [*Mammals of the Soviet Union.* Volume 2]. Moscow: Soviet Academy of Sciences.

Herbert, J. 1989. Light as a multiple control system on reproduction in mustelids. In: Seal, U. S., Thorne, E. T., Bogan, M. A., and Anderson, S. H., editors. *Conservation biology and the black-footed ferret.* New Haven: Yale University Press. pp. 138–159.

Herman, D. G. 1973. Olfaction as a possible mechanism for prey selection in the least weasel, *Mustela nivalis.* MS thesis. East Lancing, MI: Michigan State University.

Herter, K. 1939. [Psychological investigations on a mouseweasel (*Mustela nivalis* L.)]. *Zeitschrift für Tierpsychologie* 3: 249–263.

Herzog, C. 2003. The use of track plates to identify individual free-ranging fishers. MS thesis: Prescott College, New York.

Hewitt, G. M. 1999. Post-glacial re-colonization of European biota. *Biological Journal of the Linnean Society* 68: 87–112.

Hewson, R. 1972. Changes in the number of stoats, rats, and little owls in Yorkshire as shown by tunnel trapping. *Journal of Zoology, London* 168: 427–429.

Hewson, R., and Healing, T. D. 1971. The stoat *Mustela erminea* and its prey. *Journal of Zoology (London)* 164: 239–244.

Hewson, R., and Watson, A. 1979. Winter whitening of stoats (*Mustela erminea*) in Scotland and north-east England. *Journal of Zoology (London)* 187: 55–64.

Hildebrand, M. 1974. *Analysis of vertebrate structure.* New York: John Wiley & Sons.

Hill, M. 1939. The reproductive cycle of the male weasel (*Mustela nivalis*). *Proceedings of the Zoological Society of London, Series B* 109: 481–512.

Hill, S., and Hill, J. 1987. *Richard Henry of Resolution Island*. Dunedin: John McIndoe Ltd.

Hirschi, R. 1985. A mother's work is never done. *BBC Wildlife* 3: 222–226.

Hirzel, D. J., Wang, J., Das, S. K., Dey, S. K., and Mead, R. A. 1999. Changes in uterine expression of leukemia inhibitory factor during pregnancy in the western spotted skunk. *Biology of Reproduction* 60: 484–492.

Holdaway, R. N. 1999. Introduced predators and avifaunal extinction in New Zealand. In: MacPhee, R. D. E., editor. *Extinctions in near time*. New York: Kluwer Academic/Plenum Publishers. pp. 189–238.

Holland, O. J., and Gleeson, D. M. 2005. Genetic characterisation of blastocysts and the identification of an instance of multiple paternity in the stoat (*Mustela erminea*). *Conservation Genetics* 6: 855–858.

Holmes, T. 1980. Locomotor adaptations in the limb skeletons of North American mustelids. MA thesis. Arcata, California: Humboldt State University.

———. 1987. Sexual dimorphism in North American weasels, with a phylogeny of Mustelidae. Ph.D. thesis. Lawrence, Kansas, USA: University of Kansas.

Holmes, T., and Powell, R. A. 1994. Morphology, ecology and the evolution of sexual dimorphism in North American *Martes*. In: Buskirk, S. W., Harestad, A. S., Raphael, M. G., and Powell, R. A., editors. *Martens, sables and fishers: biology and conservation*. Ithaca, New York: Cornell University Press. pp. 72–84.

Hörnfeldt, B. 2004. Long-term decline in numbers of cyclic voles in boreal Sweden: analysis and presentation of hypotheses. *Oikos* 107: 376–392.

Hosey, G., and Jacques, M. 1998. Lesser black-backed gull apparently picking up and dropping live stoat. *British Birds* 91: 199.

Hosoda, T., Suzuki, H., Harada, M., Tsuchiya, K., Han, S., Zhang, Y., Kryukov, A. P., and Lin, L. 2000. Evolutionary trends of the mitochondrial lineage differentiation in species of genera *Martes* and *Mustela*. *Genes and Genetic Systems* 75: 259–267.

Huff, J. N., and Price, E. O. 1968. Vocalizations of the least weasel, *Mustela nivalis*. *Journal of Mammalogy* 49: 548–550.

Huggett, A. S. G., and Widdas, W. F. 1951. The relationship between mammalian foetal weight and conception age. *Journal of Physiology* 114: 306–317.

Hunter, T. S. 1969. Aggressive sparrows. *Country Life* 25 September 1969.

Hussell, D. J. T. 1974. Photographic records of predation at Lapland longspur and snow bunting nests. *The Canadian Field-Naturalist* 88: 503–506.

Hutchinson, G., and Parker, P. 1978. Sexual dimorphism in the winter whitening of the stoat *Mustela erminea*. *Journal of Zoology (London)* 186: 560–563.

Hutchinson, G. E. 1959. Homage to Santa Rosalia, or, why are there so many kinds of animals? *American Naturalist* 93: 145–159.

Innes, J., and Barker, G. 1999. Ecological consequences of toxin use for mammalian pest control in New Zealand—an overview. *New Zealand Journal of Ecology* 23: 111–127.

Innes, J., Warburton, B., Williams, D., Speed, H., and Bradfield, P. 1995. Large-scale poisoning of ship rats (*Rattus rattus*) in indigenous forests of the North Island, New Zealand. *New Zealand Journal of Ecology* 19: 5–17.

Innes, J. G., and Hay, J. R. 1991. The interactions of New Zealand forest birds with introduced fauna. *International Ornithological Congress* 20: 2523–2530.

Innes, J. G., Hay, R., Flux, I., Bradfield, P., Speed, H., and Jansen, P. 1999. Successful recovery of North Island kokako *Callaeas cinerea wilsoni* populations, by adaptive management. *Biological Conservation* 87: 201–214.

Jaarola, M., and Tegelström, H. 1995. Colonization history of north European field voles (*Microtus agrestis*) revealed by mitochondrial DNA. *Molecular Ecology* 4: 299–310.

Jaarola, M., Tegelström, H., and Fredga, K. 1999. Colonization history in Fennoscandian rodents. *Biological Journal of the Linnean Society* 68: 113–127.

Jarvis, B. D. W. 1999. Rabbit control RCD: dilemmas and implications. *Royal Society of New Zealand, Miscellaneous Series* 55: 1–120.

Jędrzejwska, B. 1987. Reproduction in weasels *Mustela nivalis* in Poland. *Acta Theriologica* 32: 493–496.

Jędrzejwska, B., and Jędrzejwski, W. 1989. Seasonal surplus killing as hunting strategy of the weasel *Mustela nivalis*–test of a hypothesis. *Acta Theriologica* 34: 347–359.

Jędrzejwska, B., and Jędrzejwski, W. 1998. *Predation in vertebrate communities. The Bialowieza Primeval Forest as a case study.* Berlin: Springer-Verlag.

Jędrzejwski, W., and Jędrzejwska, B. 1993. Predation on rodents in Bialowieza primeval forest, Poland. *Ecography* 16: 47–64.

———. 1996. Rodent cycles in relation to biomass and productivity of ground vegetation and predation in the Palearctic. *Acta Theriologica* 41: 1–34.

Jędrzejwski, W., Jędrzejwska, B., and McNeish, E. 1992. Hunting success of the weasel *Mustela nivalis* and escape tactics of forest rodents in Bialowieza National Park. *Acta Theriologica* 37: 319–328.

Jędrzejwski, W., Jędrzejwska, B., and Szymura, L. 1995. Weasel population response, home range, and predation on rodents in a deciduous forest in Poland. *Ecology* 76: 179–195.

Jędrzejwski, W., Jędrzejwska, B., Zub, K., and Nowakowski, W. K. 2000. Activity patterns of radio-tracked weasels *Mustela nivalis* in Bialowieza National Park (E. Poland). *Annales Zoologici Fennici* 37: 161–168.

Jędrzejwski, W., Rychlik, L., and Jędrzejwska, B. 1993. Responses of bank voles to odours of seven species of predators: experimental data and their relevance to natural predator-vole relationships. *Oikos* 68: 251–257.

Jennings, D. H., Threlfall, W., and Dodds, D. G. 1982. Metazoan parasites and food of short-tailed weasels and mink in Newfoundland, Canada. *Canadian Journal of Zoology* 60: 180–183.

Jensen, B. 1978. [Catch results with cage traps]. *Natura Jutlandica* 20: 129–136.

Johnson, D. D. P., Macdonald, D. W., and Dickman, A. J. 2000a. Analysis and review of models of the sociobiology of the Mustelidae. *Mammal Review* 30: 171–196.

Johnson, D. R., Swanson, B. K., and Eger, J. L. 2000b. Cyclic dynamics of eastern Canadian ermine populations. *Canadian Journal of Zoology* 78: 835–839.

Jones, C., Moller, H., and Hamilton, W. 2004. A review of potential techniques for identifying individual stoats (*Mustela erminea*) visiting control or monitoring stations. *New Zealand Journal of Zoology* 31: 193–203.

Kaitala, V., Mappes, T., and Ylönen, H. 1997. Delayed female reproduction in equilibrium and chaotic populations. *Evolutionary Ecology* 11: 105–126.

Kauhala, K., Helle, P., and Helle, E. 2000. Predator control and the density and reproductive success of grouse populations in Finland. *Ecography* 23: 161–168.

Kavanau, J. L., and Ramos, J. 1975. Influences of light on activity and phasing of carnivores. *The American Naturalist* 100: 391–418.

Keedwell, R. J., and Brown, K. P. 2001. Relative abundance of mammalian predators in the upper Waitaki Basin, South Island, New Zealand. *New Zealand Journal of Zoology* 28: 31–38.

Kelly, D., Brindle, C., Ladley, J. J., Robertson, A. W., Maddigan, F. W., Butler, J., Ward-Smith, T., Murphy, D. J., and Sessions, L. 2005. Can stoat (*Mustela erminea*) trapping increase bellbird (*Anthornis melanura*) populations and benefit mistletoe (*Peraxilla tetrapetala*) pollination? *New Zealand Journal of Ecology* 29: 69–82.

Kildemoes, A. 1985. The impact of introduced stoats (*Mustela erminea*) on an island population of the water vole, *Arvicola terrestris*. *Acta Zoologica Fennica*: 193–195.

King, C. M. 1975a. The sex ratio of trapped weasels (*Mustela nivalis*). *Mammal Review* 5: 1–8.

————, editor. 1975b. *Biology of mustelids: some Soviet research.* Boston Spa, Yorkshire, England: British Library Lending Division.

————. 1975c. The home range of the weasel (*Mustela nivalis*) in an English woodland. *Journal of Animal Ecology* 44: 639–668.

————. 1976. The fleas of a population of weasels in Wytham Woods, Oxford. *Journal of Zoology (London)* 180: 525–535.

————. 1977. The effects of the nematode parasite *Skrjabingylus nasicola* on British weasels (*Mustela nivalis*). *Journal of Zoology (London)* 182: 225–249.

————. 1979. Moult and colour change in English weasels (*Mustela nivalis*). *Journal of Zoology (London)* 189: 127–134.

————. 1980a. Field experiments on the trapping of stoats *(Mustela erminea).* *New Zealand Journal of Zoology* 7: 261–266.

————. 1980b. The weasel *Mustela nivalis* and its prey in an English woodland. *Journal of Animal Ecology* 49: 127–159.

————. 1980c. Population biology of the weasel *Mustela nivalis* on British game estates. *Holarctic Ecology* 3: 160–168.

————, editor. 1980d. *Biology of mustelids. Some Soviet research.* Volume 2. Wellington, New Zealand: New Zealand Department of Scientific and Industrial Research, Bulletin 227.

————. 1980e. Age determination in the weasel (*Mustela nivalis*) in relation to the development of the skull. *Zeitschrift für Säugetierkunde* 45: 153–173.

————. 1981a. The reproductive tactics of the stoat (*Mustela erminea*) in New Zealand forests. In: Chapman, J. A., and Pursley, D., editors. *Proceedings of the First Worldwide Furbearer Conference.* Frostburg, Maryland: Worldwide Furbearer Conference Inc. pp. 443–468.

————. 1981b. The effects of two types of steel traps upon captured stoats (*Mustela erminea*). *Journal of Zoology (London)* 195: 553–554.

————. 1982. Stoat observations. *Landscape (Wellington)* 12: 12–15.

————. 1983a. The life history strategies of *Mustela nivalis* and *M. erminea. Acta Zoologica Fennica* 174: 183–184.

————. 1983b. The relationships between beech (*Nothofagus* sp.) seedfall and populations of mice (*Mus musculus*), and the demographic and dietary responses of stoats (*Mustela erminea*), in three New Zealand forests. *Journal of Animal Ecology* 52: 141–166.

————. 1983c. Factors regulating mustelid populations. *Acta Zoologica Fennica* 174: 217–220.

————. 1984a. The origin and adaptive advantages of delayed implantation in *Mustela erminea. Oikos* 42: 126–128.

————. 1984b. *Immigrant killers: introduced predators and the conservation of birds in New Zealand.* Auckland: Oxford University Press.

————. 1989a. The advantages and disadvantages of small size to weasels, *Mustela* species. In: Gittleman, J. L., editor. *Carnivore behaviour, ecology and evolution.* Ithaca, New York: Cornell University Press. pp. 302–334.

————. 1989b. *The natural history of weasels and stoats.* London: Christopher Helm.

————. 1991a. A review of age determination methods for the stoat *Mustela erminea. Mammal Review* 21: 31–49.

————. 1991b. Body size-prey size relationships in European stoats (*Mustela erminea*): a test case. *Holarctic Ecology* 14: 173–185.

————. 1991c. Age-specific prevalence and a possible transmission route for skrjabingylosis in New Zealand stoats, *Mustela erminea. New Zealand Journal of Ecology* 15: 23–30.

————. 1991d. Weasel roulette. *Natural History* 11/91: 34–41.

————. 2002. Cohort variation in the life-history parameters of stoats *Mustela erminea* in relation to fluctuating food resources: a challenge to boreal ecologists. *Acta Theriologica* 47: 225–244.

————, editor. 2005a. *The handbook of New Zealand mammal,* 2nd edition. Melbourne: Oxford University Press.

————. 2005b. Stoat. In: King, C. M., editor. *The handbook of New Zealand mammals,* 2nd edition. Melbourne: Oxford University Press. pp 261–287.

————. 2005c. A bibliographic database on weasels and stoats. *DoC Research and Development Series* 217: 1–21.

King, C. M., Davis, S. A., Purdey, D. C., and Lawrence, B. 2003a. Capture probability and heterogeneity of trap response in stoats, *Mustela erminea. Wildlife Research* 30: 611–619.

King, C. M., and Edgar, R. L. 1977. Techniques for trapping and tracking stoats (*Mustela erminea*); a review, and a new system. *New Zealand Journal of Zoology* 4: 193–212.

King, C. M., Flux, M., Innes, J. G., and Fitzgerald, B. M. 1996. Population biology of small mammals in Pureora Forest Park: 1. Carnivores (*Mustela erminea, M.furo, M nivalis* and *Felis catus*). *New Zealand Journal of Ecology* 20: 241–251.

King, C. M., Griffiths, K., and Murphy, E. C. 2001. Advances in New Zealand mammalogy 1990–2000: stoat and weasel. *Journal of the Royal Society of New Zealand* 31: 165–183.

King, C. M., and McMillan, C. D. 1982. Population structure and dispersal of peak-year cohorts of stoats (*Mustela erminea*) in two New Zealand forests, with especial reference to control. *New Zealand Journal of Ecology* 5: 59–66.

King, C. M., and Moody, J. E. 1982. The biology of the stoat (*Mustela erminea*) in the national parks of New Zealand. *New Zealand Journal of Zoology* 9: 49–144.

King, C. M., and Moors, P. J. 1979a. On co-existence, foraging strategy and the biogeography of weasels and stoats (*Mustela nivalis* and *M. erminea*) in Britain. *Oecologia (Berl.)* 39: 129–150.

―――. 1979b. The life-history tactics of mustelids, and their significance for predator control and conservation in New Zealand. *New Zealand Journal of Zoology* 6: 619–622.

King, C. M., and Moller, H. 1997. Distribution and response of rats *Rattus rattus*, *R. exulans* to seedfall in New Zealand beech forests. *Pacific Conservation Biology* 3: 143–155.

King, C. M., and White, P. C. L. 2004. Decline in capture rate of stoats at high mouse densities in New Zealand *Nothofagus* forests. *New Zealand Journal of Ecology* 28: 251–258.

King, C. M., O'Donnell, C. F. J., and Phillipson, S. M. 1994. Monitoring and control of mustelids on conservation lands. Part 2: Field and workshop guide. Wellington: New Zealand Department of Conservation Reports, Technical Series 4: 1–36.

King, C. M., White, P. C. L., Purdey, D. C., and Lawrence, B. 2003b. Matching productivity to resource availability in a small predator, the stoat (*Mustela erminea*). *Canadian Journal of Zoology* 81: 662–669.

Klemola, T., Koivula, M., Korpimäki, E., and Norrdahl, K. 1997. Small mustelid predation slows population growth of *Microtus* voles–a predator reduction experiment. *Journal of Animal Ecology* 66: 607–614.

Klemola, T., Korpimäki, E., Norrdahl, K., Tanhuanpää, M., and Koivula, M. 1999. Mobility and habitat utilization of small mustelids in relation to cyclically fluctuating prey abundances. *Annales Zoologica Fennici* 36: 75–82.

Kohn, M., York, E. C., Kamradt, D. A., Haught, G., Sauvajot, R. M., and Wayne, R. K. 1999. Estimating population size by genotyping faeces. *Proceedings of the Royal Society of London* 266B: 657–663.

Kopein, K. I. 1965. The biology of reproduction of ermine in Yamal. In: King, C. M., editor. *Biology of mustelids. Some Soviet research.* Volume 2. Wellington, New Zealand: New Zealand Department of Scientific and Industrial Research, Bulletin 227 (1980). pp. 62–69.

―――. 1969. The relationship between age and individual variation in the ermine. In: King, C. M., editor. *Biology of mustelids. Some Soviet research.* Volume 2. Wellington, New Zealand: New Zealand Department of Scientific and Industrial Research, Bulletin 227 (1980). pp. 132–138.

Korpimäki, E., and Norrdahl, K. 1989a. Avian predation on mustelids in Europe 1: occurrence and effects on body size variation and life traits. *Oikos* 55: 205–215.

―――. 1989b. Avian predation on mustelids in Europe 2: impact on small mustelid and microtine dynamics–a hypothesis. *Oikos* 55: 273–276.

―――. 1998. Experimental reduction of predators reverses the crash phase of small-rodent cycles. *Ecology* 79: 2448–2455.

Korpimäki, E., Brown, P. R., Jacob, J., and Pech, R. P. 2004. The puzzles of population cycles and outbreaks of small mammals solved? *BioScience* 54: 1071–1079.

Korpimäki, E., Koivunen, V., and Hakkarainen, H. 1996. Microhabitat use and behaviour of voles under weasel and raptor predation risk: predator facilitation? *Behavioural Ecology* 7: 30–34.

Korpimäki, E., Norrdahl, K., Klemola, T., Pettersen, T., and Stenseth, N. C. 2002. Dynamic effects of predators on cyclic voles: field experimentation and model extrapolation. *Proceedings of the Royal Society of London* 269B: 991–997.

Korpimäki, E., Norrdahl, K., and Rinta-Jaskari, T. 1991. Responses of stoats and least weasels to fluctuating food abundances: is the low phase of the vole cycle due to mustelid predation? *Oecologia (Berl.)* 88: 552–561.

Korpimäki, E., Norrdahl, K. and Valkama, J. 1994. Reproductive investment under fluctuating predation risk: microtine rodents and small mustelids. *Evolutionary Ecology* 8: 357–368.

Koskela, E., Horne, T. J., Mappes, T., and Ylönen, H. 1996. Does risk of small mustelid predation affect the oestrous cycle in the bank vole, *Clethrionomys glareolus*? *Animal Behaviour* 51: 1159–1163.

Kraft, V. A. 1966. Influence of temperature on the activity of the ermine in winter. In: King, C. M., editor. *Biology of mustelids. Some Soviet research*. Boston Spa, Yorkshire, England: British Library Lending Division. pp. 104–107.

Kratochvil, J. 1977. Studies on *Mustela erminea* (Mustelidae, Mamm.) I. Variability of metric and mass traits. *Folia Zoologica* 26: 291–304.

Krebs, C. J., Boutin, S., and Boonstra, R., editors. 2001. Ecosystem dynamics of the boreal forest: the Kluane project. New York: Oxford University Press.

Krebs, C. J., Boutin, S., Boonstra, R., Sinclair, R. E., Smith, J. N. M., Dale, M. R. T., Martin, K., and Turkington, R. 1995. Impact of food and predation on the snowshoe hare cycle. *Science* 269: 1112–1115.

Kruuk, H. 1972. Surplus killing by carnivores. *Journal of Zoology (London)* 166: 233–244.

Kukarcev, V. A. *The structure of an ermine (Mustela erminea) population at different densities*. In: Obrtel, R., Folk, C. and Pellantova, J., editors. Abstracts, First International Theriological Congress, Brno, Czechoslovakia (June 1978).

Kurose, N., Abramov, A., and Masuda, R. 2000. Intrageneric diversity of the cytochrome b gene and phylogeny of Eurasian species of the genus *Mustela* (Mustelidae, Carnivora). *Zoological Science* 17: 673–679.

Kurose, N., Abramov, A. V., and Masuda, R. 2005. Comparative phylogeny between the ermine *Mustela erminea* and the least weasel *M. nivalis* of Palaearctic and Nearctic regions, based on analysis of mitochondrial DNA control region sequences. *Zoological Science* 22: 1069–1078.

Kurtén, B. 1960. Chronology and faunal evolution of the earlier European glaciations. *Commentationes Biologicae Societas Scientiarum Fennica* 21: 2–62.

———. 1966. Pleistocene mammals and the Bering bridge. *Commentationes Biologicae Societas Scientiarum Fennica* 29: 1–7.

———. 1968. *Pleistocene Mammals of Europe*. London: Weidenfeld & Nicolson.

Kurtén, B., and Anderson, E. 1980. *Pleistocene mammals of North America*. New York: Columbia University Press.

LaChapelle, E. R. 1969. *Field guide to snow crystals*. Seattle: University of Washington Press.

Lambin, X., Ims, R. A., Steen, H., and Yoccoz, N. G. 1995. Vole cycles. *Trends in Ecology and Evolution* 10: 204.

Lambin, X., Petty, S. J., and Mackinnon, J. L. 2000. Cyclic dynamics in field vole populations and generalist predation. *Journal of Animal Ecology* 69: 106–118.

Latham, R. M. 1952. The fox as a factor in the control of weasel populations. *Journal of Wildlife Management* 16: 516–517.

Lavrov, N. P. 1944. Effect of helminth invasions and infectious diseases on variations in numbers of the ermine. In: King, C. M., editor. *Biology of mustelids. Some Soviet research.* Boston Spa, Yorks: British Library Lending Division. pp. 170–187.

Lawrence, B. 1999. Live rats and mice as lures for stoats. *Conservation Advisory Science Notes* 234: 1–2.

Lawrence, B. L., and Dilks, P. J. 2000. Effectiveness of diphacinone to control stoat populations. *Conservation Advisory Science Notes* 306: 1–12.

Lawrence, B. L., and O'Donnell, C. F. J. 1999. Trap spacing and layout: experiments in stoat control in the Dart Valley, 1992–95. *Science for Conservation* 118: 1–13.

Lee, W. G., and Jamieson, I. G., editors. 2001. The Takahe: fifty years of conservation management and research. Dunedin: University of Otago Press.

Levins, R. 1966. The strategy of model building in population ecology. *American Scientist* 54: 421–431.

———. 1979. Coexistence in variable environments. *American Naturalist* 114: 765–783.

Lewis, J. W. 1967. Observations on the skull of Mustelidae infected with the nematode, *Skrjabingylus nasicola. Journal of Zoology (London)* 153: 561–564.

———. 1978. A population study of the metastrongylid nematode *Skrjabingylus nasicola* in the weasel *Mustela nivalis. Journal of Zoology (London)* 184: 225–229.

Lindenfors, P., Dalen, L., and Angerbjorn, A. 2003. The monophyletic origin of delayed implantation in carnivores and its implications. *Evolution* 57: 1952–1956.

Lindquist, E. E., Wu, K. W., and Redner, J. H. 1999. A new species of the tick genus *Ixodes* (Acari:Ixodidae) parasitic on mustelids (Mammalia: Carnivora) in Canada. *Canadian Entomologist* 131: 151–170.

Lindström, E. 1988. Reproductive effort in the red fox, *Vulpes vulpes*, and supply of fluctuating prey. *Oikos* 52: 115–119.

Linduska, J. P. 1947. Longevity of some Michigan farm game mammals. *Journal of Mammalogy* 28: 126–129.

Linn, I., and Day, M. G. 1966. Identification of individual weasels *Mustela nivalis* using the ventral pelage pattern. *Journal of Zoology (London)* 148: 583–585.

Lippincott, J. W. 1940. I trap no more weasels. *Pennsylvania Game News* March: 6 and 25.

Lisgo, K. A. 1999. Ecology of the short-tailed weasel (*Mustela erminea*) in the mixedwood boreal forest of Alberta. MS thesis. Vancouver: University of British Columbia.

Littin, K. E., Mellor, D. J., Warburton, B., and Eason, C. T. 2004. Animal welfare and ethical issues relevant to the humane control of vertebrate pests. *New Zealand Veterinary Journal* 52: 1–10.

Lloyd, A. R. 1982. *Kine.* Feltham, UK: Hamlyn Paperbacks.

———. 1989. *Witchwood.* London: Century Hutchinson.

Lloyd, B. D. 2001. Advances in New Zealand mammalogy 1990–2000: short-tailed bats. *Journal of the Royal Society of New Zealand* 31: 59–81.

Lockie, J. D. 1966. Territory in small carnivores. *Symposia of the Zoological Society of London* 18: 143–165.

Lockie, J. D., and Day, M. G. 1964. The use of anaesthesia in the handling of stoats and weasels. *Small Animal Anaesthesia.* Oxford: Pergamon Press. pp.187–189.

Lokemoen, J. T., and Higgins, K. F. 1972. Population irruption of the least weasel (*Mustela nivalis*) in east central North Dakota. *Prairie Naturalist* 4: 96.

Lopez, B. 1990. *Crow and Weasel.* Albany, CA: North Point Press.

Lynch, J. M. 1996. Postglacial colonisation of Ireland by mustelids, with particular reference to the badger (*Meles meles* L.). *Journal of Biogeography* 23: 179–185.

Lyver, P. O. B. 2000. Identifying mammalian predators from bite marks: a tool for focusing wildlife protection. *Mammal Review* 30: 31–43.

Macdonald, D. W., editor. 2001. The new encyclopaedia of mammals. Oxford: Oxford University Press.

Macdonald, D. W., Tew, T. E., and Todd, I. A. 2004. The ecology of weasels *(Mustela nivalis)* on mixed farmland in southern England. *Biologia (Bratislava)* 59: 235–241.

MacLean, S. F., Jr., Fitzgerald, B. M., and Pitelka, F. A. 1974. Population cycles in arctic lemmings: winter reproduction and predation by weasels. *Arctic and Alpine Research* 6: 1–12.

Macpherson, A. H. 1965. The origin of diversity in mammals of the Canadian arctic tundra. *Systematic Zoology* 14: 153–173.

Madison, D. M. 1984. Group nesting and its ecological and evolutionary significance in overwintering microtine rodents. In: Merritt, J. F., editor. *Winter ecology of small mammals.* pp. 267–274.

Maher, W. J. 1967. Predation by weasels on a winter population of lemmings, Banks Island, Northwest Territories. *The Canadian Field-Naturalist* 81: 248–250.

Maloney, R. F., and McLean, I. G. 1995. Historical and experimental learned predator recognition in free-living New Zealand robins. *Animal Behaviour* 50: 1193–1201.

Mandahl, N., and Fredga, K. 1980. A comparative chromosome study by means of G-, C-, and NOR-bandings of the weasel, the pygmy weasel and the stoat (*Mustela,* Carnivora, Mammalia). *Hereditas* 93: 75–83.

Mappes, T., Koskela, E., and Ylönen, H. 1998. Breeding suppression in voles under predation risk of small mustelids: laboratory or methodological artefact? *Oikos* 82: 365–369.

Mardon, D. K., and Moors, P. J. 1977. Records of fleas collected from weasels (*Mustela nivalis* L.) in north-east Scotland (Siphonatera: Hystrichopsyllidae and Ceratophyllidae). *Entomologist's Gazette* 28: 277–280.

Marshall, W. H. 1963. The ecology of mustelids in New Zealand. *New Zealand Department of Scientific and Industrial Research, Information Series* 38: 1–32.

Martin, R. D., and Potter, M. A. Unpublished. Seasonal and local weather effects on stoat activity.

Martinoli, A., Preatoni, D. G., Chiarenzi, B., Wauters, L. A., and Tosi, G. 2001. Diet of stoats (*Mustela erminea*) in an Alpine habitat: the importance of fruit consumption in summer. *Acta Oecologica-International Journal of Ecology* 22: 45–53.

Mayer, M. V. 1957. A method of determining the activity of burrowing mammals. *Journal of Mammalogy* 38: 531.

Maynard Smith, J. 1979. Game theory and the evolution of behaviour. *Proceedings of the Royal Society of London* 205B: 475–488.

McCleery, R. H., Clobert, J., Julliard, R., and Perrins, C. M. 1996. Nest predation and delayed cost of reproduction in the great tit. *Journal of Animal Ecology* 65: 96–104.

McDonald, R. A. 2000. Live fast–die young. The life history of stoats and weasels. *Biologist* 47: 120–124.

———. 2002. Resource partitioning among British and Irish mustelids. *Journal of Animal Ecology* 71: 185–200.

McDonald, R. A., and Birks, J. D. S. 2003. Effects of farming practice and wildlife management on small mustelid carnivores. In: Tattersall, F. H., and Manley, W., editors. *Conservation and conflict: mammals and farming in Britain.* London: Westbury Publishing, for Linnean Society of London, Occasional Publications 4. pp. 106–119.

McDonald, R. A., Day, M. J., and Birtles, R. J. 2001. Histological evidence of disease in wild stoats (*Mustela erminea*) in England. *Veterinary Record* 149: 671–675.

McDonald, R. A., and Harris, S. 1999. The use of trapping records to monitor populations of stoats *Mustela erminea* and weasels *M. nivalis*: the importance of trapping effort. *Journal of Applied Ecology* 36: 679–688.

McDonald, R. A., and Harris, S. 2000. The use of fumigants and anticoagulant rodenticides on game estates in Great Britain. *Mammal Review* 30: 57–64.

McDonald, R. A., and Harris, S. 2002. Population biology of stoats *Mustela erminea* and weasels *Mustela nivalis* on game estates in Great Britain. *Journal of Applied Ecology* 39: 793–805.

McDonald, R. A., Harris, S., Turnbull, G., Brown, P., and Fletcher, M. 1998. Anticoagulant rodenticides in stoats (*Mustela erminea*) and weasels (*Mustela nivalis*) in England. *Environmental Pollution* 103: 17–23.

McDonald, R. A., and Larivière, S. 2001. The diseases and pathogens of *Mustela* spp., with special reference to the biological control of stoat (*Mustela erminea*) populations in New Zealand. *Journal of the Royal Society of New Zealand* 31: 721–744.

McDonald, R. A., and Lariviere, S. 2002. Captive husbandry of stoats *Mustela erminea*. *New Zealand Journal of Zoology* 29: 177–186.

McDonald, R. A., and Murphy, E. C. 2000. A comparison of the management of stoats and weasels in Great Britain and New Zealand. In: Griffiths, H. I., editor. *Mustelids in a modern world. Management and conservation aspects of small carnivore: human interactions.* Leiden: Backhuys Publishers. pp. 21–40.

McDonald, R. A., and Vaughan, N. 1999. An efficient way to prepare mammalian skulls and bones. *Mammal Review* 29: 265–266.

McDonald, R. A., Webbon, C., and Harris, S. 2000. The diet of stoats (*Mustela erminea*) and weasels (*Mustela nivalis*) in Great Britain. *Journal of Zoology, London* 252: 363–371.

McLennan, J. A., Dew, L., Miles, J., Gillingham, N., and Waiwai, R. 2004. Size matters: predation risk and juvenile growth in North Island brown kiwi (*Apteryx mantelli*). *New Zealand Journal of Ecology* 28: 241–250.

McLennan, J. A., Potter, M. A., Robertson, H. A., Wake, G. C., Colbourne, R., Dew, L., Joyce, L., McCann, A. J., Miles, J., Miller, P. J., and Reid, J. 1996. Role of predation in the decline of kiwi, *Apteryx* spp., in New Zealand. *New Zealand Journal of Ecology* 20: 27–35.

Mead, R. A. 1981. Delayed implantation in mustelids, with special emphasis on the spotted skunk. *Journal of Reproduction and Fertility Supplement* 29: 11–24.

———. 1989. The physiology and evolution of delayed implantation in carnivores. In: Gittleman, J. L., editor. *Carnivore behavior, ecology and evolution.* New York: Cornell University Press. pp. 437–464.

———. 1993. Embryonic diapause in vertebrates. *Journal of Experimental Zoology* 266: 629–641.

———. 1994. Reproduction in *Martes.* In: Buskirk, S. W., Harestad, A. S., Raphael, M. G., and Powell, R. A., editors. *Martens, sables and fishers: biology and conservation.* Ithaca, New York: Cornell University Press. pp. 404–422.

Meia, J.-S. 1990. Étude de la variation de taille de l'hermine (*Mustela erminea* L.) et de la belette (*Mustela nivalis* L.) en Europe. *Bulletin de la Société neuchâteloise des Sciences naturelles* 113: 307–318.

Meia, J.-S., and Mermod, C. 1992. Taxonomie et variation de taille des hermines et des belettes en Europe: revue de littérature et commentaire. *Revue suisse Zoologie* 99: 109–118.

Mellett, J. S. 1981. Mammalian carnassial function and the "Every Effect." *Journal of Mammalogy* 62: 164–166.

Middleton, A. D. 1967. Predatory mammals and the conservation of game in Great Britain. *Annual Report of the Game Research Association* 6: 14–21.

Miller, C., Elliot, M., and Alterio, N. 2001. Home range of stoats (*Mustela erminea*) in podocarp forest, South Westland, New Zealand: implications for a control strategy. *Wildlife Research* 28: 165–172.

Moller, H., and Alterio, N. 1999. Home range and spatial organisation of stoats (*Mustela erminea*) ferrets (*Mustela furo*) and house cats (*Felis catus*) on coastal grasslands, Otago Peninsula, New Zealand: implications for yellow-eyed penguin (*Megadyptes antipodes*) conservation. *New Zealand Journal of Zoology* 26: 165–174.

Montague, T. L. 2002. Rabbit meat and rodent-scented lures as attractants for stoats (*Mustela erminea*). *DoC Science Internal Series* 45: 1–14.

Moorhouse, R., Greene, T., Dilks, P., Powlesland, R., Moran, L., Taylor, G., Jones, A., Knegtmans, J., Wills, D., Pryde, M., Fraser, I., August, A., and August, C. 2003. Control of introduced mammalian predators improves kaka *Nestor meridionalis* breeding success: reversing the decline of a threatened New Zealand parrot. *Biological Conservation* 110: 33–44.

Moors, P. J. 1974. The annual energy budget of a weasel (*Mustela nivalis* L.) population in farmland. Ph.D. thesis. Aberdeen: University of Aberdeen.

———. 1975. The food of weasels (*Mustela nivalis*) on farmland in north-east Scotland. *Journal of Zoology (London)* 177: 455–461.

———. 1977. Studies of the metabolism, food consumption and assimilation efficiency of a small carnivore, the weasel (*Mustela nivalis* L.). *Oecologia (Berl.)* 27: 185–202.

———. 1980. Sexual dimorphism in the body size of mustelids (Carnivora): the roles of food habits and breeding systems. *Oikos* 34: 147–158.

———. 1983a. Predation by stoats (*Mustela erminea*) and weasels (*M. nivalis*) on nests of New Zealand forest birds. *Acta Zoologica Fennica* 174: 193–196.

————. 1983b. Predation by mustelids and rodents on the eggs and chicks of native and introduced birds in Kowhai Bush, New Zealand. *Ibis* 125: 137–154.

Morozova-Turova, L. G. 1965. [Geographical variation in weasels in the Soviet Union]. In: King, C. M., editor. *Biology of mustelids. Some Soviet research.* Boston Spa, Yorkshire, England: British Library Lending Division. pp 7–29.

Morton Boyd, J. 1958. Mole and stoat on Eilean Molach, Loch Ba, Argyll. *Proceedings of the Zoological Society of London* 131: 327–328.

Mulder, J. L. 1990. The stoat *Mustela erminea* in the Dutch dune region, its local extinction, and a possible cause: the arrival of the fox *Vulpes vulpes. Lutra* 33: 1–21.

Müller, H. 1970. Beiträge zur biologie des hermelins, *Mustela erminea* Linné, 1758. *Saugetierkundliche Mitteilungen* 18: 293–380.

Mumford, R. E. 1969. Longtailed weasel preys on big brown bats. *Journal of Mammalogy* 50: 360.

Murie, A. 1935. Weasel goes hungry. *Journal of Mammalogy* 16: 321–322.

Murphy, B. D. 1989. Reproductive physiology of female mustelids. In: Seal, U. S., Thorne, E. T., Bogan, M. A., and Anderson, S. H., editors. *Conservation biology and the black-footed ferret.* New Haven: Yale University Press. pp. 107–123.

Murphy, E. C., Clapperton, B. K., Bradfield, P. M. F., and Speed, H. J. 1998. Effects of rat-poisoning operations on abundance and diet of mustelids in New Zealand podocarp forests. *New Zealand Journal of Zoology* 25: 315–328.

Murphy, E. C., and Dowding, J. E. 1994. Range and diet of stoats (*Mustela erminea*) in a New Zealand beech forest. *New Zealand Journal of Ecology* 18: 11–18.

————. 1995. Ecology of the stoat in *Nothofagus* forest: home range, habitat use and diet at different stages of the beech mast cycle. *New Zealand Journal of Ecology* 19: 97–109.

Murphy, E. C., and Fechney, L. 2003. What's happening with stoat research? Fifth report on the five year stoat-research programme. Wellington: Department of Conservation.

Murphy, E. C., Robbins, L., Young, J. B., and Dowding, J. E. 1999. Secondary poisoning of stoats after an aerial 1080 poison operation in Pureora Forest, New Zealand. *New Zealand Journal of Ecology* 23: 175–182.

Musgrove, B. F. 1951. Weasel foraging patterns in the Robinson Lake area, Idaho. *The Murrelet* 32: 8–11.

Myrberget, S. 1972. Fluctuations in a north Norwegian population of willow grouse. *International Ornithological Congress* 15: 107–120.

Nams, V. 1981. Prey selection mechanisms of the ermine (*Mustela erminea*). In: Chapman, J. A., and Pursley, D., editors. *Worldwide Furbearer Conference.* Frostburg, Maryland, USA (3–11 August 1980): Worldwide Furbearer Conference Inc. pp. 861–882.

Nasimovich, A. A. 1949. The biology of the weasel in Kola Peninsula in connection with its competitive relations with the ermine. *Zoologicheskii Zhurnal* 28: 177–182.

Nelson, L., Jr., and Clark, F. W. 1973. Correction for sprung traps in catch/effort calculations of trapping results. *Journal of Mammalogy* 54: 295–298.

Noback, C. V. 1935. Observations on the seasonal hair molt in a New York State weasel (*Mustela noveboracensis*). *Bulletin of the New York Zoological Society* 38: 25–27.

Norbury, G. 2000. The potential for biological control of stoats (*Mustela erminea*). *New Zealand Journal of Zoology* 27: 145–163.

Norbury, G. L., Norbury, D. C., and Heyward, R. P. 1998. Behavioral responses of two predator species to sudden declines in primary prey. *Journal of Wildlife Management* 62: 45–58.

Norrdahl, K. 1995. Population cycles in northern small mammals. *Biological Reviews* 70: 621–637.

Norrdahl, K., and Korpimäki, E. 1995a. Does predation risk constrain maturation in cyclic vole populations? *Oikos* 72: 263–272.

———. 1995b. Effects of predator removal on vertebrate prey populations: birds of prey and small mammals. *Oecologia (Berl.)* 103: 241–248.

———. 1998. Does mobility or sex of voles affect risk of predation by mammalian predators? *Ecology* 79: 226–232.

———. 2000a. The impact of predation risk from small mustelids on prey populations. *Mammal Review* 30: 147–156.

———. 2000b. Do predators limit the abundance of alternative prey? Experiments with vole-eating avian and mammalian predators. *Oikos* 91: 528–540.

Northcott, T. H. 1971. Winter predation of *Mustela erminea* in northern Canada. *Arctic* 24: 142–144.

Nyholm, E. S. 1959a. About the ermine and the weasel and their winter territory. *Suomen Riista* 13: 106–116.

———. 1959b. Stoats and weasels in their winter habitat. In: King, C. M., editor. *Biology of mustelids. Some Soviet research.* 1975 ed. Boston Spa, Yorkshire, England: British Library Lending Division. pp. 118–131.

O'Connor, C. E., Turner, J., Scobie, S., and Duckworth, J. 2004. Stoat reproductive biology. Lincoln: Landcare Research. Report LC0304/135.

O'Donnell, C. F. J. 1996. Predators and the decline of New Zealand forest birds: an introduction to the hole-nesting bird and predator programme. *New Zealand Journal of Zoology* 23: 213–219.

O'Donnell, C. F. J., Dilks, P. J., and Elliott, G. P. 1996. Control of a stoat (*Mustela erminea*) population irruption to enhance mohua (yellowhead) (*Mohoua ochrocephala*) breeding success in New Zealand. *New Zealand Journal of Zoology* 23: 279–286.

O'Donnell, C. F. J., and Phillipson, S. M. 1996. Predicting the incidence of mohua predation from the seedfall, mouse, and predator fluctuations in beech forests. *New Zealand Journal of Zoology* 23: 287–293.

Obara, Y. 1985. Karyological relationship between two species of mustelids, the Japanese ermine and the least weasel. *Japanese Journal of Genetics* 60: 157–160.

O'Regan, R. 1966. The effect of stoats on native birds in Westland. *Forest and Bird* November 1966: 4–10.

Oksanen, T., Oksanen, L., and Fretwell, S. D. 1985. Surplus killing in the hunting strategy of small predators. *The American Naturalist* 126: 328–346.

Oksanen, T., Oksanen, L., Jedrzejewski, W., Jedrzejewska, B., Korpimaki, E., and Norrdahl, K. 2000. Predation and the dynamics of the bank vole, *Clethrionomys glareolus*. *Polish Journal of Ecology* 48: 197–217.

Ondrias, J. 1960. Secondary sexual variation and body skeletal proportions in European Mustelidae. *Arkiv Für Zoologi* 12: 577–583.

———. 1962. Comparative osteological investigations on the front limbs of European Mustelidae. *Arkiv Für Zoologi* 13 (15): 311–320.

Osborn, D. J., and Helmy, I. 1980. The contemporary land mammals of Egypt (including Sinai). *Fieldiana: Zoology, New Series* 5: 406–409.

Osgood, F. L. 1936. Earthworms as a supplementary food of weasels. *Journal of Mammalogy* 17: 64.

Ostfeld, R. S., and Keesing, F. 2000. Pulsed resources and community dynamics of consumers in terrestrial ecosystems. *Trends in Ecology and Evolution* 15: 232–237.

Packer, J. J., and Birks, J. D. S. 1999. An assessment of British farmers' and gamekeepers' experiences, attitudes and practices in relation to the European polecat *Mustela putorius. Mammal Review* 29: 75–92.

Parkes, J., and Murphy, E. C. 2003. Management of introduced mammals in New Zealand. *New Zealand Journal of Zoology* 30: 335–359.

Parovshchikov, V. Y. 1963. A contribution to the ecology of *Mustela nivalis* Linnaeus 1766, of the Arkhangel'sk North. In: King, C. M., editor. *Biology of mustelids. Some Soviet research.* Boston Spa, Yorkshire, England: British Library Lending Division. pp. 84–97.

Pascal, M., Delattre, P., Damange, J. P., Haffner, P., and Poser, S. 1990. Identification des classes d'âge dans les populations de belettes (*Mustela nivalis* L. 1766). *Review Ecology (Terre Vie)* 45: 93–112.

Pearce, J. 1937. A captive New York weasel. *Journal of Mammalogy* 18: 483.

Pearson, O. P. 1966. The prey of carnivores during one cycle of mouse abundance. *Journal of Animal Ecology* 35: 217–233.

———. 1985. Predation. In: Tamarin, R. H., editor. *Biology of New World Microtus.* American Society of Mammalogists. pp. 535–566.

Pekkarinen, P., and Heikkila, J. 1997. Prey selection of the least weasel *Mustela nivalis* in the laboratory. *Acta Theriologica* 42: 179–188.

Peters, R. P., and Mech, L. D. 1975. Scent-marking in wolves. *American Scientist* 63: 628–637.

Petrov, O. V. 1962. The validity of Bergmann's Rule as applied to intraspecific variation in the ermine. In: King, C. M., editor. *Biology of mustelids. Some Soviet research.* Boston Spa, Yorkshire, England: British Library Lending Division. pp. 30–38.

Pierce, R. J. 1986. Differences in susceptibility to predation during nesting between pied and black stilts *Himantopus* spp. *The Auk* 103: 273–280.

Pierce, R. J., and Westbrooke, I. M. 2003. Call count responses of North Island brown kiwi to different levels of predator control in Northland, New Zealand. *Biological Conservation* 109: 175–118.

Poddubnaya, N. Y. 1992. Asynchronism of population dynamics of various mustelids in response to changes in numbers of murine rodents. *Soviet Journal of Ecology* 23: 34–38.

Polder, E. 1968. Den and cover usage by northeast Iowa spotted skunk and weasel populations [published as "Spotted skunk and weasel populations den and cover usage by northeast Iowa"]. *Proceedings of the Iowa Academy of Sciences* 75: 142–146.

Polderboer, E. B. 1942. Habits of the least weasel (*Mustela rixosa*) in northeastern Iowa. *Journal of Mammalogy* 23: 145–147.

Polderboer, E. B., Kuhn, L. W., and Hendrickson, G. O. 1941. Winter and spring habits of weasels in central Iowa. *Journal of Wildlife Management* 5: 115–119.

Polkanov, A. 2000. Aspects of the biology, ecology and captive breeding of stoats. *Conservation Advisory Science Notes* 307: 1–15.

Popov, V. A. 1943. Numerosity of *Mustela erminea* as affected by *Skrjabingylus* worms invasion. *Comptes rendus, Academie des Sciences, Paris* 39: 160–162.

Potter, M. A., and King, C. M. Unpublished. Acceptance by stoats of 1080 in small-volume baits.

Potts, G. R. 1979. *Will grey partridges suvive?* Fordingbridge, England: The Game Conservancy Annual Review for 1978. pp. 48–52.

Potts, G. R. 1980. The effects of modern agriculture, nest predation and game management on the population ecology of partridges (*Perdix perdix* and *Alectoris rufa*). *Advances in Ecological Research* 11: 2–79.

Potts, G. R. 1986. *The partridge: pesticides, predation and conservation.* London: Collins.

Potts, G. R., and Vickerman, G. P. 1974. Studies on the cereal ecosystem. *Advances in Ecological Research* 8: 107–197.

Pounds, C. J. 1981. Niche overlap in sympatric populations of stoats (*Mustela erminea*) and weasels (*Mustela nivalis*) in northeastern Scotland. D. Phil. thesis. Aberdeen, Scotland: University of Aberdeen.

Powell, C. B. 1995. *A bold carnivore.* Boulder, Colorado: Roberts Reinhart.

Powell, C. B., and Powell, R. A. 1982. The predator-prey concept in elementary education. *Wildlife Society Bulletin* 10: 238–244.

Powell, R. A. 1973. A model for raptor predation on weasels. *Journal of Mammalogy* 54: 259–263.

———. 1978a. Zig! Zag! Zap! A weasel on the prowl for prey is nature's original energy conservationist. *Animal Kingdom* 81: 20–25.

———. 1978b. A comparison of fisher and weasel hunting behavior. *Carnivore* 1: 28–34.

———. 1979a. Mustelid spacing patterns: variations on a theme by *Mustela. Zeitschrift für Tierpsychologie* 50: 153–165.

———. 1979b. Ecological energetics and foraging strategies of the fisher (*Martes pennanti*). *Journal of Animal Ecology* 48: 195–212.

———. 1982. Evolution of black-tipped tails in weasels: predator confusion. *The American Naturalist* 119: 126–131.

———. 1985a. Fisher pelt primeness. *Wildlife Society Bulletin* 13: 67–70.

———. 1985b. Possible pathways for the evolution of reproductive strategies in weasels and stoats. *Oikos* 44: 506–508.

———. 1987. Black bear home range overlap in North Carolina and the concept of home range applied to black bears. *International Conference on Bear Research and Management* 7: 235–242.

———. 1989. Effects of resource productivity, patchiness and predictability on mating and dispersal strategies. In: Standen, V., and Foley, R. A., editors. *Comparative socioecology: the behavioural ecology of humans and other mammals.* Oxford: Blackwell Scientific Publications. pp. 101–123.

———. 1993. *The fisher: life history, ecology and behavior.* Minneapolis: University of Minnesota Press.

———. 1994. Structure and spacing of *Martes* populations. In: Buskirk, S. W., Harestad, A. S., Raphael, M. G., and Powell, R. A., editors. *Martens, sables and fishers: biology and conservation.* Ithaca, New York: Cornell University Press. pp. 101–121.

———. 2000. Animal home ranges and territories and home range estimators. In: Boitani, L., and Fuller, T., editors. *Research techniques in animal ecology: controversies and consequences.* New York: Columbia University Press. pp. 65–110.

———. 2001. Who limits whom: predators or prey? *Endangered Species Update* 18: 98–102.

Powell, R. A., and Fried, J. J. 1992. Helping by juvenile pine voles (*Microtus pinetorum*), growth and survival of younger siblings, and the evolution of pine vole sociality. *Behavioral Ecology* 4: 325–333.

Powell, R. A., and King, C. M. 1997. Variation in body size, sexual dimorphism and age-specific survival in stoats, *Mustela erminea* (Mammalia: Carnivora), with fluctuating food supplies. *Biological Journal of the Linnean Society* 62: 165–194.

Powell, R. A., and Proulx, G. 2003. Trapping and marking terrestrial mammals for research: integrating ethics, performance criteria, techniques, and common sense. *Institute of Laboratory Animal Research Journal* 44: 259–276.

Powell, R. A., and Zielinski, W. J. 1983. Competition and coexistence in mustelid communities. *Acta Zoologica Fennica* 223–227.

———. 1989. Mink response to ultrasound in the range emitted by prey. *Journal of Mammalogy* 70: 637–638.

Powell, R. A., Zimmerman, J. W., and Seaman, D. E. 1997. *Ecology and behaviour of North American black bears.* London: Chapman & Hall.

Powlesland, R. G. 1983. Breeding and mortality of the South Island robin in Kowhai Bush, Kaikoura. *Notornis* 30: 265–282.

Price, E. O. 1971. Effect of food deprivation on activity of the least weasel. *Journal of Mammalogy* 52: 636–640.

Proulx, G. 2000. The impact of human activities on North American mustelids. pp. 53–75. In Griffiths, H. I., editor. *Mustelids in a modern world: management and conservation aspects of small carnivore: human interactions.* Leiden: Backhuys Publishers.

Pruitt, W. O. J. 1978. *Boreal ecology.* London: Arnold.

Pryde, M. A., O'Donnell, C. F. J., and Barker, R. J. 2005. Factors influencing survival and long-term population viability of New Zealand long-tailed bats (*Chalinolobus tuberculatus*): implications for conservation. *Biological Conservation* 126: 175–185.

Purdey, D. C., King, C. M., and Lawrence, B. 2004. Age structure, dispersion and diet of a population of stoats (Mustela erminea) in southern Fiordland during the decline phase of the beechmast cycle. *New Zealand Journal of Zoology* 31: 205–225.

Purdey, D. C., Petcu, M., and King, C. M. 2003. A simplified protocol for detecting two systemic bait markers (Rhodamine B and iophenoxic acid) in small mammals. *New Zealand Journal of Zoology* 30: 175–184.

Pyle, E. 2003. The alarming discovery about stoats. *Forest and Bird* 309: 31–32.

Quick, H. F. 1944. Habits and economics of the New York weasel in Michigan. *Journal of Wildlife Management* 8: 71–78.

─────. 1951. Notes on the ecology of weasels in Gunnison County, Colorado. *Journal of Mammalogy* 32: 281–290.

Radinsky, L. B. 1981. Evolution of skull shape in carnivores 2. Additional modern carnivores. *Biological Journal of the Linnean Society* 16: 337–355.

Raffaelli, D., and Moller, H. 2000. Manipulative field experiments in animal ecology: do they promise more than they can deliver? *Advances in Ecological Research* 30: 299–338.

Ragg, J. R., Moller, H., and Waldrup, K. A. 1995. The prevalence of bovine tuberculosis (*Mycobacterium bovis*) infections in feral populations of cats (*Felis catus*), ferrets (*Mustela furo*) and stoats (*Mustela erminea*) in Otago and Southland, New Zealand. *New Zealand Veterinary Journal* 43: 333–337.

Ralls, K., and Harvey, P. H. 1985. Geographic variation in size and sexual dimorphism of North American weasels. *Biological Journal of the Linnean Society* 25: 119–167.

Ratz, H. 1997. Ecology, identification and control of introduced mammalian predators of yellow-eyed penguins (*Megadyptes antipodes*). Ph.D. thesis. Dunedin, New Zealand: University of Otago.

─────. 2000. Movements by stoats (*Mustela erminea*) and ferrets (*M. furo*) through rank grass of yellow-eyed penguin (*Megadyptes antipodes*) breeding areas. *New Zealand Journal of Zoology* 27: 57–69.

Raymond, M., and Bergeron, J.-M. 1982. Réponse numérique de l'hermine aux fluctuations d'abondance de *Microtus pennsylvanicus*. *Canadian Journal of Zoology* 60: 542–549.

Raymond, M., Bergeron, J.-M., and Plante, Y. 1984. Dimorphisme sexuel et régime alimentaire de l'hermine dans un agrosystème du Québec. *Canadian Journal of Zoology* 62: 594–600.

Raymond, M., Robitaille, J.-F., Lauzon, P., and Vaudry, R. 1990. Prey-dependent profitability of foraging behaviour of male and female ermine, *Mustela erminea*. *Oikos* 58: 323–328.

Reichstein, H. 1957. [Skull variability in European weasels (*Mustela nivalis* L.) and stoats (*Mustela erminea* L.) in relation to distribution and sex]. *Zeitschrift für Säugetierkunde* 22: 151–182.

Reid, D. G., Krebs, C. J., and Kenney, A. J. 1997. Patterns of predation on noncyclic lemmings. *Ecological Monographs* 67: 89–108.

Reig, S. 1997. Biogeographic and evolutionary implications of size variation in North American least weasels (*Mustela nivalis*). *Canadian Journal of Zoology* 75: 2036–2049.

Renfree, M. B., and Shaw, G. 2000. Diapause. *Annual Review of Physiology* 62: 353–375.

Reynolds, J. C., and Tapper, S. C. 1996. Control of mammalian predators in game management and conservation. *Mammal Review* 26: 127–155.

Rickard, C. G. 1996. *Introduced small mammals and invertebrate conservation in a lowland podocarp forest, South Westland, New Zealand*. M. For. Sc. thesis. Christchurch: University of Canterbury.

Robinson, R. A., and Sutherland, W. J. 2002. Post-war changes in arable farming and biodiversity in Great Britain. *Journal of Applied Ecology* 39: 157–176.

Robitaille, J.-F., and Baron, G. 1987. Seasonal changes in the activity budget of captive ermine, *Mustela erminea* L. *Canadian Journal of Zoology* 65: 2864–2871.

Robitaille, J.-F., and Raymond, M. 1995. Spacing patterns of ermine, *Mustela erminea* L., in a Quebec agrosystem. *Canadian Journal of Zoology* 73: 1827–1834.

Robson, G. 1998. The breeding ecology of the curlew *Numenius arquata* on north Pennine moorland. Ph.D. thesis. Sunderland: University of Sunderland.

Ronkainen, H., and Ylönen, H. 1994. Behaviour of cyclic bank voles under risk of mustelid predation———do females avoid copulations? *Oecologia (Berl.)* 97: 377–381.

Rothschild, M., and Lane, C. 1957. Note on change of pelage in the stoat (*Mustela erminea* L.). *Proceedings of the Zoological Society of London* 128: 602.

Rowlands, I. 1972. Reproductive studies in the stoat. *Journal of Zoology (London)* 166: 574–576.

Rubina, M. A. 1960. [Some features of weasel (*Mustela nivalis* L.) ecology based on observations in the Moscow region]. *Byulleten' Moskovskogo Obshchestva Ispytatelei Prirody: (Otdel Biologicheskii)* 65: 27–33.

Ruscoe, W. A., Choquenot, D., Heyward, R., Yockney, I., Young, N., and Drew, K. 2003. Seed production, predators and house mouse population eruptions in New Zealand beech forests. In: Singleton, G. R., Hinds, L. A., Krebs, C. J., and Spratt, D. M., editors. *Rats, mice and people: rodent biology and management.* Canberra: ACIAR. pp. 334–337.

Rust, C. C. 1962. Temperature as a modifying factor in the spring pelage change of short-tailed weasels. *Journal of Mammalogy* 43: 323–328.

———. 1965. Hormonal control of pelage cycles in the short tailed weasel (*Mustela erminea bangsi*). *General and Comparative Endocrinology* 5: 222–231.

———. 1968. Procedure for live trapping weasels. *Journal of Mammalogy* 49: 318.

Rust, C. C., and Meyer, R. K. 1969. Hair color, molt, and testis size in male, short-tailed weasels treated with melatonin. *Science* 165: 921–922.

Salomonsen, F. 1939. Moults and sequence of plumages in the Rock Ptarmigan (*Lagopus mutus* (Montin)). *Videnskabelige Meddelelser fra Danske Naturhistorisk Forening* 103: 1–491.

Samson, C., and Raymond, M. 1995. Daily activity pattern and time budget of stoats (*Mustela erminea*) during summer in southern Quebec. *Mammalia* 59: 501–510.

Sandell, M. 1984. To have or not to have delayed implantation: the example of the weasel and the stoat. *Oikos* 42: 123–126.

———. 1986. Movement patterns of male stoats *Mustela erminea* during the mating season: differences in relation to social status. *Oikos* 47: 63–70.

———. 1988. Stop-and-go stoats. *Natural History* 97: 55–65.

———. 1989. Ecological energetics, optimal body size and sexual size dimorphism: a model applied to the stoat, *Mustela erminea* L. *Functional Ecology* 3: 315–324.

———. 1990. The evolution of seasonal delayed implantation. *The Quarterly Review of Biology* 65: 23–42.

Sandell, M., and Liberg, O. 1992. Roamers and stayers: a model on mating tactics and mating systems. *American Naturalist* 139: 177–189.

Sanderson, G. C. 1949. Growth and behavior of a litter of captive long-tailed weasels. *Journal of Mammalogy* 30: 412–415.

Sato, J. J., Hosoda, T., Wolsan, M., Tsuchiya, K., Yamamo, M., and Suzuki, H. 2003. Phylogenetic relationships and divergence times among mustelids (Mammalia: Carnivora) based on nucleotide sequences of the nuclear interphotoreceptor retinoid binding protein and mitochondrial cytochrome b genes. *Zoological Science* 20: 243–264.

Saunders, A. 2000. A review of Department of Conservation mainland restoration projects and recommendations for further action. Wellington: Department of Conservation.

Schmidt, K. 1992. Skull variability of *Mustela nivalis* Linnaeus, 1766 in Poland. *Acta Theriologica* 37: 141–162.

Schmidt-Neilsen, K. 1984. *Scaling: why is animal size so important?* Cambridge: Cambridge University Press.

Seaman, D. E. 1993. Home range and male reproductive activity in black bears. Ph.D. thesis. Raleigh, NC: North Carolina State University.

Seaman, D. E., and Powell, R. A. 1996. Accuracy of kernel estimators for animal home range analysis. *Ecology* 77: 2075–2085.

Segal, A. N. 1975. Postnatal growth, metabolism, and thermoregulation in the stoat. *Ekologia* 1: 38–44.

Seton, E. T. 1926. *Lives of game animals.* Garden City, New York: Doubleday, Doran & Co Inc.

Shackelford, R. M. 1952. Superfetation in the ranch mink. *American Naturalist* 86: 311–319.

Shelden, R. M. 1972. The fate of short-tailed weasel, *Mustela erminea*, blastocysts following ovariectomy during diapause. *Journal of Reproduction and Fertility* 31: 347–352.

Shore, R. F., Birks, J. D. S., and Freestone, P. 1999. Exposure of non-target vertebrates to second-generation rodenticides in Britain, with particular reference to the polecat *Mustela putorius*. *New Zealand Journal of Ecology* 23: 199–206.

Short, H. L. 1961. Food habits of a captive least weasel. *Journal of Mammalogy* 42: 273–274.

Short, M. J., and Reynolds, J. C. 2001. Physical exclusion of non-target species in tunnel-trapping of mammalian pests. *Biological Conservation* 98: 139–147.

Sidorovich, V. E., Polozov, A. G., Solovej, I. A., and Ivanova, N. V. Unpublished. Resource separation between the weasel (*Mustela nivalis*) and the stoat (*M. erminea*) in Belarus.

Simms, D. A. 1978. Spring and summer food habits of an ermine (*Mustela erminea*) in the central arctic. *The Canadian Field-Naturalist* 92: 192–193.

———. 1979a. North American weasels: resource utilization and distribution. *Canadian Journal of Zoology* 57: 504–520.

———. 1979b. Studies of an ermine population in southern Ontario. *Canadian Journal of Zoology* 57: 824–832.

Sittler, B. 1995. Response of stoats (*Mustela erminea*) to a fluctuating lemming (*Dicrostonyx groenlandicus*) population in north-east Greenland: preliminary results from a long term study. *Annales Zoologici Fennici* 32: 79–92.

Sleeman, D. P. 1987. The ecology of the Irish stoat. Ph.D. thesis. Cork, Eire: National University of Ireland (University College Cork).

———. 1988a. Irish stoat road casualties. *Irish Naturalists' Journal* 22: 527–529.

————. 1988b. *Skrjabingylus nasicola* (Leuckhart) (Metastrongyloidae) as a parasite of the Irish stoat. *Irish Naturalists' Journal* 22: 525–527.

————. 1989a. Ectoparasites of the Irish stoat. *Medical and Veterinary Entomology* 3: 213–218.

————. 1990. Dens of Irish stoats. *Irish Naturalists' Journal* 23: 202–203.

————. 1992. The diet of Irish stoats. *Irish Naturalist's Journal* 24: 151–153.

————. 2004. Reproduction in the Irish stoat, *Mustela erminea hibernica*. *Irish Naturalist's Journal* 27: 344–348.

Sleeman, P. 1989b. *Stoats and weasels, polecats and martens*. London, England: Whittet Books.

Smith, D. H. V. 2001. *The movement, diet and relative abundance of stoats Mustela erminea in the Murchison Mountains (Special Takahe Area), Fiordland*. MSc thesis thesis. Dunedin: University of Otago.

Sommer, R., and Benecke, N. 2004. Late- and post-glacial history of the Mustelidae in Europe. *Mammal Review* 34: 249–284.

Sonerud, G. A. 1988. What causes extended lows in microtine cycles? Analysis of fluctuations in sympatric shrew and microtine populations in Fennoscandia. *Oecologia (Berl.)* 76: 37–42.

Song, J. H., Houde, A., and Murphy, B. D. 1998. Cloning of leukemia inhibitory factor (LIF) and its expression in the uterus during embryonic diapause and implantation in the mink (*Mustela vison*). *Molecular Reproduction and Development* 51: 13–21.

Southern, H. N. 1970. The natural control of a population of tawny owls (*Strix aluco*). *Journal of Zoology, London* 162: 197–285.

Southern, H. N., and Lowe, V. P. W. 1982. Predation by tawny owls (*Strix aluco*) on bank voles (*Clethrionomys glareolus*) and wood mice (*Apodemus sylvaticus*). *Journal of Zoology (London)* 198: 83–102.

Spurr, E. B. 1997. Assessment of the effectiveness of Transonic™ ESP and Yard Gard™ ultrasonic devices for repelling stoats (*Mustela erminea*). *Conservation Advisory Science Notes* 151.

————. 1999. Developing a long-life toxic bait and lures for mustelids. *Science for Conservation* 127: 1–24.

————. 2000. Hen eggs poisoned with sodium monofluoroacetate (1080) for control of stoats (*Mustela erminea*) in New Zealand. *New Zealand Journal of Zoology* 27: 165–172.

————. 2002a. Rhodamine B as a systemic hair marker for assessment of bait acceptance by stoats (Mustela erminea). *New Zealand Journal of Zoology* 29: 187–194.

————. 2002b. Iophenoxic acid as a systemic blood marker for assessment of bait acceptance by stoats (Mustela erminea) and weasels (Mustela nivalis). *New Zealand Journal of Zoology* 29: 135–142.

Spurr, E. B., and O'Connor, C. E. 1999. Sound lures for stoats. *Science for Conservation* 127: 25–38.

Spurr, E. B., O'Connor, C. E., Airey, A. T., and Kerr, J. H. 2002. FeraCol[(R)] for the control of stoats (*Mustela erminea*). *DoC Science Internal Series* 61: 1–15.

St-Pierre, C., Ouellet, J.-P., and Crete, M. In press-a. Do competitive intraguild interactions affect space and habitat use by small carnivores in forested landscapes? *Ecography*.

St-Pierre, C., Ouellet, J.-P., Dufresne, F., Chaput-Bardy, A., and Hubert, F. In press-b. Morphological and molecular discrimination of ermines (*Mustela erminea*) and long-tailed weasels (*M. frenata*) in eastern Canada. *Northeastern Naturalist.*

Stake, M. M., and Cimprich, D. A. 2003. Using video to monitor predation at black-capped vireo nests. *Condor* 105: 348–357.

Steadman, D. W. 1995. Prehistoric extinctions of Pacific Island birds: biodiversity meets zooarchaeology. *Science* 267: 1123–1131.

Stenseth, N. C. 1985. Optimal size and frequency of litters in predators of cyclic preys: comments on the reproductive biology of stoats and weasels. *Oikos* 45: 293–296.

———. 1995. Snowshoe hare populations: squeezed from below and above. *Science* 269: 1061–1062.

Stephen, D. 1969. Miniature maneaters. *The Scotsman*; 25 October 1969.

Stewart, C. L., Kaspar, P., Brunet, L. J., Bhatt, H., Gadi, I., Kontgen, F., and Abbondanzo, S. J. 1992. Blastocyst implantation depends on maternal expression of leukemia inhibitory factor. *Nature* 359: 76–79.

Stoate, C. 2002. Multifunctional use of a natural resource on farmland: wild pheasant (*Phasianus colchicus*) management and the conservation of farmland passerines. *Biodiversity and Conservation* 11: 561–573

Stoddart, D. M. 1976. Effect of the odour of weasels (*Mustela nivalis* L.) on trapped samples of their prey. *Oecologia (Berl.)* 22: 439–441.

Stolt, B.-O. 1979. Colour pattern and size variation of the weasel *Mustela nivalis* L. in Sweden. *Zoon* 7: 55–61.

Strann, K.-B., Yoccoz, N. G. and Ims, R. A. 2002. Is the heart of Fennoscandian rodent cycle still beating? A 14–year study of small mammals and Tengmalm's owls in northern Norway. *Ecography* 25: 81–87.

Stroganov, S. U. 1937. [A method of age determination and an analysis of the age structures of ermine populations (*Mustela erminea* L)]. In: King, C. M., editor. *Biology of mustelids. Some Soviet research.* Volume 2. Bulletin 227 (1980) ed. Wellington, New Zealand: New Zealand Department of Scientific and Industrial Research. pp 93–108.

Stuart, A. J. 1982. *Pleistocene vertebrates in the British Isles.* London: Longman.

Stubbe, M. 1970. [On the evolution of anal marking organs of Mustelids]. *Biologisches Zentralblatt* 89: 213–223.

———. 1972. Die analen markierungsorgane der *Mustela*-Arten. *Zoologische Garten N. F., Leipzig* 42: 176–188.

Sullivan, T. P., and Sullivan, D. S. 1980. The use of weasels for natural control of mouse and vole populations in a coastal coniferous forest. *Oecologia (Berl.)* 47: 125–129.

Sullivan, T. P., Sullivan, D. S., Ransome, D. B., and Lindgren, P. M. F. 2003. Impact of removal-trapping on abundance and diversity attributes in small-mammal communities. *Wildlife Society Bulletin* 31: 464–474.

Sumption, K. J., and Flowerdew, J. R. 1985. The ecological effects of the decline in rabbits (*Oryctolagus cuniculus*) due to myxomatosis. *Mammal Review* 15: 151–186.

Sundell, J. 2003. Reproduction of the least weasel in captivity: basic observations and the influence of food availability. *Acta Theriologica* 48: 59–72.

Sundell, J., Eccard, J. A., Tiilikainen, R., and Ylonen, H. 2003. Predation rate, prey preference and predator switching: experiments on voles and weasels. *Oikos* 101: 615–623.

Sundell, J., and Norrdahl, K. 2002. Body size-dependent refuges in voles: an alternative explanation of the Chitty effect. *Annales Zoologici Fennici* 39: 325–333.

Sundell, J., Norrdahl, K., Korpimaki, E., and Hanski, I. 2000. Functional response of the least weasel, *Mustela nivalis nivalis*. *Oikos* 90: 501–508.

Svendsen, G. E. 1976. Vocalizations of the long-tailed weasel (*Mustela frenata*). *Journal of Mammalogy* 57: 398–399.

———. 2003. Weasels and black-footed ferret, *Mustela* species. In: Feldhamer, G. A., Thompson, B. C., and Chapman, J. A., editors. *Wild mammals of North America: biology, management, and conservation*, 2nd edition. Baltimore: Johns Hopkins University Press. pp. 650–661.

Swanson, G. and Fryklund, P. O. 1935. The least weasel in Minnesota and its fluctuation in numbers. *The American Midland Naturalist* 16: 120–126.

Tapper, S. C. 1976. The diet of weasels, *Mustela nivalis*, and stoats, *Mustela erminea*, during early summer, in relation to predation on gamebirds. *Journal of Zoology (London)* 179: 219–224.

———. 1979. The effect of fluctuating vole numbers (*Microtus agrestis*) on a population of weasels (*Mustela nivalis*) on farmland. *Journal of Animal Ecology* 48: 603–617.

———. 1982. Using estate records to monitor population trends in game and predator species, particularly weasels and stoats. *International Congress of Game Biologists* 14: 115–120.

———. 1992. *Game heritage: an ecological review from shooting and gamekeeping records.* Fordingbridge, Hants., UK: Game Conservancy.

———. 1999. *A question of balance.* Fordingbridge, Hants: The Game Conservancy Trust.

Tapper, S. C., Green, R. E., and Rands, M. R. W. 1982. Effects of mammalian predators on partridge populations. *Mammal Review* 12: 159–167.

Tapper, S. C., Potts, G. R., and Brockless, M. H. 1996. The effects of an experimental reduction in predation pressure on the breeding success and population density of grey partridges *Perdix perdix*. *Journal of Applied Ecology* 33: 965–978.

Tattersall, F., and Manley, W., editors. 2003. Conservation and conflict: mammals and farming in Britain. London: Westbury Publishing, for the Linnean Society of London.

Taylor, R., and Smith, I. 1997. *The state of New Zealand's environment.* Wellington: Ministry for the Environment.

Taylor, R. H., and Tilley, J. A. V. 1984. Stoats (*Mustela erminea*) on Adele and Fisherman Islands, Abel Tasman National Park, and other offshore islands in New Zealand. *New Zealand Journal of Ecology* 7: 139–145.

Teer, J. G. 1964. Predation by long-tailed weasels on eggs of blue-winged teal. *Journal of Wildlife Management* 28: 404–406.

Tenquist, J. D., and Charleston, W. A. G. 1981. An annotated checklist of ectoparasites of terrestrial mammals in New Zealand. *Journal of the Royal Society of New Zealand* 11: 257–285.

Teplov, V. P. 1948. The problem of sex ratio in ermine. In: King, C. M., editor. *Biology of mustelids. Some Soviet research.* Boston Spa, Yorkshire, England: British Library Lending Division. pp. 98–103.

————. 1952. Taking a census of otter, sable, marten and small mustelids. In: King, C. M., editor. *Biology of mustelids. Some Soviet research.* Volume 2. Bulletin 227 (1980) ed. Wellington, New Zealand: New Zealand Department of Scientific and Industrial Research. pp. 173–179.

Ternovsky, D. V. 1983. [The biology of reproduction and development of the stoat *Mustela erminea* (Carnivora, Mustelidae)]. *Zoologischeskii Zhurnal* 62: 1097–1105.

Theobald, S., and Coad, N. 2002. Den control of stoats (*Mustela erminea*) in Trounson Kauri Park, Northland. *DOC Science Internal Series* 90: 1–15.

Thom, M. D., Johnson, D. D. P., and MacDonald, D. W. 2004. The evolution and maintenance of delayed implantation in the mustelidae (Mammalia: Carnivora). *Evolution* 58: 175–183.

Thompson, F. R., and Burhans, D. E. 2003. Predation of songbird nests differs by predator and between field and forest habitats. *Journal of Wildlife Management* 67: 408–416.

Thompson, I. D., and Colgan, P. W. 1987. Numerical responses of martens to a food shortage in northcentral Ontario. *Journal of Wildlife Management* 51: 824–835.

Towns, D. R., and Broome, K. G. 2003. From small Maria to massive Campbell: forty years of rat eradications from New Zealand islands. *New Zealand Journal of Zoology* 30: 377–398.

Towns, D. R., Simberloff, D., and Atkinson, I. A. E. 1997. Restoration of New Zealand islands: redressing the effects of introduced species. *Pacific Conservation Biology* 3: 99–124.

Trout, R. C., and Tittensor, A. M. 1989. Can predators regulate wild rabbit *Oryctolagus cuniculus* population density in England and Wales? *Mammal Review* 19: 153–173.

Tumanov, I. L., and Levin, V. G. 1974. Age and seasonal changes in some physiological characters in the weasel (*Mustela nivalis* L.) and ermine (*Mustela erminea* L.). In: King, C. M., editor. *Biology of mustelids. Some Soviet research.* Volume 2. Wellington: New Zealand Department of Scientific and Industrial Research, Bulletin 227 (1980). pp. 192–196.

Vaisfeld, M. A. 1972. Ecology of the ermine in the cold season in the European north. In: King, C. M., editor. *Biology of mustelids. Some Soviet research.* Volume 2. Bulletin 227 (1980) ed. Wellington: New Zealand Department of Scientific and Industrial Research. pp. 1–10.

van Soest, R. W. M., and van Bree, P. J. H. 1969. On the moult of the stoat, *Mustela erminea* Linnaeus, 1758, from the Netherlands. *Beaufortia* 20: 85–97.

————. 1970. Sex and age composition of a stoat population (*Mustela erminea* Linnaeus, 1758) from a coastal dune region of the Netherlands. *Beaufortia* 17: 51–77.

van Soest, R. W. M., van der Land, J., and van Bree, P. J. H. 1972. *Skrjabingylus nasicola* (Nematoda) in skulls of *Mustela erminea* and *Mustela nivalis* (Mammalia) from the Netherlands. *Beaufortia* 20: 85–97.

van Wijngaarden, A., and Mörzer Bruijns, M. F. 1961. [The ermine *Mustela erminea* L., on the island of Terschelling]. *Lutra* 3: 35–42.

van Zyll de Jong, C. G. 1992. A morphometric analysis of cranial variation in Holarctic weasels (*Mustela nivalis*). *Zeitschrift für Säugetierkunde* 57: 77–93.

Vaudry, R., Raymond, M., and Robitaille, J. F. 1990. The capture of voles and shrews by male and female ermine *Mustela erminea* in captivity. *Holarctic Ecology* 13: 265–278.

Venables, L. S. V., and Venables, U. M. 1955. Stoat (*Mustela erminea*). In: Venables, L. S. V., and Venables, U. M., editors. *Birds and mammals of Shetland*. Edinburgh: Oliver & Boyd. pp. 69–70.

Vernon-Betts, R. J. 1967. Weasels' running battle. *The Field* 15 June 1967.

Vershinin, A. A. 1972. The biology and trapping of the ermine in Kamchatka. In: King, C. M., editor. *Biology of mustelids. Some Soviet research.* Volume 2. Bulletin 227 (1980) ed. Wellington, New Zealand: New Zealand Department of Scientific and Industrial Research. pp. 11–23.

Vik, R. 1955. Invasion of *Skrjabingylus* (Nematoda) in Norwegian mustelidae. *Nyt Magasin for Zoologi* 3: 70–78.

Walters, E. L., and Miller, E. H. 2001. Predation on nesting woodpeckers in British Columbia. *Canadian Field-Naturalist* 115: 413–419.

Watzka, M. 1940. [Microscopic-anatomical research into the breeding seasons and length of pregnancy period of the ermine (*Putorius ermineus*)]. *Zeitschrift für Mikroskopisch-Anatomische Forschung* 48: 24–37.

Weber, J.-M. 1986. Aspects quantitatifs du cycle de *Skrjabingylus nasicola* (Leukart, 1842), nematode parasite des sinus frontaux des mustelides. Ph.D. thesis. Neuchatel: University of Neuchatel.

Weber, J.-M., and Mermod, C. 1983. Experimental transmission of *Skrjabingylus nasicola*, a parasitic nematode of mustelids. *Acta Zoologica Fennica* 174: 237–238.

———. 1985. Quantitative aspects of the life cycle of *Skrjabingylus nasicola*, a parasitic nematode of the frontal sinuses of mustelids. *Zeitschrift fur Parasitenkunde* 71: 631–638.

Wederlin, L., and Turner, A.. 1996. Turnover in the guild of larger carnivores in Euroasia across the Miocene-Pliocene boundary. *Acta Zoologica Cracoviensia* 39: 585–592.

Whitaker, J. 1970. The biological subspecies: an adjunct of the biological species. *The Biologist* 52: 12–15.

White, P. C. L., and King, C. M. In press. Predation on native birds in New Zealand beech forests: the role of functional relationships between stoats and rodents. *Ibis.*

White, T. C. R. 1993. *The inadequate environment: nitrogen and the abundance of animals.* Berlin: Springer-Verlag.

———. 2001. Opposing paradigms: regulation or limitation of populations? *Oikos* 93: 148–152.

Willans, M. 2000. Annual review Te Kakahu O Tamatea stoat eradication project. Te Anau: New Zealand Department of Conservation.

Willey, R. L. 1970. Sound location of insects by the dwarf weasel. *The American Midland Naturalist* 84: 563–564.

Wilson, K-J. 2005. The state of New Zealand's birds 2005. *Wingspan* 15 (4, Suppl.).

Wilson, D. J., Krebs, C. J., and Sinclair, T. 1999. Limitation of collared lemming populations during a population cycle. *Oikos* 87: 382–398.

Wilson, P. R., Karl, B. J., Toft, R. J., Beggs, J. R., and Taylor, R. H. 1998. The role of introduced predators and competitors in the decline of kaka (*Nestor meridionalis*) populations in New Zealand. *Biological Conservation* 83: 175–185.

Winder, S. J. 2003. The stoat (*Mustela erminea*): the role of behaviour in management. MSc thesis. Hamilton, New Zealand: University of Waikato.

Wittmer, H., King, C. M., and Powell, R. A. Unpublished.Using elasticity values from matrix models to develop management strategies for fluctuating populations: the importance of cohort effects.

Wójcik, M. 1974. [Remains of mustelidae (Carnivora, Mammalia) from the Late Pleistocene deposits of Polish caves. *Acta Zoologica Cracoviensia* 19: 75–93.

Wolff, J. O., and Davis, B. R. 1997. Response of gray-tailed voles to odours of a mustelid predator: a field test. *Oikos* 79: 543–548.

Wood, K. 1946. Five o'clock killer. *Fauna, Philadelphia* 8: 44–46.

Woods, J. G., Paetkau, D., Lewis, D., McNellan, B. N., Proctor, M., and Strobeck, C. 1999. Genetic tagging of free-ranging black and brown bears. *Wildlife Society Bulletin* 27: 616–627.

Worthy, T. H., and Holdaway, R. N. 2002. *The lost world of the moa.* Bloomington: Indiana University Press.

Wright, L. W. 1980. Decision making and the logging industry: an example from New Zealand. *Biological Conservation* 18: 101–115.

Wright, P. L. 1942a. Delayed implantation in the long-tailed weasel (*Mustela frenata*), the short-tailed weasel (*Mustela cicognani*), and the marten (*Martes americana*). *Anatomical Record* 83: 341–353.

———. 1942b. A correlation between the spring molt and spring changes in the sexual cycle in the weasel. *Journal of Experimental Zoology* 91: 103–110.

———. 1947. The sexual cycle of the male long-tailed weasel (*Mustela frenata*). *Journal of Mammalogy* 28: 343–352.

———. 1948. Breeding habits of captive long-tailed weasels (*Mustela frenata*). *The American Midland Naturalist* 39: 338–344.

———. 1950. Development of the baculum of the longtailed weasel. *Proceedings of the Society for Experimental Biology and Medicine* 75: 820–822.

———. 1963. Variations in reproductive cycles in North American mustelids. In: Enders, A. C., editor. *Delayed implantation.* Chicago, Illinois, USA: University of Chicago Press. pp. 77–97.

Yalden, D. 1999. *The history of British mammals.* London: T & AD Poyser.

Yamaguchi, and Macdonald, D. W. 2003. The burden of occupancy: intraspecific resource competition and spacing patterns in American mink, *Mustela vison. Journal of Mammalogy* 84: 1341–1355.

Ylönen, H. 1989. Weasels *Mustela nivalis* suppress reproduction in cyclic bank voles *Clethrionomys glareolus. Oikos* 55: 138–140.

———. 1994. Vole cycles and anti-predatory behaviour. *Trends in Ecology and Evolution* 9: 426–430.

Ylönen, H., and Ronkainen, H. 1994. Breeding suppression in the bank vole as anti-predatory adaptation in a predictable environment. *Evolutionary Ecology* 8: 658–666.

Ylönen, H., Sundell, J., Tiilikainen, R., Eccard, J. A., and Horne, T. 2003. Weasels'

(*Mustela nivalis nivalis*) preference for olfactory cues of the vole (*Clethrionomys glareolus*). *Ecology* 84: 1447–1452.

Yom-Tov, Y., Yom-Tov, S., and Moller, H. 1999. Competition, co-existence, and adaptation amongst rodent invaders to Pacific and New Zealand islands. *Journal of Biogeography* 26: 947–958.

Zavaleta, E. S., Hobbs, R. J., and Mooney, H. A. 2001. Viewing invasive species removal in a whole-ecosystem context. *Trends in Ecology and Evolution* 16: 454–459.

Zielinski, W. J. 1986. Circadian rhythms of small carnivores and the effects of restricted feeding on daily activity. *Physiology and Behavior* 38: 613–620.

———. 1988. The influence of daily variation in foraging cost on the activity of small carnivores. *Animal Behaviour* 36: 239–249.

———. 1995. Track plates. In: Zielinski, W. J., and Kucera, T. E., editors. *American marten, fisher, lynx and wolverine: survey methods for their detection*: USDA Forest Service GTR-157.

———. 2000. Weasels and martens: carnivores in northern latitudes. In: Halle and Stenseth, N. C., editors. *Activity patterns in small mammals*. Berlin: Springer-Verlag. pp. 95–118.

Zima, J., and Cenevova, E. 2002. Coat colour and chromosome variation in central European populations of the weasel (*Mustela nivalis*). *Folia Zoologica* 51: 265–274.

Index

Note: In this list, as in the text, we index general concepts applicable to any or all weasel species either without qualification, or with the word "weasel" alone to refer to weasels in general. References to particular species are listed under their common names. Entries in bold refer to illustrations.